W0051141

POLYMERS
IN MEDICINE
Biomedical and Pharmacological Applications

POLYMER SCIENCE AND TECHNOLOGY

Editorial Board:

William J. Bailey, *University of Maryland, College Park, Maryland*
J. P. Berry, *Rubber and Plastics Research Association of Great Britain,*
        Shawbury, Shrewsbury, England
A. T. DiBenedetto, *The University of Connecticut, Storrs, Connecticut*
C. A. J. Hoeve, *Texas A & M University, College Station, Texas*
Yōlchi Ishida, *Osaka University, Toyonaka, Osaka, Japan*
Frank E. Karasz, *University of Massachusetts, Amherst, Massachusetts*
Oslas Solomon, *Franklin Institute, Philadelphia, Pennsylvania*

Recent volumes in the series:

A Continuation Order Plan is available for this series. A continuation order will bring delivery of each new volume immediately upon publication. Volumes are billed only upon actual shipment. For further information please contact the publisher.

POLYMERS IN MEDICINE

Biomedical and Pharmacological Applications

Edited by

Emo Chiellini

and

Paolo Giusti

University of Pisa
Pisa, Italy

PLENUM PRESS • NEW YORK AND LONDON

Library of Congress Cataloging in Publication Data

Main entry under title:

Polymers in medicine.

(Polymer science and technology; v. 23)
"Contains the collected papers presented at the International Symposium on Polymers in Medicine: Biomedical and Pharmacological Applications, which was held at Porto Cervo, Italy, May 24–28, 1982"—Pref.
Includes bibliographical references and index.
1. Polymers in medicine—Congresses. I. Chiellini, Emo. II. Giusti, Paolo. III. International Symposium on Polymers in Medicine, Biomedical and Pharmacological Applications (1982: Porto Cervo, Sardinia) IV. Series. [DNLM: 1. Polymers—Congresses. 2. Drugs—Congresses. 3. Biocompatible materials—Congresses. QV 55 I616p 1982]
R857.P6P63   1983   610'.28   83-17711
ISBN 978-1-4615-7645-7   ISBN 978-1-4615-7643-3 (eBook)
DOI 10.1007/978-1-4615-7643-3

Proceedings of an International Symposium on Polymers in Medicine:
Biomedical and Pharmacological Applications, held May 24–28, 1982, at
Porto Cervo, Sardinia, Italy

©1983 Plenum Press, New York
Softcover reprint of the hardcover 1st edition 1983
A Division of Plenum Publishing Corporation
233 Spring Street, New York, N.Y. 10013

All rights reserved

No part of this book may be reproduced, stored in a retrieval system, or transmitted in any form or by any means, electronic, mechanical, photocopying, microfilming, recording, or otherwise, without written permission from the Publisher
recording, or otherwise, without written permission from the Publisher

PREFACE

        This book contains the collected papers presented at the Inter-
national Symposium on Polymers in Medicine, Biomedical and Pharmacological Applications, which was held at Porto Cervo, Italy, May
24-28, 1982.

        To the best of our knowledge, this symposium was the first to
be organized in Italy entirely devoted to the several aspects of
the use of synthetic and semisynthetic macromolecular materials in
the field of biomedical and pharmacological applications. The inten-
tion of the Organizing Committee of the symposium was the promotion
of a scientific and cultural initiative to gain the attention of
various experts in line research of the potential of suitably designed "man-made" polymeric materials in biomedical applications.

        With highly qualified and worldwide attendance, the above
goal was fully satisfied. Indeed the opportunity of meeting together in a well conceived and discreet corner of the world, scientists with different cultural backgrounds and objectives helped extend the meaning of the symposium far beyond the Italian borders and
the perspectives of the National Research Council of Italy (CNR),
the major sponsor of the meeting.

        The papers collected in the book represent all the contributions
presented at the symposium and very plainly reflect the general chal-
lenging atmosphere permeating over the several sessions of the symposium. According to the three major lines on which the meeting was
articulated, there are papers focused on synthetic and semisynthetic
polymers:
i). a. With biologically active functional groups imparting drug
properties to the polymeric material (Polymeric Drugs); b. able to
bind either covalently or ionically conventional drugs, that are released at a controlled rate by hydrolytic, enzymatic, and oxidative
means. Systems are also described in which the polymer acts as a
barrier for the physically trapped drug, that is released at a rate
controlled by diffusion or matrix degradation processes (Drug Delivery Systems). These aspects are described in Section I.
ii). With intrinsic structural and morphological requirements that

make them suitable for implants and prosthetic devices. In this area, Section 2 (Polymers as Biomaterials) in particular is devoted to the description of a few polymeric materials of appropriate synthetic procedures or physical modifications guaranteed for their biocompatibility in temporary and longlasting devices.
Section 3 (medical and Surgical Applications) stresses, among others, some applications in the cardiovascular field, that along with the dental and orthopaedic applications of polymers constitute one of the main and most attractive outlets for ad hoc formulated macromolecular materials.

The Editors are well aware that the information reported here merely represents a few of the desirable research investigations in the general area of the symposium topics. They do hope that this book will help sustain ongoing and stimulate new basic and applied research in the medical and pharmacological field.

Finally, the Editors wish to express their heartful and sincere thanks to all participants in the symposium, along with the National Research Council of Italy and the companies that contributed to the success of the meeting. We are also indebted to the people of the Institute of General Chemistry, Faculty of Engineering, and the Institute of Industrial Organic Chemistry, Faculty of Sciences, University of Pisa, who gave their spontaneous personal support to the organization. Particular gratitude has to be reserved to Dr. G. Galli for his efforts provided during the several steps of the organization of the symposium and for his help in setting up some of the contributed manuscripts.

Emo Chiellini
Paolo Giusti

Institute of General Chemistry
Faculty of Engineering
University of Pisa
56100 Pisa, Italy

CONTENTS

SECTION II
POLYMERS AS BIOMATERIALS

SECTION III

MEDICAL AND SURGICAL APPLICATIONS OF POLYMERS

SECTION I
POLYMERIC DRUGS AND DRUG AND DRUG DELIVERY SYSTEMS

CHARACTERISTIC BIOLOGICAL EFFECTS OF ANIONIC POLYMERS

Raphael M. Ottenbrite,* Kristine Kuus,* & Alan M. Kaplan°

*Department of Chemistry and Microbiology and the MCV/VCU
Cancer Center, Virginia Commonwealth University,
Richmond, VA 23284

°Department of Microbiology and Immunology, School of
Medicine, Kentucky University, Lexington, KY 40506

INTRODUCTION

Polyanions are polyelectrolytes with negative charges that reside on pendant groups attached to the backbone of the polymer chain. Both natural polyanions such as heparin and heparinoids and synthetic polyanions such as polyacrylic acid and pyran produce a variety of biological activities which have been discussed in detail [1,2]. These biological effects include induction of interferon and antiviral activity including inhibition of Maloney sarcoma, Rauscher leukemia, Friend leukemia, polyoma, vesicular stomatitis virus, Mengo, encephalomyocarditis, and foot-and-mouth disease.

Polyanions can enter into biological functions by distribution throughout the host and behave similar to certain proteins, glycoproteins and polynucleotides which modulate a variety of biological responses related to host defense reaction. These effects include enhanced host resistance to bacteria and fungi, enhanced immune responsiveness, inhibition of adjuvant arthritis and depending on polymer size, either depress or stimulate the functional phagocytic activity of the reticuloendothelial system. In relation to these immunologic and hormonal responses, inflammation, wound repair, blood clotting and tissue damage are subject to the action of these macromolecules.

The action of polyanions as mitotic inhibitors and their functional role in neoplastic processes have been reported as has the role of polynucleotides in immunology and virus resistance [1]. A possible mechanism for the activity of polyanions on tumor growth may be related to coupling of the polyanion to tumor antigen. However, the action of polyanions on a wide range of enzymes, such as alteration of the isoelectric point of proteins, displacement of nucleohistone and antiviral action all indicate possible alternative concepts of antitumor action. For example, immunopotentiators, or "host-resistance inducing agents," have been used for decades in an attempt to treat tumor-bearing animals and man. These attempts have been largely empirical until recently, when the interactions of the basic immune response to tumors have been partially delineated. Cytotoxic lymphocytes, cytotoxic antibody and "activated" macrophages have all been shown capable of inhibiting or destroying tumor cells [1].

Recently considerable evidence has emerged that implicates the macrophage as a major effector of tumor cytotoxicity and/or cytostasis. Both synthetic reticuloendothelial stimulants such as poly-(acrylic acid-alt-maleic anhydride), pyran and polyriboinosinic-polycytidylc acid as well as biologic reticuloendothelial stimulants are known to enhance macrophage function as well as to induce resistance to tumor growth [1]. Moreover, macrophages from animals treated with polyanionic stimulants have been demonstrated to be cytostatic and/or cytotoxic for tumor cells while demonstrating quantitatively less cytotoxicity for normal cells.

Synthetic polyanions that have received considerable attention include pyran, polyacrylic acid, and poly(acrylic acid-alt-maleic anhydride) copolymer. More recently we have made an indepth study of several other copolymers of maleic anhydride [2,3].

PYRAN COPOLYMER

The synthetic polymer which has received the most interest is divinyl ether-maleic anhydride copolymer, commonly referred to as pyran copolymer due to the tetrahydropyran ring that was reported to form during polymerization [4]. In the literature it is also referred to by the acronym DIVEMA (divinyl ether-maleic anhydride copolymer) and, more recently, as MVE (maleic anhydride-vinyl ether copolymer). Pyran was first reported by Butler in 1960 [4]. It was submitted to the NIH screen independently by Butler (University of Florida) and by Breslow (Hercules Corporation) [5]. Pyran showed activity and was designated as NSC 46015 by the National Cancer Institute. It has been under investigation for use in cancer chemotherapy and has also exhibited a wide range of other biological activities [1]. Pyran has been reported to have interferon inducing capacity [6-9] and to be active against a number of viruses [6-14] including Friend leukemia, Rauscher leukemia, Maloney sarcoma, polyoma, vesicular stomatitis, Mengo, encephalomyocarditis, and foot-and-mouth disease. It has antibacterial [15-17] and antifungal activity [15]; it stimulates immune response [15-28] and is a blood anticoagulant [22]. Pyran inhibits adjuvant disease [5,23], a hypersensitive reaction to mycrobacterial antigens, similar to rheumatoid arthritis and also shows potential for removing plutonium from the liver [24] (see Figure 1 for antitumor activity).

Pyran NSC 46015, similar to poly(ethylene sulfonate), was found to be too toxic in human patients. Specifically, it caused thromocytopenia [25], a condition characterized by a decrease in the absolute number of thrombocytes in the blood circulation. Its other toxic effects involved cytoplasmic inclusions throughout the blood as well as in liver and spleen cells of the reticuloendothelial system (RES) [25]; inhibition of microsomal enzymes; sensitization to endotoxin and enlargement of the liver (hepatomegaly) and spleen (splenomegaly) [23]. At high dosages, 12 mg/kg/day, pyran NSC 46015 induced fever and blocked the conversion of fibrinogen to fibrin. Interesting, no patient suffered hemorrhage despite significant increases in clotting time [25]. Initially, the

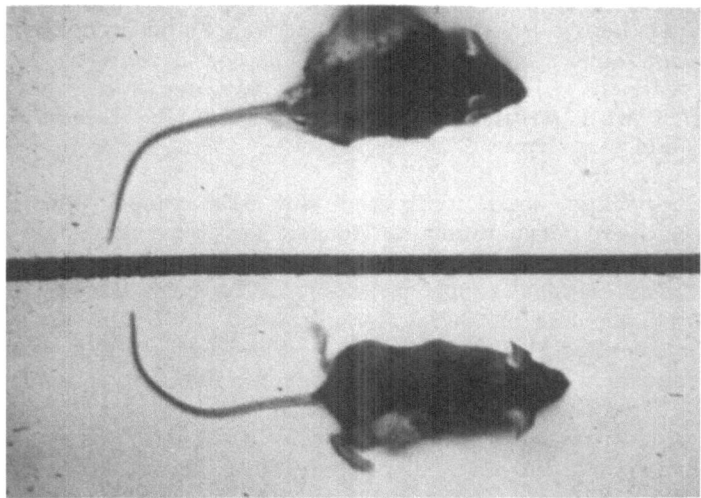

Figure 1. This is fifteen days after both mice were implanted with Lewis Lung Carcinoma subcutaneously. The top animal was untreated and the bottom animal was given pyran copolymer on days 1 and 5.

toxicity of pyran NSC 46015 was too high for further clinical investigations [1,26]. It has since been shown [1,27] that toxicity
is less prevalent in lower molecular weight fractions and when the
calcium salt is used instead of the sodium salt of pyran. These
findings have prompted another phase I clinical study of the drug
and phase II studies are presently being completed.

EFFECT OF MOLECULAR WEIGHT AND STRUCTURE ON BIOLOGICAL ACTIVITY

A number of investigations have clearly demonstrated that the
structure and molecular weight of synthetic polyanions are directly
related to biological activity and toxicity. Breslow [5] showed
that the acute toxicity caused by pyran in mice increased with increased molecular weight. Pyran fractions prepared with narrow
polydispersity and molecular weights ranging from 2,500-32,200 were
evaluated for biological activity. The data indicated that molecular weights up to 15,000 stimulated RES whereas higher molecular
weight fractions suppressed RES, resulting in biphasic response. It
was found that the level of serum glutamic pyruvate transaminase
(SGPT), which is a measure of liver damage, also increased with molecular weight, as did inhibition of drug metabolism and sensitization to endotoxin. However, the activities against Lewis lung and
Ehrlich ascites tumor were shown to be independent of molecular
weight.

Kaplan found [28] that a low molecular weight pyran sample of
narrow polydispersity (X18571-31) activated macrophages and was
active against Lewis lung carcinoma. It was further noted that this
sample did not sensitize the mice to endotoxin. A very low molecular weight sample (X19543-27) was not only non-toxic but inactive
whereas very high molecular weight samples were active against
Lewis lung but very toxic.

These findings were confirmed and elaborated upon by Ottenbrite et al. [29]. Two lower molecular weight fractions (PM10 and
PM30) were isolated from Hercules XA124-177 pyran copolymer, (a
broad molecular weight range polymer). The toxicologic, antitumor
and antiviral properties were investigated and results are listed
in Table 1. As expected, the higher molecular weight parent pyran
XA124-177 caused hepatosplenomegaly, sensitization to endotoxin and
inhibition of microsomal enzymes as measured by increased hexobar-
bitol sleeping times. However, the lower molecular weight PM 10 and
and PM 30 fractions showed much higher LD_{50} (lower toxicity values)
than the parent polymer, caused less hepatosplenomegaly and did not
significantly sensitize to bacterial endotoxin. Inhibition of
microsomal enzymes became almost insignificant. The activity
against Lewis lung carcinoma was comparable to the parent polymer.
However, the antiviral activity was greatly decreased; the lowest
molecular weight (PM 10) fraction showed no antiviral activity and

TABLE I. Pharmacologic Effects and Pyran Molecular Weight.

[a]	Control	Pyran XA124-177	PM-10	PM-30
Intrinsic Viscosity	--	0.21	0.05	0.06
Toxicologic Properties				
LD_{50} [b]		74.0	120.0	115.0
Liver Weight [c]	5.4	7.8	5.1	5.9
Spleen Weight	0.4	1.1	0.4	0.4
LD_{50} [d] (Endotoxin)	25.0	0.1	15.0	15.0
Hexobarbital Sleeping Time [e] (min)	36.8	97.6	42.8	48.6
Antitumor and Antiviral Properties				
Antitumor Activity [f]				
Lewis Lung % Inhibition	0.0	76.0	69.0	64.0
Antiviral (EMC) % Protection	0.0	89.0	0.0	30.0

[a] PM 10 Filtrate of XA124-177 passed through Amicon PM 10 filter and PM 30 Filtrate of XA124-177 passed through Amicon PM 30 filter.

[b] Polymers (mg/kg) administered intravenously. Mortality recorded after 24 hours.

[c] Polymers administered in a dose of 25 mg/kg intravenously. Organ weights determined 7 days after drug injection and expressed as percent of total animal weight.

[d] Mg/kg of S. typhosa 0904 lipopolysaccharide administered 24 hours after a single dose of 24 mg/kg of polymer.

[e] Mice were inoculated with 25 mg/kg i.v. of the polymer; 24 hrs later an anesthetic dose (80 mg/kg) of sodium hexobarbital was administered i.v. and duration of anesthesia recorded.

[f] Mice were inoculated with 10^6 Lewis lung cells. Polymers were administered daily i.p. for 10 consecutive days following tumor implantation. Primary tumor size was determined on day 17 (% inhibition) and increased life span (% ILS) calculated for mean time to death.

the PM 30 fraction was only 30% effective compared to the whole polymer.

For further investigations into the effect of molecular weight and polyanionic structures on biological activity [1,2], several different molecular weight fractions of pyran, poly(acrylic acid-alt-maleic acid) (PAAMA), poly(maleic acid) (PMA), and poly(acrylic acid-co-3,6-endoxo-1,2,3,6-tetrahydrophthalic acid) (BCEP) were prepared. These were evaluated for activity against Lewis lung carcinoma and encephalomyocarditis virus (Table II). Except for BCEP which inhibited tumor growth by only 30%, the other three polymers reduced tumor growth by approximately 75%. However, it was shown that inhibition of tumor growth does not necessarily lead to a con-

TABLE II. Antitumor and Antiviral Activity
With Polymer Molecular Weight.

Polymer [a]	PYRAN (%)	PAAMA (%)	PMA (%)	BCEP (%)
Inhibition of Tumor Size [b]				
Whole polymer	78	74	75	30
(Increased life span)	(40)	(33)	(15)	(<10)
1,000- 10,000	89	--	74	22
10,000- 30,000	84	80	72	--
30,000- 50,000	74	--	76	--
50,000-100,000	70	78	74	30
Antiviral Encephalomyocarditis Protection [c]				
Whole Polymer	89	90	30	25
1,000- 10,000	0	0	0	0
10,000- 30,000	30	29	0	--
30,000- 50,000	58	54	<10	--
50,000-100,000	86	80	26	38

[a] Poly(divinylether-co-maleic anhydride) (Pyran); Poly(acrylic acid-co-maleic anhydride) (PAAMA); Poly(maleic anhydride) (MA); Poly(acrylic acid-co-3,6-endoxo-1,2,3,6-tetrahydrophthalic anhydride) (BCEP).

[b] BDF, mice were inoculated with 10^6 Lewis' lung cells into right-hind gluteus muscle. Polymers were administered daily by the intraperitoneal route for 10 consecutive days following tumor implantation. Primary tumor size was determined on day 14 (% inhibition) and increased life span (% ILS) calculated from mean time of death.

[c] Mice were inoculated intravenously with 10 LD_{50} of encephalomyocarditis virus 24 hours after administration (i.v.) of 25 mg/kg of polymer. Percent protection based on 20 mice/group calculated from eman time to death.

commitant increased life span (ILS). The greatest ILS was obtained with pyran copolymer (40%) and the least ILS was obtained with BCEP (10%). It appears that the polymeric structure has significant effect on biological activity. Only the higher molecular weight fractions (>30,000) showed antiviral activity whereas the lower molecular weight fractions were ineffective.

Similar to previous studies, it was shown that as the molecular weight of the polymer fraction decreased, so did the acute toxicity (Table III). Endotoxin sensitization showed a dependence on molecular weight distribution particularly with the more active polymers such as pyran and PAAMA.

TABLE III. Toxicologic Effects and Polymer Molecular Weight.

POLYMER	PYRAN	PAAMA	PMA	BCEP
Polymer Molecular Weight and Acute Toxicity $(LD_{50})^a$				
Whole Polymer	74	110	120	150
1,000- 10,000	120		160	>200
10,000- 30,000	115	>200	150	---
30,000- 50,000	95		140	---
50,000-100,000	84	180	135	180
Polymer Molecular Weight and Endotoxin Sensitization^b				
Whole Polymer	0.12	1.0	15	3.0
1,000- 10,000	15	---	>20	>10
10,000- 30,000	15	---	>20	>10
30,000- 50,000	---	>20	>20	>3
50,000-100,000	---	>20	>20	>3

^{a} LD_{50} was determined by administering the polymer in mg/kg intraveneously and the mortality recorded 24 hours later.

^{b} LD_{50} of S. typhosa 0904 lipopolysaccharide 24 hours after single administration of 24 mg/kg of polymer. The dosage of endotoxin was administered in mg/kg.

Recently, Munson, et al. [27], reported that the calcium salts of pyran were less toxic than sodium salts. It was also demonstrated that toxicity and sensitization to bacterial endotoxin were molecular weight dependent (Table IV). While the two salts showed differences in acute toxicity, little difference was seen in the sensitization to bacterial endotoxin.

TABLE IV. Effect of Molecular Weight and Counterions of Pyran on Acute Toxicity and Sensitivity to Bacterial Endotoxin.

Polymer Molecular Weight	LD_{50}^{a}		Endotoxim LD_{50}^{b}	
	Na$^+$ Salt	Ca^{++} Salt	Na$^+$ Salt	Ca^{++} Salt
12,500	112	--	24.0	--
15,500	98	190	7.0	10
21,300	94	--	0.8	--
52,600	86	170	0.5	0.5

^{a} Mice were injected i.v. with sodium and calcium salts of Pyran. Mortality was observed over a 14 day period.

^{b} Mice were injected i.v. with 25 mg/kg of pyran followed 24 hours later by i.s. administration of S. typhosa lysopolysaccharide.

These studies suggest that: (a) the biological toxicity of polyanions increases with molecular weight (significant toxicity increases occur above 15,000 MW and becomes serious over 50,000 MW); (b) antitumor activity does not appear to be molecular weight dependent; (c) antiviral activity is related to molecular weight (it is diminished below 50,000 MW); and (d) sensitization to bacterial endotoxin is related to molecular weight in the case of pyran and seems to be affected by polydispersity of the polyanionic polymer.

TARGET CELL SPECIFICITY OF MACROPHAGES ACTIVATED BY ANIONIC POLYMERS

A major deterrent in the search for effective antineoplastic agents is the problem of specificity. Presently, most of the commonly used chemotherapeutic agents are as deleterious to normal host cells as they are to neoplastic cells. Consequently scientists and clinicians work within a very narrow margin between lethal toxicity versus clinical efficacy when administering chemothera-peutic agents for the control and eradication of tumors.

Synthetic polyanionic immunomodulators present another ap-proach for the control of neoplasia. These agents activate host elements to become cytotoxic to tumor cells thus eliminating or re-ducing the tumor size while the activated elements do not appear to be harmful to normal cells. These polyanionic polymers are not, however, void of toxic effects but it appears that these deleter-ious effects are indeed side-effects and not directly caused by the interaction of the immunomodulator with normal cells.

Macrophages may be "activated" in vivo to become cytotoxic to tumor cells in vitro. Activation is accomplished by elicitation of peritoneal macrophage by agents such as pyran, C. Parvum and Bacillus Calmuette Gueirain (BCG). Normal peritoneal macrophages and those elicited with thioglycollate or protease peptone are not cytotoxic (Figure 2). More recently we have been able to activate macrophages to tumoricidal capacity with several polyanionic co-polymers [3]. A common finding among all these activated macrophage populations is that they are cytotoxic or cytostatic for tumor cells such as Lewis lung and Ehrlich ascites while they have no ap-parent effect on normal cell populations such as newborn mouse fi-broblasts and fetal mouse fibroblasts [30,31] (Figure 3). There-fore, polyanion activated macrophages possess the unique capability of discriminating between a normal cell and a neoplastic cell.

At the present time the basis for this discrimination is unknown, but it must be assumed that the activated macrophage can recognize a feature of a neoplastic cell that is inconsistent with that found on a normal cell. For several decades researchers have

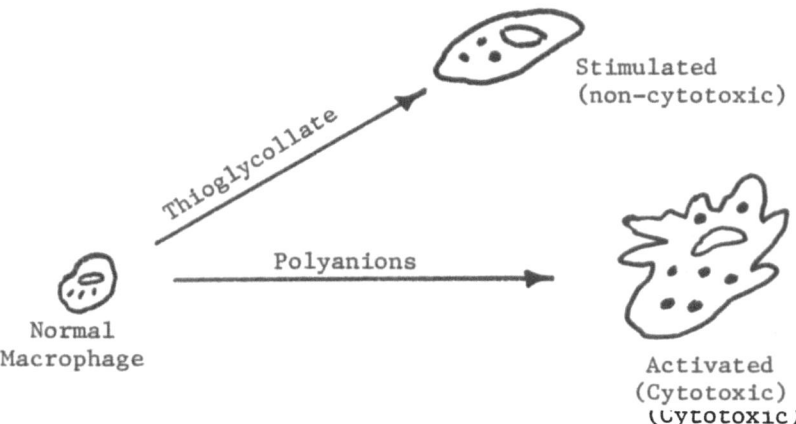

Figure 2. "Stimulated" and "Activated" macrophage to non-cytotoxic and tumor-cytotoxic states.

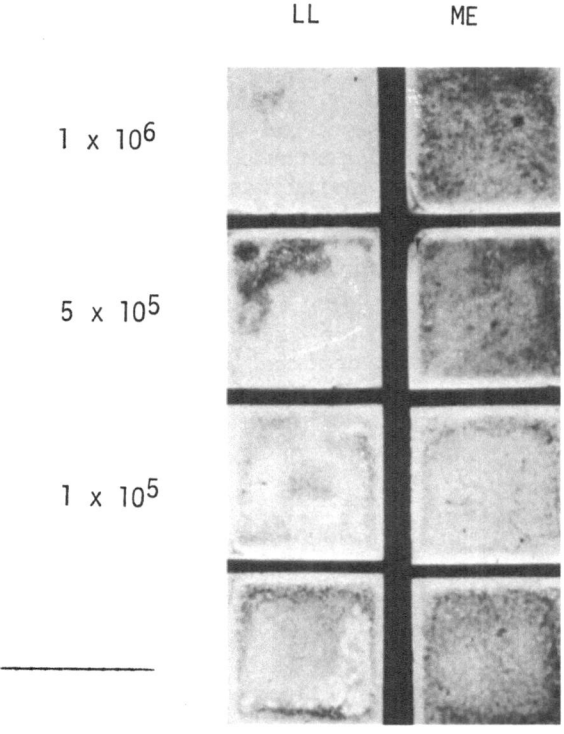

Figure 3. The effect of concentration of pyran activated peritoneal macrophage cells (PMC) on Lewis Lung (LL) carcinoma and normal mouse embryo (ME) cell cultures. (At 10^6 PMC all the LL cells are killed while the ME cells are unaffected.

been trying to delineate singular differences between normal and neoplastic cells in a given system. Tumor cells exhibit differences in lectin binding and agglutination, membrane micro-viscosity, enzyme activity and cytochemical structure when compared to normal cells [32,33]. It appears, however, that in every instance these apparent aberrations could be duplicated in completely normal cells under the appropriate conditions. At the present time the mechanism by which activated macrophages recognize tumor cells remains to be elucidated. However, the consistency of the phenomenon and the potential value in the management of cancer renders it unique and worthy of further study.

DISTRIBUTION OF C^{14}-PYRAN IN THE HOST

The site of injection of an agent may profoundly effect its action based on the barriers and organs that it encounters via each specific route. It is therefore of interest to compare studies on the organ distribution of C^{14}-labeled pyran administered intravenously (i.v.) by Munson et al. (unpublished observations) and intraperitoneally (i.p.) by Papamatheakis et al. [34].

C^{14}-pyran was administered i.v. and i.p. at dosages of 44 mg/kg and 25 mg/kg, respectively. The organ distribution of the radioactivity was follwed for 9 months after i.v. injection and 3 weeks after i.p. injection. C^{14}-pyran was cleared quickly from the blood after both i.v. and i.p. injection. Administration by i.p. resulted in peak levels after 2 hours and was completely cleared in 6 hours while 0.2% of the residual activity remained in the blood 24 hours after i.v. injection. The concentration of radiolabeled pyran in the brain paralleled that found in the blood after i.v. injection while no activity was detectable in the brain after i.p. injection, at any time. The patterns were also similar for lung distribution since i.v. injection gave 4% of the residual C^{14} activity in the lung after 24 hours and this level decreased to 1% after 7 days, 0.38% at 5 months and 0.05% at 9 months. However, i.p. injection resulted in a constant residual level of activity in the lung (500 cpm vs. 5500 and 8000 cpm in the liver and spleen respectively at the height of observed activity). The high activity after 24 hours in the lung after i.v. injection is probably due to the fact that the lung capillary beds are among those first encountered after i.v. injection.

The bulk of the C^{14}-pyran administered at either site had a reticuloendothelial distribution with most of the activity residing in the liver and spleen even though the kinetics of distribution differed. Peak levels of activity in the liver were found 5 days after i.v. injection and 2 days after i.p. injection. Hepatic levels after i.v. injection declined to 11% after 2 weeks, 16% after 5 weeks, 4.7% after 5 months and 4.1% after 9 months. The levels after i.p. injection declined about 20-25% from the peak at 3-4 days and remained relatively constant for 21 days.

Peak levels of C^{14}-pyran in the spleen were 6% at day 5 and 5500 cpm at day 1 for i.v. and i.p. administration, respectively. At all time points, in both cases, the splenic levels were greatly reduced compared to those found in the liver. Levels of activity in the spleen after i.v. injection were 4% at 2 weeks, 9% at 5 weeks, 0.90% at 5 months and 0.74% at 9 months. I.p. administration resulted in a drop from the peak activity of 5500 cpm to relatively constant levels at 2500 cpm at day 3 for up to 3 weeks.

These patterns demonstrated that while the kinetics of the distribution of C^{14}pyran by i.p. or i.v. injection vary somewhat in the observed peak levels, the polymer has a reticuloendothelial distribution and is concentrated primarily in the liver and spleen. Pyran is persistent in these organs and significant levels are detectable even 9 months after administration. Whether or not the measured activity at these later time points may be attributed to intact pyran is not known.

CELLULAR UPTAKE OF PYRAN COPOLYMER

The uptake and location of pyran into the macrophage is of central importance in understanding the mechanism(s) by which these polyanionic polymers activate the macrophage to tumoricidal capacity. Very few studies have been addressed to this question with the exception of those performed by Pratten et al. [35].

There are two general mechanisms for the pinocytosis of a substance by a macrophage. Fluid phase pinocytosis requires cellular energy and is represented by the uptake of I^{125}-polyvinyl pyrrolidine (PVP). Absorptive pinocytosis requires negligible amounts of cellular energy and is represented by the uptake of ^{198}-Au by the macrophage [35]. The rate and mechanism of pinocytosis is influenced by the serum content of the media used in the culturing of macrophages in vitro and the chemical nature and form of the substance that is pinocytized. Enhanced positive charge, increased hydrophobicity and the inclusion of pinocytic substrates into a liposome all enhance pinocytic uptake [35,36]. An inverse relationship has been observed on the effect of serum concentration and the mechanism of pinocytic uptake. Serum in the medium used to culture cells for the assessment of pinocytosis enhances fluid-phase pinocytosis and inhibits absorptive pinocytosis [37]. In addition, macrophages demonstrate a preferential uptake of higher molecular weight fractions of polyvinyl pyrolidine [38].

In vitro, ^{14}C-pyran and ^{125}-I labelled pyran were rapidly accumulated by rat peritoneal macrophages. The uptake of labeled

pyran is one hundred times more rapid than I^{125}-PVP. Uptake is inhibited when the reaction is performed at 4°C and when 2,4-dini-
trophenol, an inhibitor of cellular metabolism, is included in the
culture system. These data indicate that pyran was internalized and
not merely binding to the cell surface. In addition, the rate of
uptake was consistent with an adsorptive pinocytic mechanism which
may be receptor mediated [35].

Further studies on the uptake and localization of other polyanionic polymers into cells should provide us with a clearer understanding of the mechanism(s) by which these agents activate the
macrophage to become specifically cytotoxic to tumor cells.

POLYMER STRUCTURAL EFFECTS ON MACROPHAGE ACTIVATION

We have recently evaluated a series of polyanionic compounds
(Figure 4a,b) which differ in molecular weight, lipophilicity,
chain rigidity and surface charge for their ability to induce
macrophages to specific states of activation. There are generally
three levels of activation attributed to peritoneal macrophage
(1) normal/ resident/ unstimulated; (2) inflammatory/stimulated--
these cells are elicited with inflammatory agents such as thioglycollate and protease peptone; and (3) activated/tumoricidal--these
cells are elicited with agents such as C. Parvum, BCG and pyran.
These different states of macrophage activation may be distin-
guished by morphological, functional and biochemical criteria
[39,40]. In order to determine the level of activation which was
induced by our test polymers we evaluated peritoneal exudate cells
(PEC) which were elicited by intraperitoneal injection of test
polymers for their ability to kill Lewis lung tumor cells. In
addition, we performed ecotoenzyme analysis on lysates of these
cell populations. The ectoenzyme profile consisted of
5'-nucleotidase (5'-N), a marker enzyme for resident/unstimulated
macrophages, alkaline phosphodiesterase (APD), a marker enzyme for
thioglycollate/inflammatory macrophages and leucine aminopeptidase
(LAP), a marker enzyme for activated/tumoricidal macrophages [41].

The ability of macrophages elicited with test polymers to
become activated to tumoricidal capacity was evaluated in ^{3}H
release and morphological assays. Results from these experiments
are shown in Figure 5. PEC from mice which received 50 mg/kg test
polymers of pyran, or 17.25 mg/kg of C. Parvum, i.p., were
collected on day 7. PEC from mice receiving 1 ml of a 10% solution
of Brewers thioglycollate were collected on day 3. A generally good
correlation was obtained for the two assays of the test polymers.
The polymers poly(itaconic acid-alt-styrene) (IAS), poly(cyclohexyl-
1,3-dioxepin-alt-maleic anhydride) (CDA-MA), and poly(4-methyl-2-
pentenone-alt-maleic anhydride) (MP-MA) demonstrated the greatest
capacity for induction of tumoricidal macrophages in both assays.

NAME	APPROXIMATE MOLECULAR WEIGHT	STRUCTURE
ETHYLENE-MALEIC ANHYDRIDE COPOLYMER (E-MA)	~5,000	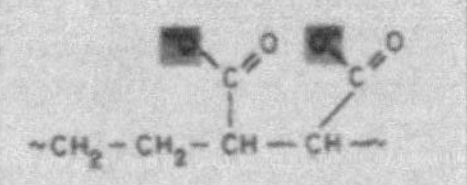
VINYL ACETATE-MALEIC ANHYDRIDE COPOLYMER (MAVA)	UM2 1,000 to 10,000 PM10 10,000 to 20,000	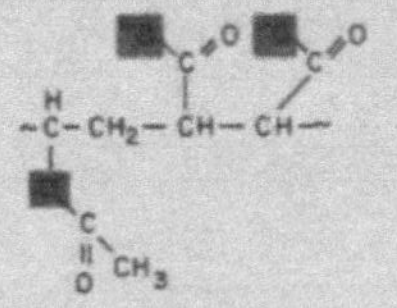
ALLYL-UREA-MALEIC ANHYDRIDE COPOLYMER (AV-MA)	~20,000	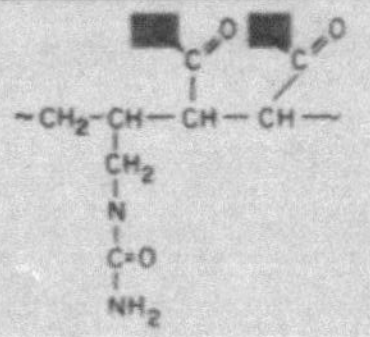
4-METHYL-2-PENTONYL-MALEIC ANHYDRIDE COPOLYMER (MP-MA)	~20,000	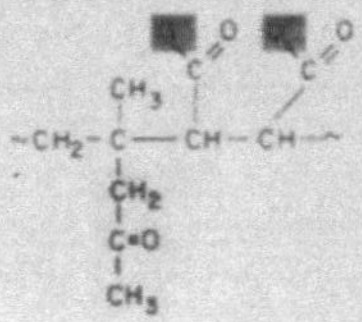

NAME	APPROXIMATE MOLECULAR WEIGHT	STRUCTURE
CYCLOHEXYL-1,3-DIOXEPIN-MALEIC ANHYDRIDE COPOLYMER (CDA-MA)	~30,000	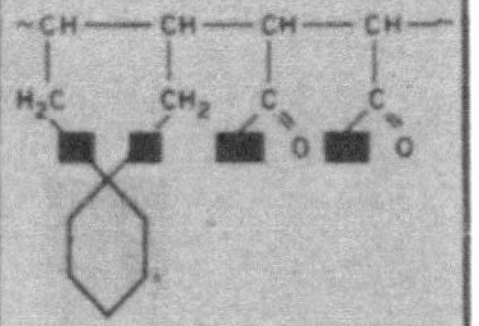
STYRENE-MALEIC ANHYDRIDE COPOLYMER (LS-MA)	~1500	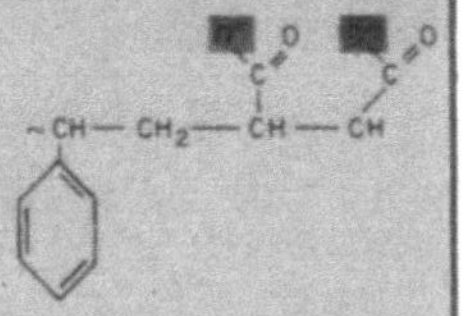
ITACONIC ACID-STYRENE-COPOLYMER (IAS)	UM2 1,000 to 10,000 PM10 10,000 to 20,000	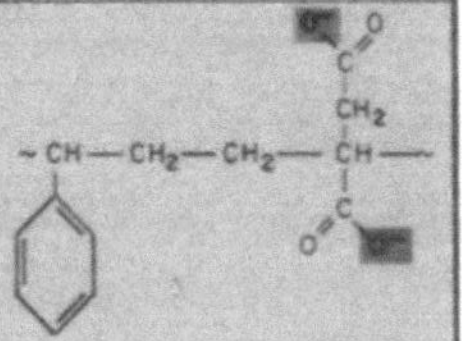
α-ETHACRYLIC ACID HOMOPOLYMER (EA)	~5,000	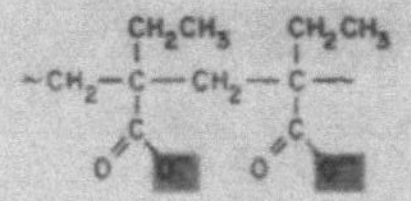

DENOTES ATOMS WITH POTENTIAL NEGATIVE CHARGE AT PHYSIOLOGICAL pH

DENOTES LIPOPHILIC GROUPS

FIGURE 4a. Lipophilic type polymers evaluated for macrophage activation.

FIGURE 4b. Polar type polymers evaluated for macrophage activation

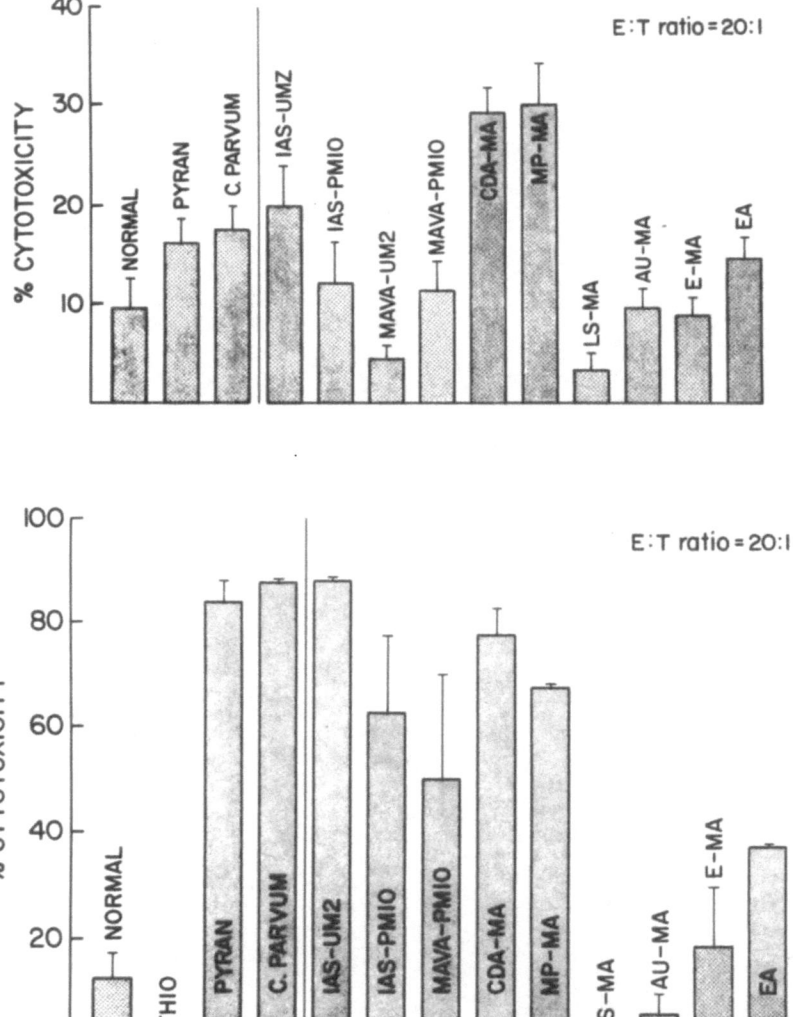

Figure 5. Introduction of cytotoxic macrophages with polyanionic compounds. Data represents the mean +SEM of triplicate experiments Macrophages were harvested seven days after administration of eliciting agents. A) tritium release assay; B) morphological assay.

Each of these polymers possess a lipophilic group, enhanced surface charge properties and chain rigidity.

The ectoenzyme profiles of test polymer elicited macrophages are shown in Figure 6. It can be seen that 5'-N activity is elevated in all the test polymer elicited-PEC populations compared to thioglycollate, pyran or C. Parvum elicited PEC. The 5'-N levels detected in CDA-MA and MP-MA elicited macrophages which also possessed enhanced tumor cytotoxicity, approached the level of resident populations. It may be concluded that the cell populations which are activated to tumoricidal capacity by test polymers do not possess the low 5'-N phenotype of the conventional activating agents, pyran and C. Parvum. Test polymers elicited PEC do not share the elevated APD phenotype of inflammatory populations. The LAP activity were not significantly enhanced in any "activated" populations. We have since stopped using this assay for the initial screenings of test polymer activities.

The elevated 5'-N activity observed in test polymer elicited macrophages also possess good tumoricidal activity. This may be attributed to the possibility that these agents are activating resident macrophage populations in situ and are not recruiting macrophages from peripheral monocyte pools, which is thought to occur upon administration of pyran [41]. These data indicate a fundamental difference in the mechanism of activation which occurs with these new polymers as compared to pyran.

Survival studies performed in mice which had received 2×10^4 Lewis lung cells subcutaneously indicate that CDA-MA and MP-MA are the most effective antitumor agents in vivo (Table V). CDA-MA enhanced survival at all dosages and MP-MA was most effective at 100 mg/kg. Mice which received either of these agents did not exhibit any overt toxic effects. IAS administration results in a reversal of the dose response. This is probably due to direct toxic effects of the compound. Only 2 out of 5 mice which received 100 mg/kg of IAS survived 3 days after administration. LS-MA was ineffective in prolonging the survival time of tumor innoculated mice. CDA-MA (50 mg/kg and 100 mg/kg) and MP-MA (100 mg/kg) were more effective in inhibiting tumor growth than pyran or C. Parvum at previously determined optimal doses.

These data point to basic differences between the cellular response and pharmacologic efficacy between the structurally modified anionic polymers and pyran. These data give precedence for the synthesis and evaluation of other polyanionic polymers.

TUMOR CELL CYCLE CHANGES EFFECTED BY "ACTIVATED" MACROPHAGES

All dividing cells progress through a well defined sequence of events during replication. These events are collectively termed the

Ectoenzyme Activity of Polymer Induced M∅

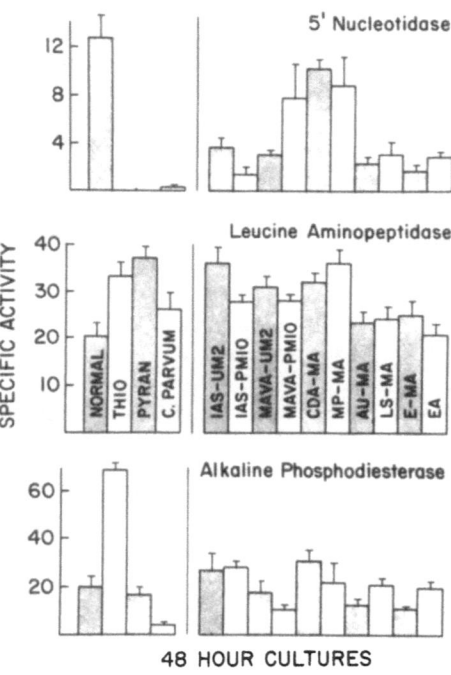

FIGURE 6. Ectoenzyme activity of polymer-induced macrophages.

TABLE V. Enhanced Survival of Test Polymer Treated Mice (%).

Dosage (mg/kg)	25	50	100
AGENT[a]			
CDA-MA	>130[b]	>160	>173
MP-MA	95	115	>158
IAS	>130	>120	> 80
LS-MA	102	107	98
PYRAN	--	>142	--
C. parvum[c]	>130		--

Mice received 2×10^4 Lewis lung tumor cells SQ in the right hind flank on day 0. Agents were administered i.p. at the indicated dosages on days 0, 1 and 2. Data are expressed as a function of mice which received tumor cells alone.

[a] CDA-MA: Poly(cyclohexyl-1,3-dioxepin-alt-maleic anhydride); MP-MA: poly(4-methyl-2-propenone-alt-maleic anhydride); IAS: poly(itaconic acid-alt-styrene); LS-MA: poly(styrene-alt-maleic anhydride).

[b] > Indicates groups where mice survived more than 90 days.

[c] C. parvum was administered at a dosage of 17.5 mg/kg.

cell cycle. The major phases of this cycle are depicted in Figure 7. During the normal replicative cycle cells will duplicate their DNA (S phase) before dividing (G_2M) thereby causing a doubling of their DNA content prior to division (diploid). After division each daughter cell is left with the normal compliment of DNA (haploid). Tumor cells do not abide by these rules of cellular division and may have a much higher concentration of DNA per cell than do normal cells. Tumor cells are therefore considered to be aneuploid.

It has been observed that when pyran activated macrophages were cultured with Lewis lung tumor cells, the tumor cells' DNA content was reduced by 50% but no tumor cell division took place. The decrease in DNA content resulted in tumor cell death as information essential for cell metabolism and/or replication was no longer available to the cell.

Cell cycle analysis of Lewis lung tumor cells cultured with pyran activated macrophages (Table VI) shows a shift in the cellular population with time from $G_2M \rightarrow S \rightarrow G_1$. Thus, the population is shifting from states of mitotic activity to less active, senescent states with increased time of incubation with activated macrophages. If the tumor cells are not dividing it is unlikely that they can do very much harm. Control studies indicate that normal or thioglycollate elicited macrophages to not cause this shift in tumor cell activity.

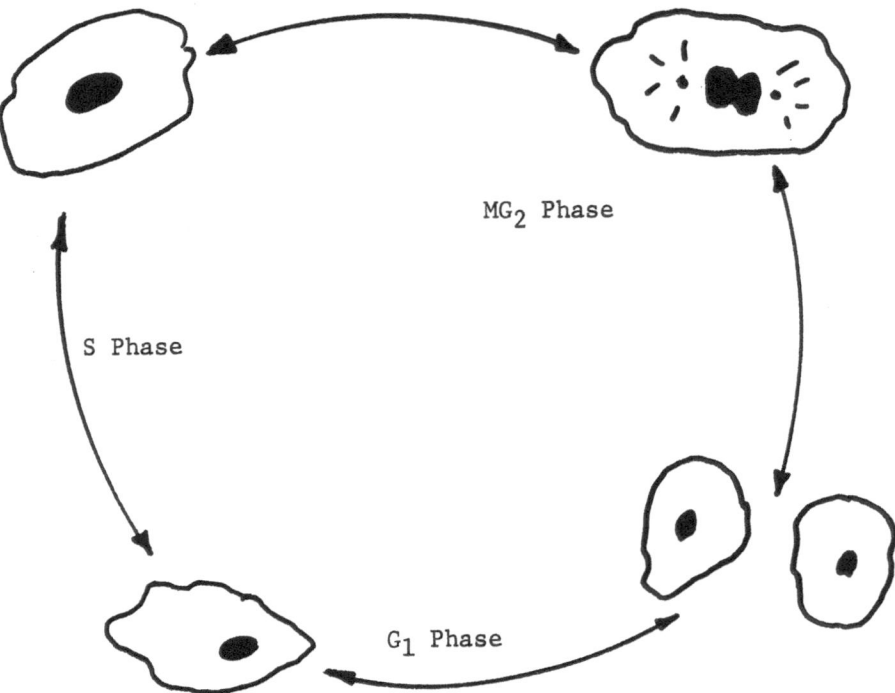

Figure 7. Phases of cell cycle. G_1 (resting) phase lasts about 24 hours. S (DNA synthesis) phase lasts about 8 hours. G_2M (mitosis) phase lasts about 2-3 hours.

TABLE VI. DNA Distribution of Lewis Lung Cell Cultivated With Pyran Activated Macrophages.

Lewis Lung Cultured[a] with AMØ for	Percent in Phase[b]		
	G_1	S	G_2M
0 hr	15	52	33
4 hr	16	52	32
8 hr	32	47	20
16 hr	62	34	4
24 hr	74	22	4

[a] Activated macrophage were obtained from mice injected i.p. with 25 mg/kg of pyran 7 days before peritoneal lavage.
[b] A pattern recognition program was used to determine G_1, S and G_2M phase.

In vivo studies illustrate a similar phenomenon (Table VII). Pyran treatment of mice resulted in a marked decrease in mitotic activity (G_2M) with a concommitant shift to the resting state (G_1). These results indicate that a possible mechanism of activated macrophage tumoricidal activity involves the induction of tumor cells with a reduced DNA content.

TABLE VII. Cell Cycle Analysi of Ehrlich Ascites Tumor Cells Before and After in Vivo Exposure to Pyran.

Days Following Tumor Inoculation[a]	Pyran Treatment[b]	Percent in Phase[c]		
		G_1	S	G_1M
2	-	20	58	22
2	+	98	2	0
6	-	36	47	17
6	+	98	2	0

[a] Mice were inoculated i.p. with 1×10^6 EA cells and killed on various days thereafter.

[b] Pyran was injected i.p. (23 mg/kg) 2 days prior to tumor inoculation.

[c] Cell cycle analysis of peritoneal EA cells was determined with a cell sorter utilizing a pattern recognition program.

Acknowledgments

The authors wish to thank the NIH for grant Al-15612 which supported this work in part as well as the MCV/VCU Cancer Center. We also wish to thank Jeff Jones and Gene Dunaway for helping prepare this manuscript.

REFERENCES
1. L. G. Donaruma, R. M. Ottenbrite and O. Vogl, eds., "Anionic Polymeric Drugs," John Wiley and Sons, New York (1980).
2. R. M. Ottenbrite, The antitumor and antiviral effects of polycarboxylic acid polymers, in: "Bilogical Activites of Polymers, "C. E. Carraher, Jr. and C. G. Gebbelein, eds., American Chemical Society Symposium Series 186 (1982).
3. K. Kuus, R. M. Ottenbrite and A. M. Kaplan, Fed. Proc. 42:3364 (1982).
4. G. B. Butler, J. Poly. Sci. 48:279 (1960).
5. D. S. Breslow, Pure Appl. Chem. 46:103 (1976).
6. T. C. Merigan, Nature 214:416 (1967).
7. T. C. Merigan, New Engl. J. Med. 277:1283 (1967).
8. T. C. Merigan, "Ciba Foundation Symposium on Interferon," G. W. Wolstenholme, M. J. O'Conner and A. Churchill, eds., London, p 50 (1967).
9. E. DeClereq and T. C. Merigan, Arch. Intera. Med. 126:94 (1970).
10 S. J. Mohr, M. A. Chirigos, F. S. Fuhrman, J. W. Pryor, Cancer Res. 35:3750 (1975).

11   P. S. Morahan and A. M. Kaplan, Int. J. Cancer 17:82 (1976).

12. W. Regelson, Adv. Exp. Med. Biol. 1:315 (1967).

13. T. C. Merigan, M. S. Finkelstein, Virology 35:363 (1968).

14. E. DeClereq and T. C. Merigan, J. Gen. Virol. 5:359 (1969).

15. W. Regelson, A. Munson, W. Wooles, "International Symposium on Standards of Interferon and Interferon Inducers," London (1969); "Symposium Series Immunobiological Standards," vol. 14, Karger, Basel, New York (1970), pp. 227-236.

16. F. F. Pindak, Infec. Immun. 1:217 (1970).

17. D. J. Givon, J. P. Schmidt, R. J. Ball and F. F. Pindak, Antimicrob. Agents and Chemother. 1:80 (1972).

18. J. Y. Richmond, Infec. Immun. 3:249 (1971) and Arch. Ges. Vi. Rusfersch. 36:232 (1972).

19. G. B. Schuller, P. S. Morahan and M. J. Snodgrass, "Tenth National Meeting of the Reticulo Society," abstract 28 (1973).

20. C. H. Campbell and J. Y. Richmond, Infec Immun. 7:199 (1973).

21. W. Regelson, A. E. Munson, Ann. N.Y. Acad. Sci. 173:831 (1970).

22. Y. Shamash, B. Alexander, Biochim. Biophys. Acta. 1:449 (1969).

23. M. A. Kapusta and J. Mendelson, Arthitis Rheum. 12:463 (1969).

24. D. W. Baxter, M. W. Rosenthal and A. Lindenbaum, "Abstracts of the Twenty-first Annual Meeting of the Radiation Research Society," St. Louis, MO, April 29, 1973; N. E. Egan, G. S. Kalesperus, E. S. Moretti and J. J. Russel, "Annual Report," Division of Biological Medical Research, Argonne National Laboratory (1972), pp 121-125.

25. W. Regelson, Biologically Active Water-Soluble Polymers, in: "Polymer Science and Technology," vol. 2, N. M. Bikales, ed., Plenum Press, New York (1973), pp 161-177.

26. T. J. Leavitt, T. C. Merigan, J. M. Freeman, Am. J. Dis. Child. 121:43 (1971).

27. A. E. Munson, White, P. Klykken, Cancer Res. 16:329 (1981).

28. Daniel N. Lapedes, Editor-in-Chief, "Dictionary of Scientific and Technical Terms," McGraw Hill Book Co., New York (1974), p 1257.

29. R. Ottenbrite, E. Goodell and A. Munson, Polymer 18:461 (1977).

30. A. M. Kaplan, P. S. Morhan and W. Regelson J.N.C.I. 54:989 (1975).

31. R. P. Harml and B. Zbar, JNCl 54:989 (1975).

32. G. B. Pierce, R. Shikes and L. M. Fink, in: "Cancer," Prentice Hall (1978).

33. H. C. Pitot, in: "Fundamentals of Oncology," Marcel Dekker, Inc. (1981).

34. J. D. Papamatheakis, R. M. Schultz, M. A. Chrigos and J. G. Massicot, Cancer Treat. Rept. 62:1845 (1978).

35. M. K. Pratten, R. Duncan, H. C. Cable, R. Schhe, H. Ringsdorf and J. B. Lloyd, Chem. Biol. Inter. 35:319 (1981).

36. M. K. Pratten and J. B. Lloyd, Biochem. J. 180:567 (1979).

37. T. Koolstra, M. K. Pratten and J. B. Lloyd, Bioscience Repts. 1:587 (1981).

38. M. K. Pratten, P. C. Millard and J. B. Lloyd, Mioscience Repts. 1:125 (1981).

39. M. L. Karngusky and J. K. Lazdins, J. Immunol. 121:809 (1978).

40. I. Carr, Clin. Invest. Med. 1:59 (1978).

41. P. S. Morahan, P. J. Edelson and K. Gass, J. Immunol. 125:1312 (1980).

42. R. vanFurth and Z. A. Cohn, J. Exp. Med. 128:415 (1968).

43. M. S. Melter, R. W. Tucker and A. C. Breuer, Cell. Immunol. 17:30 (1975).

POLYMERIC ANTITUMOUR AGENTS ON A MOLECULAR AND ON A CELLULAR LEVEL?

Leo Gros and Helmut Ringsdorf

Fachhochschule Fresenius, Dambachtal 20
D-6200 Wiesbaden, FRG and
Inst.f.Organ.Chemie, Johannes Gutenberg-Universität
J.-J.-Becher-Weg 18-20, D-6500 Mainz, FRG

INTRODUCTION

Polymers may be pharmacologically active as such. If used as carriers, they may, due to their intrinsic properties, influence body distribution, excretion or cell uptake of the drugs they carry. Considering the differences between normal and malignant cells, and taking into account several problems with low molecular weight drugs, polymeric antitumour agents have been designed and tested (1). The according model considerations (2) and a few examples will be discussed, with special emphasis on endocytosis of polymeric drugs and affinity chemotherapy. The cytostatic action of drugs of this type is ultimately a chemical effect upon cell metabolism or cell propagation on the molecular level.

Our own body, however, disposes of a perfect biological system for affinity therapy - the immune response. It is fascinating to observe what happens to a tumour cell attacked by a lymphocyte. Obviously, cell death caused by lymphocyte contact is ultimately due to a destruction of the cell membrane.

Once the cell has lost its compartment-forming structure which separated it from the environment, it can no longer maintain its internal medium and is thus killed. Substances inducing such a process may be called antitumour agents acting on a cellular level (3). Is there a chance to mimick such a process? Criteria for the design of polymeric antitumour agents on a cellular level will be discussed. First attempts to realize this concept include the synthesis of simple, stable cell models. Recent results in the

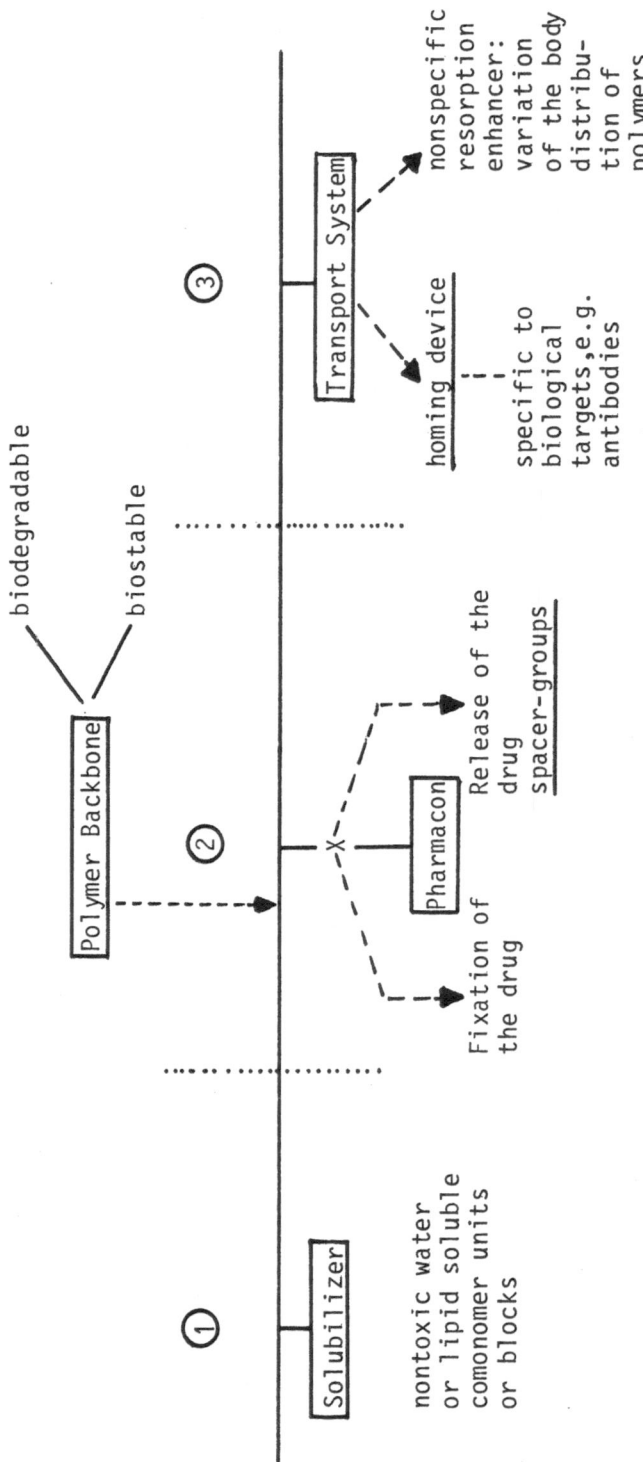

Fig.1: Model for pharmacologically active polymers

synthesis of such models stabilized by polymerization of synthetic amphiphiles in liposomal structures will be presented. Considerations and first experimental experience concerning recognition and the "cork-like" opening of such cell models capable of carrying antitumour agents will be explained.

POLYMERIC ANTITUMOUR AGENTS ON A MOLECULAR LEVEL - A MODEL, THREE EXAMPLES AND A CONCLUSION

A model of pharmacologically active polymers (2) was based on the idea that a combination of desired properties could be achieved by linking the appropriate structural units along a polymer chain (Fig. 1).

Taken into account some possible polymer specific effects of pharmacologically active polymers (4), one can conclude that such drugs may be expected to lead beyond the horizon of mere "me-too-drugs", i.e. that they offer other or more possibilities than do low molecular weight drugs. Since a review of the whole field is far beyond the scope of this article, three examples will illustrate this.

The first example is a so called lysosomotropic (5) carrier polymer(6). There are tumour cells which are resistant to methotrexate (MTX), a folate antagonist widely used in tumour chemotherapy. It was assumed that these cells lack the appropriate transport system for MTX. If the drug is covalently bound to poly-L-or D-Lysin, it is readily internalized by resistant and non-resistant tumour cells via endocytosis (a process similar to engulfment of food by amoebae). Yet, only the L-polymer acts cytostatically. In contrast to the D-polymer, it can be degraded in the lysosomes of the cell - the place where endocytosed material goes. So the "lyso-somotropic vector" poly-L-Lysin is a good example for the ability of many polymers to enter cells via endocytosis and for the importance of polymer stereochemistry.

The phenomenon of endocytosis leads us to a second example: extensive and systematic studies on endocytosis using an in vitro system (7) include the evaluation of polymers containing different peptide spacers (cf Fig. 1) which link iodotyrosine as a drug model to poly(2-hydroxypropyl-methacrylamide) (8). This study shows that by variation of peptide spacer composition the rate of intracellular drug release caused by lysosomal hydrolysis of the spacer can be controlled.

The third example is an approach to affinity chemotherapy (9) using a <u>homing device</u> (cf. Fig. 1). This polymer shown in Fig. 2 combines a <u>solubilizer</u> (a), an antitumour agent (b), and an immunoglobulin from the blood of rabbits treated with mouse lymphoma cells (c). This system showed better antitumour effect against mouse lymphoma than the appropriate control systems. Although there is some doubt whether this is a true homing or a synergistic effect, this polymer is a synthetic example of the model shown in Fig. 1.

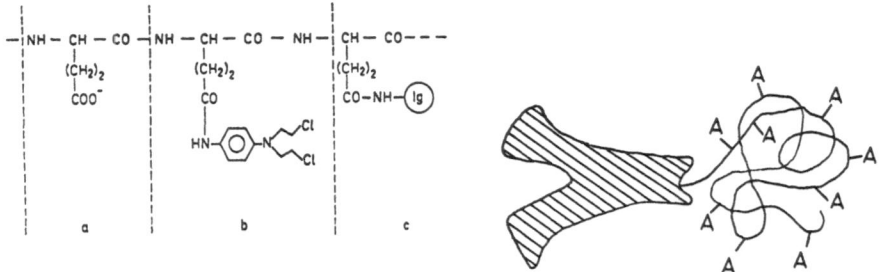

Fig.2 left: Polymeric, possibly tumour specific, alkylating anti-
            tumour agent (cf. model Fig. 1)
            a) Solubilizer, poly-(glutamic acid)
            b) pharmacon, p-phenylene diamine-Lost (A)
            c) homing device, immunoglobulin
     right: schematic picture of polymer containing A, linked to
            Y-shaped antibody molecule

Even though polymeric antitumour agents on a molecular level are far from being used clinically, the model discussion and a variety of synthetic approaches are under investigation and are improved by systematic studies. A better understanding of tumour cells and of in vivo behaviour of polymeric drugs are needed. Recent progress in the production of tumour cell specific antibodies may help to improve the homing effect.

POLYMERIC ANTITUMOUR AGENTS ON A CELLULAR LEVEL - A CONCEPT AND FIRST STEPS TO REALIZE IT

Our own body disposes of a perfect homing system: immune cells specifically recognize and destroy foreign material as well as tumour cells. Although these may escape the immune surveillance, it is fascinating to watch them being attacked by lymphocytes (a class of immune cells).

Fig. 3 shows a T-lymphocyte (small cell) before and after contact to a tumour cell. After membrane contact with the lymphocyte, the tumour cell membrane forms protrusions ("bubbles") and is destabilized. This is a perfect example of tumour cell death on a cellular level.

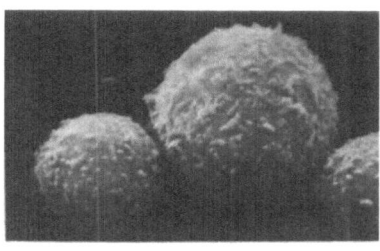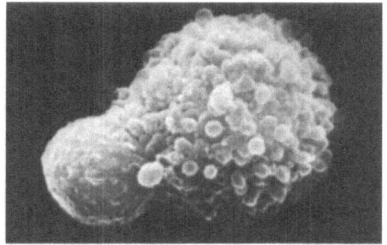

Fig. 3: Electron microscope pictures showing attack of T-Lymphocyte on tumour cell (10).
   a) Tumour cell (right) bears antigens on its surface which
      are selectively recognized by activated lymphocyte
      (smaller cell, left) via membrane receptor molecules.
   b) Death of a tumour cell, after membrane contact with
      lymphocyte; "bubbles" (membrane protrusions) show
      destabilization of membrane

Without discussing possible mechanisms of this process, one can
think of how to successfully mimic it. The simplest model of a
cell membrane is a synthetic lipid vesicle of liposomes (Fig. 4),
a bilayer ball-shaped structure formed by amphiphilic lipid molecules. If we want to use such a device as an antitumour agent on
a cellular level, it has to meet three main requirements:
1. membrane stability similar to or higher than that of a tumour
cell membrane, 2. presence of a cell destroying principle, 3. presence of a tumour cell specific recognition unit.

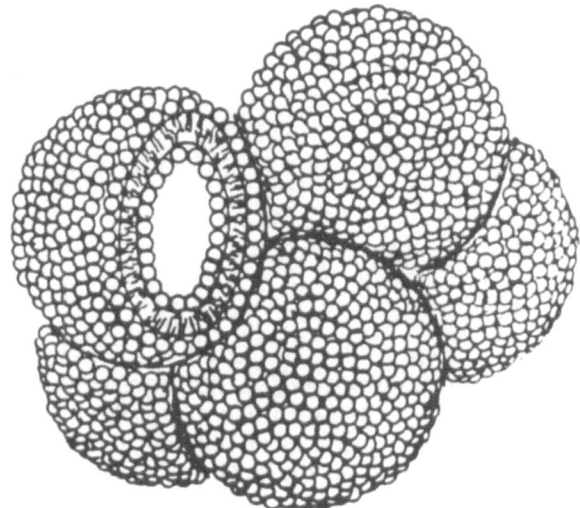

Fig. 4: Schematic representation of unilamellar vesicles
        (consisting of one lipid bilayer)

A normal liposome does not meet the first requirement: when
interacting with a tumour cell, the vesicle will be destroyed, not
the cell. Similarly, a liposome carrying membrane destroying agents
would itself be destroyed. It is therefore essential to increase
liposome stability - a first goal which seems accessible from the
point of view of a polymer chemist: if we synthesized lipid analogues containing polymerizable units and if these formed liposomes,
we should be able to prepare polymerized liposomes with increased
mechanical and biological stability.

Starting from this idea, a considerable number of polymerizable
lipids have been synthesized and investigated (11) (3). When spread
on a water surface in a Langmuir trough (a film balance used to
measure force-area relationships of monolayers of amphiphilic substances), they behave like natural lipids, forming different types
of monomolecular layers. Upon UV-irradiation, they polymerize in
the monolayer retaining the ordered structure (Fig. 5a). What can
be done in a monolayer also works in liposomal solutions (Fig. 5b):

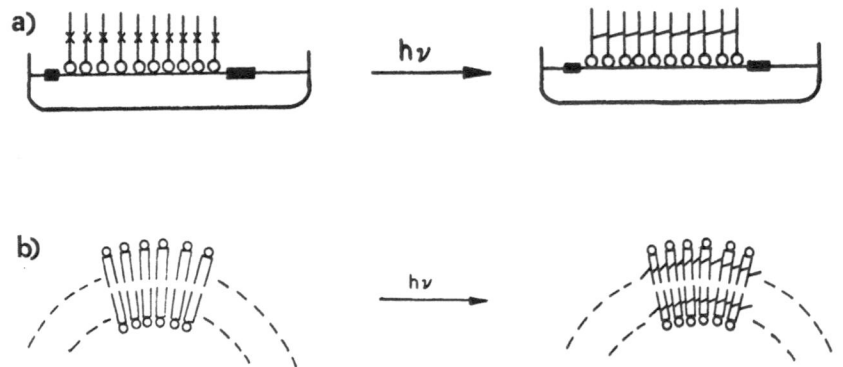

Fig.5: Schematic representation of polymerization of monomeric
       lipids
       a) in monolayer
       b) in liposomal solution

When the polymerizable lipids are sonicated in aqueous suspension,
they form liposomes. Upon UV-irradiation, clear solutions of polymeric liposomes are obtained. Scanning electron micrographs show
that the structure is retained and that the resulting polymeric
liposomes are mechanically stable under the high vacuum needed for
this technique. In contrast to monomeric liposomes, they are remarkably stable against organic solvents such as ethanol and against
detergents like sodium dodecyl sulfate. Thus, stable liposomes can

be synthesized which perhaps meet the first requirement mentioned above. Yet, the release of a cell destroying principle requires some mechanism which we might compare with the uncorking of a wine bottle - i.e. a destabilizable domain in the polymeric membrane. (Fig.6)

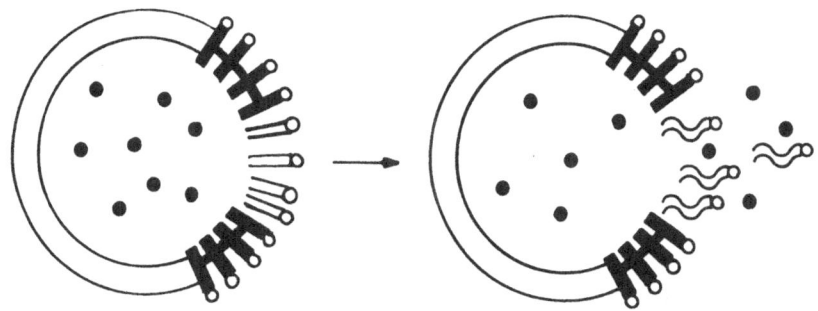

Fig. 6: Destabilization of labile domain in polymeric liposome
leading to release of entrapped substance ("cork-model")

Besides photochemical, thermal or pH-dependent opening mechanisms, one could use enzymic cleavage of lipid molecules for this purpose. To demonstrate that this effect can be achieved, mixed monolayers consisting of polymerized lipids with domains of natural lipids were spread on a Langmuir trough. After injection of Phospholipase A, an enzyme cleaving lipids and thus destroying the amphiphilic nature of the molecules, a decrease of monolayer area at constant pressure indicated that the natural lipid domains were lysed. The resulting cleavage products are water soluble and dissolve in the aqueous subphase. Within a few minutes, the area of a 1:1 mixture of polymerized and natural lipids decreases to 50% of the initial value. Without discussing the nature of possible cell destroying principles (12), one can say that these systems (13) present a tool for controlled release of substances from stable liposomes.

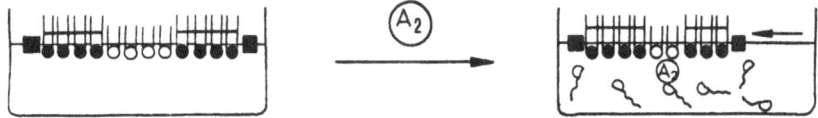

Fig.7: Schematic representation of the enzymatic hydrolysis of the
natural lipid ($\rightleftharpoons$⊙) in mixed monolayers with polymeric
lipid ($\rightleftharpoons$●) (13)

Finally, the targeting of liposomal antitumour agents on a cellular level requires recognition units. As a simple and readily accessible model system, we chose the interaction of lectins (plant proteins) which have specific recognition sites for certain sugars. The well known lectin Concanavalin A is able to bind four

units of D-Glucose or D-Mannopyranose per molecule of Con A.
Lipids with the appropriate sugars as headgroups and with polymerizable units in the hydrophobic chain were synthesized, and liposomes
from these monomers were formed and polymerized. Upon addition of
Con A to the liposome solution, a precipitate formed that contained
intact liposomes (as controlled by e.m.). If low molecular weight
sugars which can compete with the liposomes for the binding sites
of Con A were added, the liposomes re-dissolve (Fig. 8). Thus,
in the case of a model system, it could be shown that recognition
of polymeric liposomes by specific protein structures can be
achieved (14).

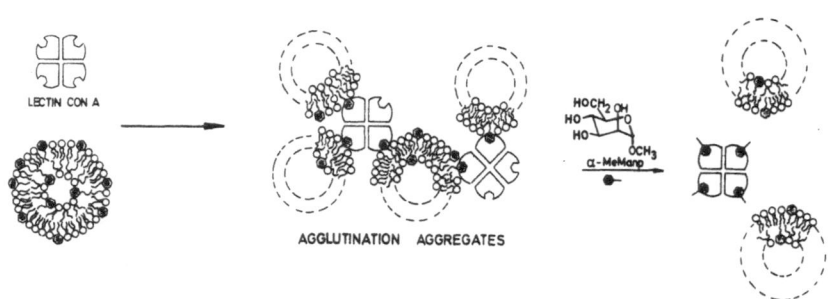

Fig. 8: Principle of recognition of sugar-containing liposomes by
        Con A. Agglutinated liposomes (middle) re-dissolve when
        ∝ -Methylmannopyranoside (∝⁻Me Manp) is added (14).

     Although only first attempts have been made, and though the
hypothetic concept of polymeric antitumour agents on a cellular
level is far from being realized, it seems worth while to further
investigate the potential of polymeric liposomes and to try to get
closer to the structures of which they are a simple model:
to cells (Fig. 9).

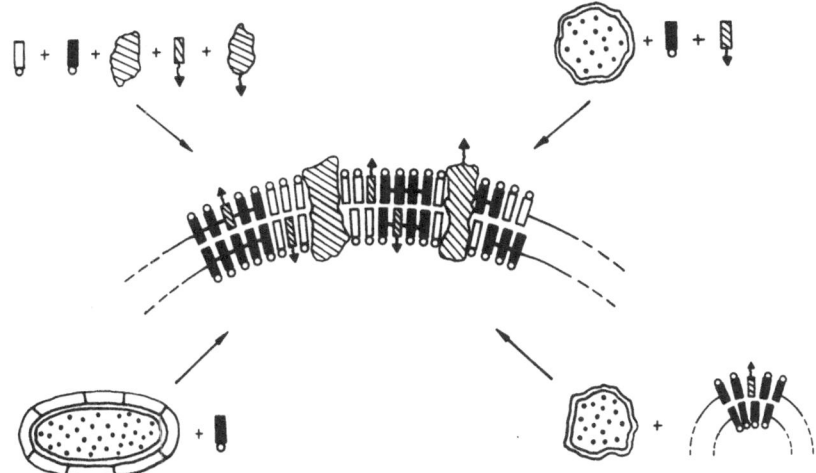

Fig. 9: Schematic representation of few possible ways to build up
stabilized cell models: synthesis of membranes from
single components (upper left) (15);
insertion of lipids into cell membranes via hemolysis (16)
(upper right); fusion of cells with monomeric liposomes
and subsequent polymerization (lower right); metabolic
uptake of fatty acid by bacteria (lower left) (17)

References

(1) D.S.Zaharko, M. Przybylski, V.T. Oliverio, Methods Cancer Res.
16, 347 (1979)
(2) H.Ringsdorf, J.Polym.Sci., Polym.Symp. 51, 135 (1975)
(3) L. Gros, H.Ringsdorf, H. Schupp, Angew.Chem.Int.Ed.Engl. 20,
305 (1981)
(4) discussed in more detail in (1) and (3)
(5) C. de Duve, Th. De Varsy, B. Poole, A. Trouet, P.Tulkens,
F. van Hoof, Biochem. Pharmacol. 23, 2495 (1974)
(6) W.C. Shen, H.J.P.Ryser, Mol.Pharmacol. 16, 614 (1979)
(7) R.Duncan, J.B. Lloyd, Biochim.Biophys.Acta 544, 647 (1978)
(8) R.Duncan, P.Rejmanova, J. Kopeček, J.B. Lloyd, Biochim.
Biophys. Acta 678, 143 (1981)
(9) M: Wilchek, Makromol.Chem.Suppl. 2, 207 (1979)

(10) L.J. Old, Scientific American 236 (5), 62 (1977)

(11) A. Akimoto, K. Dorn, L. Gros, H. Ringsdorf, H. Schupp, Angew.Chem.Int.Ed.Engl. 20, 90 (1981)

(12) This is done in ref. (3)

(13) R. Büschl, B. Hupfer, H. Ringsdorf, Makromol.Chem. Rapid Commun. 3, 589, (1982)

(14) H.Bader, H.Ringsdorf, J. Skura, Angew.Chem.Int.Ed.Engl. 20, 91 (1981) and H. Bader, Diploma Thesis Mainz 1981

(15) N. Wagner, K. Dose, H. Koch, H. Ringsdorf, FEBS Lett. 132, 313 (1981)

(16) P. Scheurich, U.Zimmermann, M. Mischel, J. Lamprecht, Z.Natur- forsch. C 35, 1081 (1980)

(17) D.S. Johnston, S. Sanghera, M. Pons, D.Chapman, Biochim. Biophys. Acta 602, 57 (1980)

UTILIZATION OF STABILIZED FORMS OF POLYNUCLEOTIDES

Hilton B. Levy and Freddie L. Riley

Laboratory of Viral Diseases
National Institute of Allergy and Infectious Diseases
National Institutes of Health
Fort Detrick, Frederick, Maryland 21701

Within a few years after the description of interferon by Isaacs and Lindemann (1957), it was recognized that interferon potentially was a broad spectrum antiviral agent of possibly high value in clinical medicine. However, the difficulty of preparing enough interferon, either non-human or human, prevented the adequate testing of this potentiality. Serious efforts were made to find non-replicating agents that would cause the host to synthesize his own interferon in large quantity. While a number of compounds were found that are capable of causing mice, and possibly humans, to make interferon, they either induced too small amounts or were too toxic (Merigan, 1973). Fields et al. (1967) reported that a number of natural and synthetic double-stranded (d.s.) RNAs are capable of inducing interferon. In particular the d.s. RNA polyinosinic-polycytidylic acid (poly I·poly C) was highly effective in rodents as an interferon inducer (Field et al., 1967), as an antiviral agent (Parks and Baron, 1968; Worthington et al., 1973) and as an antitumor agent (Zelezinck and Bhuyan, 1969; Levy et al., 1969).

Preclinical toxicity studies lead to phase I trials in cancer patients in two studies (Robinson et al., 1976; Young, 1971). Toxicity was very mild but there were only very low levels of interferon induced, and no antitumor action was found. In monkeys and in chimpanzees no interferon was induced by poly I·poly C (Levy and Gibbs, unpublished results).

Like most d.s. RNAs poly I·poly C is pyrogenic in rabbits. Figure 1 shows the fever response curve of a rabbit injected with poly I·poly C. If poly I·poly C is incubated with human serum (or ribonuclease), the ability of the compound to elicit fever in rabbits

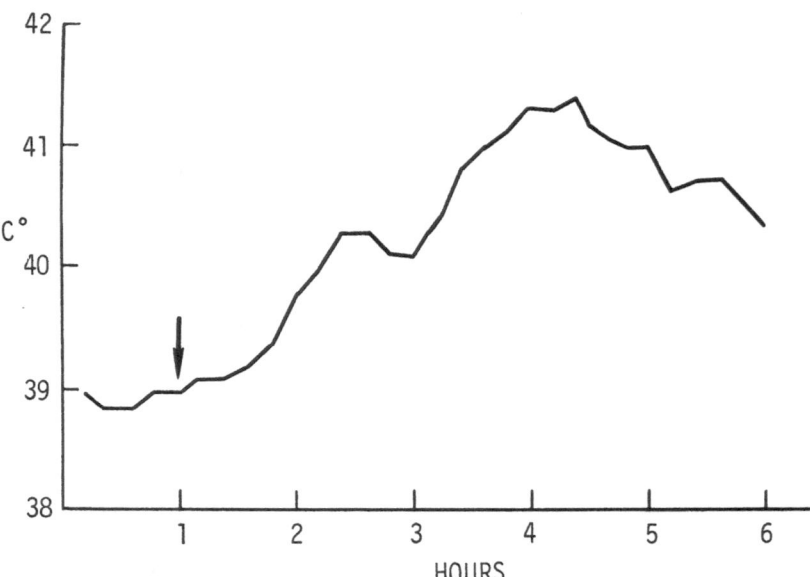

Fig. 1.   Fever response in rabbits (mean of 3) after i.v. injection
of 30 µg poly I·poly C in 0.17 ml of 0.15 M pyrogen-free saline.

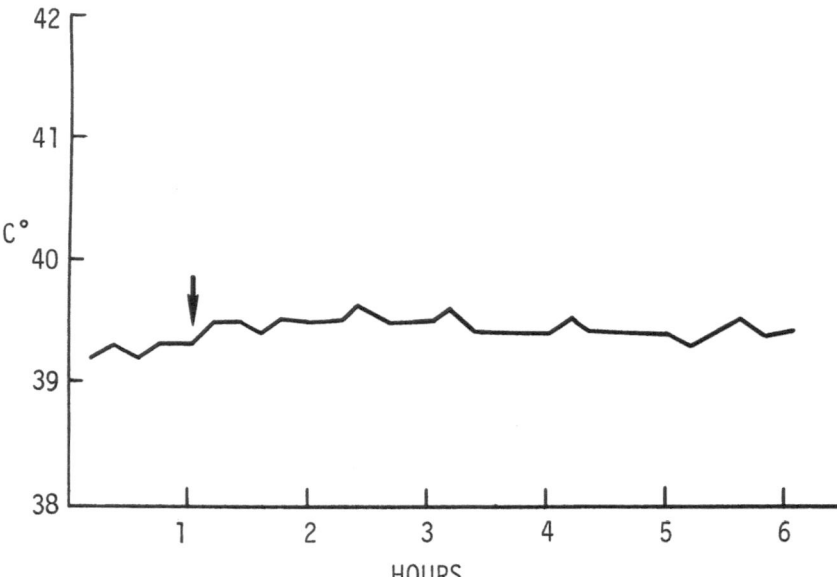

Fig. 2.   Fever response in rabbits (mean of 3) after i.v. injection
of 30 µg poly I·poly C in 0.15 ml of 0.15 M pyrogen-free saline that
had been treated with 20 µg/ml pancreatic ribonuclease plus 30 units/
ml T_1 ribonuclease.

is lost, as shown in Figure 2. Associated with this loss of pyro-
genicity is a loss of interferon-inducing capacity (Nordlund et al.,
1970). The d.s. RNA is hydrolyzed to smaller oligonucleotides, including acid soluble products.

The ability to hydrolyze poly I·poly C is much greater in human
serum than in mouse serum. In general, animal species that have high
hydrolytic capacity are poor responders to poly I·poly C, and good
responders have low hydrolytic capacity. While these observations
are consistent with the idea that there is a cause and effect relationship between hydrolytic capacity and interferon production,
they certainly do not establish such a cause and effect relationship.
One of the points that will be emphasized in this review deals with
the frequency with which hydrolysis resistance of a polynucleotide
is associated with its ability to induce interferon in primates.

In general, two approaches have been used to prepare polynucleotide inducers that are at least partially protected from hydrolysis.
One involves making the d.s. RNA from modified nucleotides with the
hope that such a d.s. RNA will be more resistant to hydrolysis. The
other is to complex the d.s. RNA with another substance with the same
goal in mind. Examples of both approaches follow.

Thiolated derivatives

There is a group of sulfur-containing d.s. RNAs that are more
resistant to hydrolysis than is plain poly I·poly C. They are of
two types. One has the sulfur as a side chain in the pyrimidine
ring, and the other has the sulfur substituting for one of the oxygens
in the phosphate backbone, yielding thiophosphate groups. The
general conclusion from these studies is that such substitution does
not appear to offer any significant clinical advantage over non-
thiolated compounds, as discussed below.

Thioketo substitution in the 2 position in uracil or cytosine
in poly U or poly C greatly increased the resistance to hydrolysis
of the d.s. RNA made by subsequent complexing with poly A or poly I,
respectively (Reuss et al., 1976). In general, thioketo substitution
also raised the T_m over that of the non-substituted polynucleotide.
There was, however, no regular alteration in the amount of interferon
induced as a result of such changes. There were both increases and
decreases, depending on 1) the test system used, tissue culture cells
in vitro or rabbit or dogs in vivo, and 2) the particular chemical
complexes studied.

In a somewhat related study, poly C with from 1 to 10% of the
poly C substituted by 5-mercaptocytosine was prepared and annealed
with poly I. In general, the larger the amount of substituted base,
the lower was the T_m. Also, in general, the poorer was the interferon

induction in tissue culture. On hydrolysis with ribonuclease, the thiolated compounds yielded larger fragments (O'Malley et al., 1975).

Somewhat more encouraging results were obtained at first with compounds in which each phosphate was replaced by a thiophosphate. Such thiophosphate derivatives were 0.1 to 0.01 as sensitive to hydrolysis by various enzymes and were significantly better as interferon inducers both in tissue culture and in rabbits than were the unsubstituted compounds (DeClercq et al., 1969). Later studies, however, showed that greater in vivo toxicity of the thiophosphates minimized any potential clinical advantage they might have offered (Merigan, personal communication). There are several difficulties inherent in all the studies with sulfur containing polynucleotides. It is apparent that many specific structural factors are important in determining whether a given compound will induce interferon in any test system (DeClercq, et al., 1969). When one substitutes a sulfur for an oxygen, one changes the chemical specificity as well as resistance to hydrolysis and thermal denaturation temperature (T_m). It is not easy to distinguish which of these factors may be responsible for any increase or decrease in the interferon-inducing capacity of the modified complexes. In addition, increased resistance to hydrolysis becomes important from a clinical viewpoint only when dealing with certain animal species, particularly primates. Primate sera have a relatively high hydrolytic capacity for d.s. RNA. Poly I·poly C is a good interferon inducer in rodents, but not in primates [see section of poly(ICLC)]. In none of the studies on the sulfur containing analogues was in vivo interferon-inducing capacity in primates tested.

DEAE-dextran

A different approach to enhancing the action of d.s. RNA was taken by Dianzani et al. (1968). It has previously been observed that treatment of cells with polybasic substances, such as diethylaminoethyl dextran (DEAE-Dx) enhanced the infectivity of viral RNA, presumably by facilitating the uptake of the infectious RNA. Working on the hypothesis that cellular uptake of poly I·poly C is necessary for it to be effective as an interferon inducer, Dianzani et al. treated mouse L-cells with 10 µg/ml of poly I· poly C with and without 400 µg/ml of DEAE-Dx added immediately prior to the inducer. The L-cells did not respond to 10 µg/ml of plain poly I·poly C, but those cultures that had DEAE-Dx in addition to the poly I·poly C yielded 300 units of interferon per ml. The mechanism of this enhancement, however, appears to be complex. One can prepare a complex of poly I·poly C with DEAE-Dx simply by mixing the two in proper proportions. Such a complex is more resistant to hydrolysis by RNase than is plain poly I·poly C. When this dextran complex is administered to mice it elicits 10 times more serum interferon than does poly I·poly C. It is also more effective in L-cells than is the uncomplexed d.s. RNA.

On the basis of these data it would appear that the enhanced resistance to hydrolysis may play an important role in increasing the effectiveness of poly I·poly C as an inducer. However, treatment of L-cells with DEAE-Dx before adding the stabilized inducer leads to still higher effectiveness. That the L-cell membranes are modified by treatment with DEAE-Dx is indicated because such treated cells are extremely resistant to swelling by water, and to disruption with a Dounce homogenizer (Levy, Riley and Dianzani, unpublished). It could indeed be that the DEAE-Dx-treated cell does allow uptake of the inducer, but it could also be that DEAE-Dx is bound at the surface of the cell and gives additional protection against nucleases on the cell membrane. Dextran sulfate, an anionic polymer that rapidly combines with DEAE-Dx, a cationic polymer, very rapidly removes the enhancing action of DEAE-Dx on the cell. However, this observation does not sharply distinguish between the two possible explanations--enhanced uptake or increased resistance to hydrolysis.

A number of basic substances in addition to DEAE-Dx, including neomycin, protamine, histone and methylated albumin all have been shown to enhance the action of poly I·poly C in tissue culture and in some instances in mice (Billiau et al., 1970). One problem with all these preparations, however, is that they are very insoluble ($\leq 10^{-5}$ M), and yield precipitates at higher concentrations. Not surprisingly such complexes act as better antiviral agents than uncomplexed poly I·poly C, both in tissue culture and in mice. They have not been studied for toxic effects, nor have they been tested in primates.

Liposomes

A procedure which augments the induction of interferon by poly I·poly C is the inclusion of the d.s. RNA into liposomes (Straub et al., 1974). Both negatively and positively charged liposomes were toxic to L-cells and, therefore, their effectiveness could not be determined. However, when given to mice, both types increased interferon production by poly I·poly C by 5- to 15-fold, with positive liposomes being somewhat more effective than negative ones. Radioactive tracing techniques revealed that both the lipid and their poly I·poly C component were rapidly cleared from the blood stream and were concentrated in the liver. The rate of degradation to acid soluble components of poly I·poly C in liposomes was slower than that of unprotected poly I·poly C. The increased inducing activity, therefore, might have been due to increased cellular uptake as well as greater resistance to hydrolyses. No studies of in vivo toxicity were reported, nor were there reports of studies in primates.

Stabilization with poly-lysine

Rice et al. (1970) showed that the addition of poly-D-lysine (M.W. ca 10^5 daltons) to poly I·poly C (6S size) increased the interferon-inducing capacity in mice by about 5- to 10-fold. However, the poly-D-lysine complex with this low molecular weight poly I·poly C was not significantly better than plain poly I·poly C made with 9S homopolymers. When poly-D-lysine was added to 9S poly I·poly C, a precipitate was formed at all concentrations greater than about 10^{-5} M.

Eight years before the role of d.s. RNA in interferon induction was appreciated, Felsenfeld and Huang described the formation of a stable complex between a d.s. RNA (A+U) and polylysine (Felsenfeld and Huang, 1970). Tsuboi and his associates (1966) later studied extensively the interaction of poly-L-lysine with poly I·poly C. These latter studies were done in low salt at very low concentrations of poly I·poly C (ca 5.0 x 10^{-5} M) and poly-L-lysine (ca 2.25 x 10^{-5} M). The complex was formed at room temperature. The poly-L-lysine used in these complexes was of a high molecular weight (60-90,000). These investigators reported the binding reaction of poly-L-lysine to polynucleotide to be quantitative, irreversible and with a definite stoichiometric ratio. The poly-L-lysine to poly I·poly C ratio at these concentrations was 0.5 NH_2:1 PO_4. The thermal denaturation profile of this complex was found to be a one-step transition with a T_m at about 89°C in a solvent which contained 0.05 M-NaCl + 0.001 M-Na Citrate (~ 0.3X SSC). Complexes formed between poly I·poly C and poly-L-lysine with less poly-L-lysine than the 1P:0·5 NH_2 ratio gave two-step thermal denaturation profiles with T_m's at 57°C and 89°C. The lower temperature (T_m) indicated that there was still free poly I·poly C.

Tsuboi et al. (1966) also proposed a model for the poly I·poly C· poly-L-lysine complex. Knowing that in the A-form of poly I·poly C, the translation distance along the helix axis per nucleotide pair is 3.0 A, they concluded that poly-L-lysine would probably wind around the double helical poly I·poly C.

It is of interest to note that the model proposed by these authors for poly-L-lysine-DNA complex is quite different from the RNA to poly-L-lysine model described above. They concluded that a molecular model similar to that of deoxyribonucleoprotamine proposed by Wilkins (1956) would be acceptable for the poly-L-lysine DNA complex. In this model, an almost fully extended polypeptide chain also winds helically around the double-stranded polynucleotide chains. Unlike the RNA model, however, the pitch of the polypeptide helix is the same as that of the polynucleotide helices: This model requires one amino acid residue per one nucleotide residue, i.e., NH_2:P = 1:1, as is experimentally observed. On the other hand, this model would not be acceptable for poly I·poly C to poly-L-lysine complex, where NH_2:P ratio is 0.5:1.

Additional studies were later done by Haynes et al. (1970) on complexes of DNA, RNA and synthetic polynucleotides (A+U) and (I+C), with poly-L-lysine prepared at high salt concentrations. They reported that the complexes containing DNA and the synthetic polynucleotides show anomalous circular dichroism, with greatly enhanced rotational strength, and an inversion of sign in DNA and poly(A+U). DNA remained in the B form, as determined by x-ray diffraction, and there was no evidence of any large degree of distortion of the helix. The poly(A+U) complex, it was determined, was evidently in the three-stranded form.

The above data were obtained in solutions approximately 10^{-5} M, or a few micrograms per ml. When attempts were made to prepare solutions of higher concentrations,a heavy gummy precipitate was obtained. However, it one first forms a complex between poly-lysine and carboxymethylcellulose and then adds poly I·poly C, a soluble material is obtained. This material, poly(ICLC), is thermodynamically more stable than poly I·poly C, as shown in thermal denaturation studies (Fig. 3). These studies were done in 0.1X SSC, as poly(ICLC) did not "melt" below 100°C in SSC. The new compound is more resistant to hydrolysis by human sera or RNase than is poly I·poly C (Fig. 4). Poly(ICLC) induces 5-10 times more interferon in mice than does poly I·poly C (Levy et al., 1975a).

It is also of interest to note that it has been reported that single-stranded poly I or poly C complexed with polylysine and carboxymethylcellulose act as antiviral agents in mice but do not induce the formation of serum interferon (Stebbing and Dawson, 1979). The mechanism of this action is obscure.

Poly(ICLC) in monkeys

The important difference between poly I·poly C and poly(ICLC) is that the latter is able to induce significant quantities of interferon in primates. Figure 5 shows representative interferon induction kinetics in two Rhesus monkeys. In studies designed to determine the lethal dose of poly(ICLC) in monkeys (20 to 40 mg/kg body weight), up to 200,000 units of interferon per ml of serum were found.

Importance of size of components of poly(ICLC)

It has been shown by several groups that the molecular size of poly I and poly C in poly I·poly C is one of the determinants of the degree of effectiveness as an interferon inducer, with the larger homopolymers giving rise to somewhat better inducers in mice (DeClercq

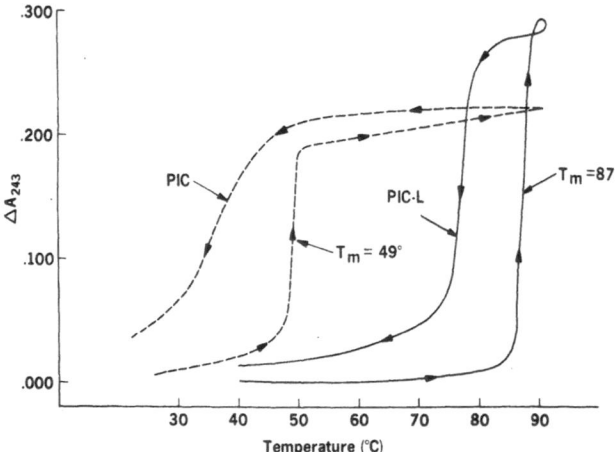

Fig. 3.   Thermal denaturation of poly I·poly C and the poly-L-lysine complex of poly I·poly C (poly [ICLC]).   The compounds, at a concentration of 50 µg poly I·poly C/ml in 0.1 x standard saline-citrate, were heated to the indicated temperatures in a recording spectrophotometer set at 260 mm (T_m).

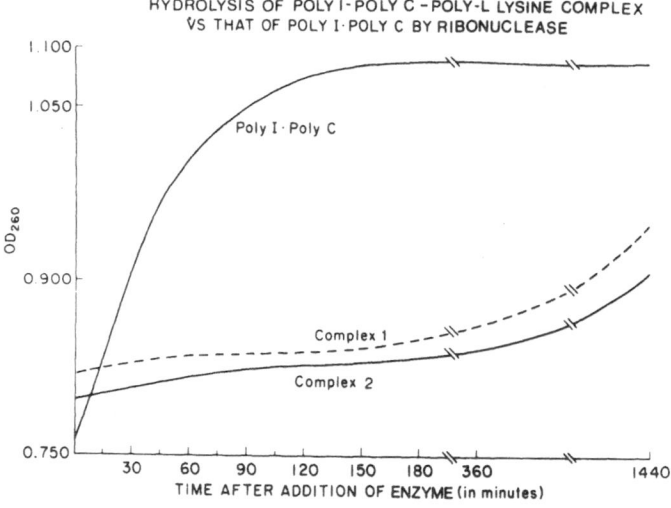

Fig. 4.   Hydrolysis of poly(ICLC) complex vs. poly I·poly C by ribonuclease.   Hydrolysis of poly I·poly C and two different lots of poly-L-lysine complex by pancreatic RNase.   The complexes, at a concentration of 50 µg poly I·poly C/ml in 0.15 m NaCl and 0.001 M phosphate buffer (pH 7.2), were exposed to 5 µg pancreatic RNase/ml at room temperature (about 25°). A_{260} mm readings were taken at 10 minute intervals.

et al., 1969). A more striking dependence on size of homopolymers and of poly-L-lysine is seen in poly(ICLC) as it relates to ability to induce interferon in monkeys. A series of poly(ICLC) preparations were prepared using poly-L-lysine of M.W. of 27,000 in a complex with polynucleotides of either 4S, 6S or 9S (Levy et al., 1981). Table 1 shows the differences among these three preparations. For convenience, these complexes are referred to as 4S, 6S and 9S poly(ICLC).

It can be seen that as the size of the d.s. RNA increases so does the T_m increase, from about 84°C for 4S to 88.5°C for the 9S complex. The resistance to hydrolysis by RNase A is also increased as the size is increased. The 4S complex is nearly totally hydrolysed by RNase, whereas 9S complex is only hydrolysed about 7.8% (increase in A_{260}). The interferon-inducing capacity can be seen to increase significantly as the size of the molecule increases. 4S poly(ICLC) induces very little interferon in monkeys whereas 9S poly(ICLC) induced greater than 1,000 units.

In another set of experiments, 9S homopolynucleotides were used to formulate complexes with poly-L-lysines of various sizes (Riley et al., 1981). The results are seen in Table 2.

As illustrated, increasing the size of poly-L-lysine increased the resistance of the complex to hydrolysis by RNase A. More importantly again, however, was the fact that   as the size of the poly-L-lysine increased so died the interferon-inducing capacity of the complex. The complex containing 2,000 M.W. lysine induced only about 10 units of interferon in monkeys, whereas the complex with

Table 1.   Correlation Between Homopolymer Size, Hydrolysis and Interferon Induction in Monkeys by poly(ICLC)

s Values of poly I and poly C in poly(ICLC)[b]	Hydrolysis of poly(ICLC)[a] (%)	T_m	Peak values of interferon ($\log_{10}$ units/ml serum)
9S	7.8	88.5	3.2
6S	39.0	86.0	1.75
4S	100.0	84.0	0.3

[a] Hydrolysis of poly I·poly C considered as 100%, 20 μg RNase/ml, 1 hour, 25°C.
[b] Poly-L-lysine used was of M.W. 27,000 daltons.

Table 2.   Correlation Between Molecular Weight of
          Poly-L-lysine, Hydrolysis and Interferon
          Induction in Monkeys by Poly(ICLC)[a]

Molecular weight of poly-L-lysine	Hydrolysis of poly(ICLC)[b]	Peak serum interferon Levels ($\log_{10}$ units/ml)
27,000	7.8	3.2
17,000	11.4	2.0
3,400	27.2	1.4
2,000	61.6	1.2

[a] Homopolynucleotides were about 9S.
[b] Hydrolysis of poly I·poly C taken as 100%, 20 µg RNase/ml, 1 hour, 25°C.

27,000 M.W. poly-L-lysine induced more than 1,000 units. Not shown is the observation that increased size of poly-L-lysine had only slight effect on the T_m of the complexes. They all "melted" at about 88°C.

    Thus, in both series of complexes the ability to induce inter-feron in primates runs parallel to the ability to resist hydrolysis. It is not unreasonable to consider that poly(ICLC) in primates is like poly I·poly C in rodents, with poly(ICLC) being effective in primates because of its ability to resist the increased hydrolytic activity of primate serum. All the subsequent studies presented here were done with poly(ICLC) made with 9S poly I·poly C, and 27,000 dalton M.W. of polylysine.

Antiviral studies in monkeys

    Levels of drug of 1 to 3 mg/kg have proven effective in con-trolling a number of viral diseases in monkeys. Rabies remains a serious problem in many underdeveloped countries. In those countries relatively crude antiserum and vaccine are used. Frequent and pain-ful injections are required. Allergic encephalitis is a sequela of importance. The combination of one dose of poly(ICLC) and one or two doses of vaccine have proven efficacious in post exposure prophylaxis of street rabies infection in monkeys (Baer et al., 1977). Table 3 summarizes some results. Studies in monkeys are necessarily restricted to small numbers of animals. In order to achieve statistically significant results, a very large challenge of virus must be used to ensure death of all the control monkeys. With this unnaturally high infection, treatment must begin within 48 hours of infection. With

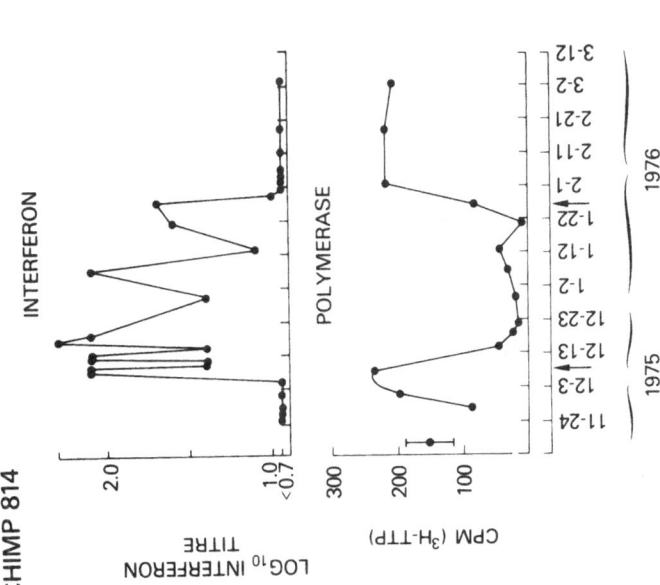

Fig. 6. Treatment of chronic hepatitis in young chimpanzees with poly I.poly C. The interferon titer is shown above and the virus associated polymerase activity is shown below. The beginning and end of the treatment period are indicated by arrows on the abscissa. The dot and bar on the graph of DNA polymerase response indicate the mean (±1 SD) of polymerase activity detected in six serum samples obtained during the 5 weeks immediately preceding the experiment.

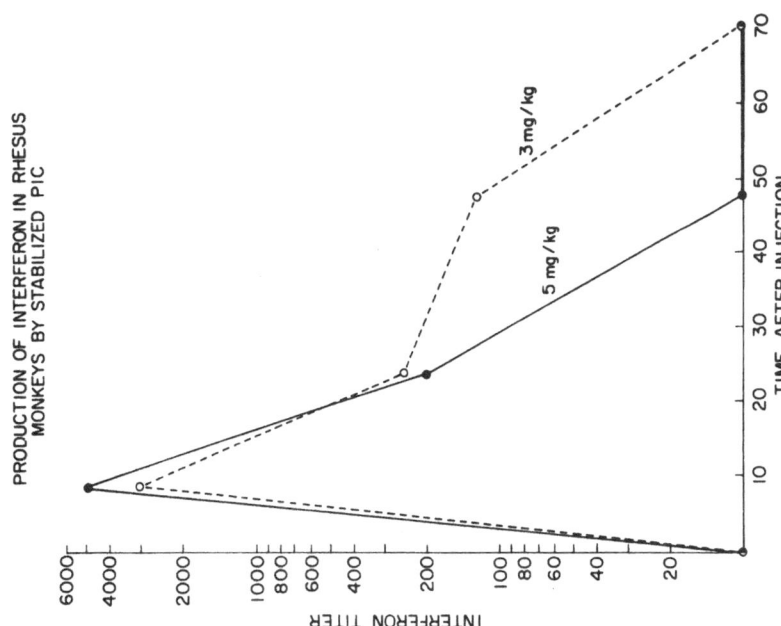

Fig. 5. Kinetics of induction of serum interferon in rhesus monkeys by i.v. administration of 3 or 5 mg/kg poly (ICLC) (one monkey per dose).

Table 3.   Effect of Poly(ICLC) and Human Diploid Cell
           Vaccine 48 Hours After Rabies Infection in
           Rhesus Monkeys

	Number dead
Untreated	8/8
Vaccine	5/8
Vaccine + poly(ICLC)	0/8

This experiment was done twice with similar results.

mice, larger numbers of animals can be used, and a less severe challenge of virus is possible. Under these conditions, treatment can be
delayed for up to a week after infection.

Table 4 gives a list of virus infections in monkeys that have
been treated systemically with poly(ICLC). With the exception of
Bolivian hemorrhagic fever and Tacaribe virus, all have been benefited.

Poly(ICLC) induces the formation of adequate levels of interferon in chimpanzees (Purcell et al., 1976). There is a model of
chronic hepatitis B in young chimpanzees. When chimpanzees were
injected with poly(ICLC) at 1 mg/kg body weight, they produced up to
750 units of interferon per ml of serum. When hepatitis B carriers
were treated with poly(ICLC) evidence of the virus disappeared during
the 16 week treatment, but reappeared after treatment was terminated
(Fig. 6). Whether or not really prolonged treatment would have a more
permanent effect has not been determined.

Adjuvant actions

Poly(ICLC) has proven to be a good immune adjuvant in primates.
When used in conjunction with a number of weak vaccines, the action
of the vaccine has been strongly augmented. Among the vaccines are
those to Venezuelan equine encephalomyelitis (Harrington et al.,
1979), swine flu (Stephen et al., 1977a), Japanese encephalitis
(Harrington, unpublished observations), herpes envelope antigen
(Levy et al., 1980), Haemophilus influenzae (Levy, et al., 1980)
and Rift Valley fever virus (Levy et al., 1980).

Monovalent influenza virus subunit vaccine, designated A/swine
X-53, prepared from ANJ/76 (New Jersey, Swine) is only moderately
to weakly effective when given as a single dose to young people.
When the vaccine was given to monkeys simultaneously with one dose

Table 4. Virus Diseases of Animals that Have Been Treated with Poly(ICLC)

Disease	Animal	Results	References
Simian hemor-rhagic fever	Monkey	Complete protection if given before virus, none if given after virus	Levy et al. (1975b)
Venezuelan equine encephalitis	Monkey	Using a nonlethal virus challenge; poly(ICLC) reduced viremia by 50%	Stephen et al. (1979)
Yellow fever	Monkey	75% protection up to 8 hr after challenge	Stephen et al. (1977b)
Japanese encephalitis	Monkey	50% protection up to 24 hr post challenge	Harrington et al. (1977)
Tacaribe virus	Mice	No effect by poly (ICLC)	Stephen, unpublished observation
Rabies	Monkey and mouse	See text	Baer et al. (1977)
Hepatitis	Chimpanzee	See text	Purcell et al. (1976)
Machupo virus	Monkey	Possible worsening of disease	Eddy, unpublished observation
Tick-borne encephalitis	Monkey	Strong beneficial effect	Burgasova et al. (1977)
Vaccinia	Monkey	Strong beneficial effect	Andzhaparidze et al. (1977)
Vaccinia skin lesions	Rabbit (topical treatment)	Spread of lesions stopped	Levy et al. (1978)

of poly(ICLC), HAI antibody titers in the serum were detectable earlier and rose to higher levels than in monkeys receiving vaccine alone. Four monkeys were used per group, each receiving 200 CCA units. Figure 7 shows results with monkeys receiving as little as 10 μg of drug per kg body weight, an amount that would induce no interferon and is associated with no physiological or pathological changes that were detected (Stephen et al., 1977a).

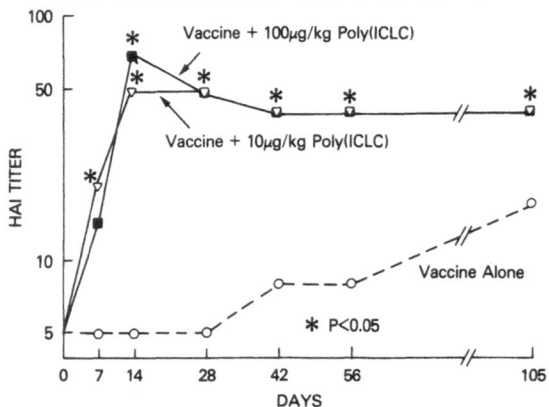

Fig. 7. Effect of one injection of poly(ICLC) on antibody production by rhesus monkeys in response to a subunit vaccine to swine flu (four monkeys per group).

Analogous results were obtained in monkeys using inactivated Venezuelan equine encephalomyelitis virus vaccine (Harrington et al., 1979). Figure 8 shows some of the data. It can be seen that antibody levels in serum were boosted about 40-fold after primary immunization when one compares levels attained after administration of vaccine along with poly(ICLC) with that attained with vaccine alone, and perhaps 200-fold after a secondary immunization. There was no alteration in the progression of IgM and IgG development. At the peak of antibody levels, most of the antibody was IgG. Polylysine complexed to carboxymethylcellulose, without poly I·poly C, had no adjuvant action.

A polysaccharide vaccine made from Haemophilus influenzae is a poor vaccine in very young children, where the disease threat is maximum. The vaccine is also poor in young monkeys. Table 5 shows this. The data presented are normalized values, obtained by radioimmune assays done by Dr. Porter Anderson (Levy et al., 1980). The value of 100 was assigned in each case to the amount of radio-activity found using the serum obtained prior to immunization. The vaccine alone caused a minimum boost, but when given with poly(ICLC) there was a more pronounced boost. Polyadenylic·polyuridylic acid complexed to carboxymethylcellulose and polylysine was not so effect-ive as poly(ICLC). Poly(ICLC) is not a universal adjuvant. With albumin and pneumococcal polysaccharide antigen, there was inhibition of antibody production (Levy, unpublished observations).

Studies in man

Levine et al. (1979) reported on a phase I study to determine what levels of drug would be acceptable and what levels of inter-feron could be induced in terminal cancer patients. Doses of poly (ICLC) ranging from 0.5 mg/m^2 to 24 mg/m^2 were administered according to the following regimen: One injection of the drug at the lowest level was given, and the patient observed for a period of a week. Then 14 daily doses were given. At least three patients were treated at each drug level before going on to the next higher level. Twenty-five patients with a wide variety of terminal cancers were studied. All had become refractory to other therapy. Peak serum interferon levels were usually found 8 to 24 hours after injection. Table 6 shows the mean peak serum interferon level found for each of the drug levels. At 18 and 24 mg/m^2 levels of serum interferon up to 15,000 units were found; but because of the toxicity noted, those doses were not considered acceptable. At 12 mg/m^2, the highest tolerated dose, peak serum interferon levels of about 2,000 were achieved. At the higher levels of drug, significant levels of interferon were found in the cerebrospinal fluid (Table 7).

There were a number of toxic manifestations associated with administration of the drug in this study. They are summarized in Table 8. Fever was always seen, although the degree tended to

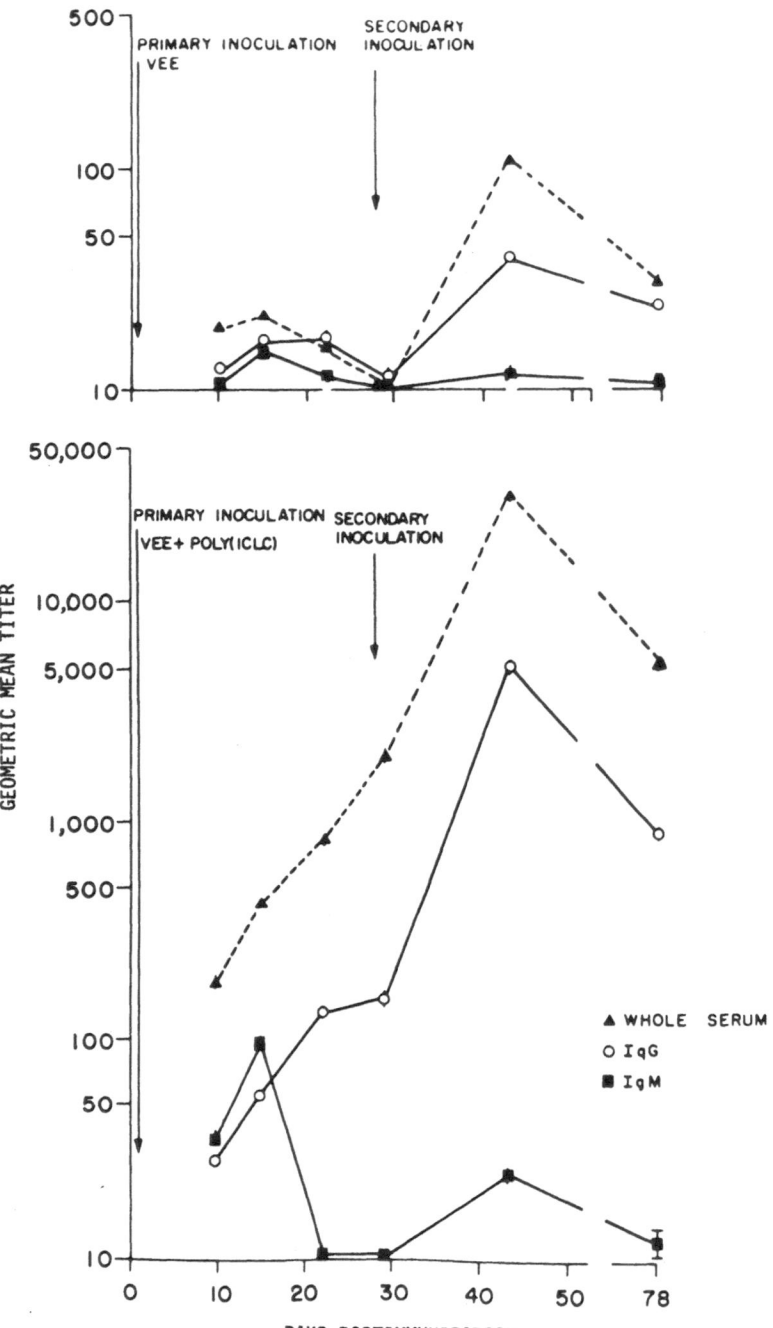

Fig. 8. Serum neutralizing antibody response by immunoglobulin class of rhesus monkeys inoculated on days 0 and 28 with (A) inactivated VEE virus vaccine (n=4), or (B) vaccine combined with 200 µg of poly(ICLC)/kg (n=4). Symbols: (▲) whole serum antibody titers, (o) antibody from IgG fraction, (■) antibody from IgM fractions.

Table 5. Effect of Low Doses of Poly(ICLC) on Antibody
          Production by Rhesus Monkeys in Response to a
          Polysaccharide Vaccine for Haemophilus influenzae

Treatment	Pre	Day 7	Day 14	Day 20	Day 28	Day 34	Day 42
Vaccine	100	590	348	386	225	187	113
Vaccine + poly(ICLC), 0.3 mg/kg	100	5,643	5,040	3,340	2,063	2,162	780
Vaccine + poly(ICLC), 0.3 mg/kg	100	6,589	3,904	1,884	1,132	839	721

Numbers represent the amount of radioactivity in the immune precipitate obtained from sera at the indicated time, normalized to the
amount of radioactivity present in the precipitate from the pretreat-
ment sera.

Table 6. Mean Peak Serum Interferon Titers After Treatment
          with Poly(ICLC)

Dose levels (mg/sq m)	Interferon titer (mean reference units/ml[a])
0.5	15 (0–25)[b]
2.5	15 (0–25)
7.5	198 (25–250)
12.0	1,940 (200–5,000)
18.0	4,473 (600–15,000)
27.0	5,820 (2,000–10,143)

[a]Assay was done in HSF4 cells, using vesicular stomatitis virus as
challenge and cytopathic effect as endpoint (international reference
units); 8 hour sample following first dose; ≥ 3 trials at each level.
[b]Numbers in parentheses, range.

Table 7. Mean Peak Cerebrospinal Fluid Interferon Titers
After Treatment with Poly(ICLC)

Dose level (mg/sq m)	Mean reference units/ml[a]
0.5	5 (0-10)[b]
12.0	34 (5-63)
18.0	55 (0-115)
27.0	515 (79-1000)

[a]Assay was done in HSF4 cells, using vesicular stomatitis virus as challenge and cytopathic effect as endpoint (international reference units); 8 hour sample following first dose; $\geq$ 3 trials at each level.
[b]Numbers in parentheses, range.

Table 8. Poly(ICLC) Toxicity

Manifestation	No./total
Fever	25/25 (100)[a]
Nausea	11/25 (44)
Thrombocytopenia and leukopenia	17/25 (68)
Hypotension[b]	7/25 (28)
Syndrome of erythema, polyarthralgia, and polymyalgia[b]	4/25 (16)
Renal failure[b]	1/25 (4)
Trial aborted[c]	5/25 (20)

[a]Numbers in parentheses, percentage of total.
[b]Related to dose level and/or magnitude of interferon induction.
[c]3, hypotension; 1, renal failure (27 mg/m^2); 1, serum sickness (?). Maximum tolerated dose, 12 mg/m^2.

decrease with repeated administration. Hyporesponsiveness for interferon production was manifest upon repeated dosage, unless 3 to 4 days elapsed between doses. Fever, myalgia and leukopenia are reminiscent of what is seen with administration of exogenous interferon. Hypotension is sometimes seen and appears to be dose related. Variations from this pattern of toxicity have been seen in some of the other clinical studies and will be mentioned later.

The Children's Hospital Cancer Testing Group (Lampkin and Levy, unpublished observations) has looked at the effect of poly(ICLC) on

far advanced cases of the null type of acute lymphoblastic leukemia
in children. No remissions have been induced, although anti-leukemic
effects, as evidenced by a marked decrease in the absolute number of
lymphoblasts in the blood of four patients and in the bone marrow of
at least one. These children were very ill and showed unacceptable
levels of toxicity at lower levels of drug than seen in the study by
Levine.

        Similarly, a phase I toxicity study by Krown et al. (1980) at
the Sloan-Kettering Institute on patients with a wide variety of
malignancies also revelaed unacceptable levels of toxicity at lower
levels of drug than reported by Levine.

        In contrast, in studies done by Leventhal, Whiznant and Levy
(unpublished observations) at Johns Hopkins and the University of
North Carolina on healthy children with juvenile laryngopapilloma,
toxicities were less than those seen by Levine. Fever, mild myalgia
and occasional rises in liver enzymes were seen. These patients
also were able to tolerate even higher levels of drug than did the
patients seen by Levine. There was also marked improvement in the
clinical condition of seven out of seven patients. A typical course
is seen in Figure 9. Each dot represents a surgical intervention.
It can be seen that surgery was required less and less frequently
after treatment with poly(ICLC) began. The regimen for these patients
consisted of initiation at 4 mg/m^2 and working up to 12 or even
15 mg/m^2 over several weeks.

        Less dramatic, but of some interest, were the results obtained
by Durie and Salmon on patients with multiple myeloma (Durie, Salmon
and Levy, unpublished observations), summarized in Figure 10. These
were patients who had been refractory to all other therapy. There
was subjective and objective improvement in these people but only of
a modest degree. They received about 4 to 6 mg/m^2 of the drug, once
or twice a week, and produced interferon levels ranging from 100 to
1,000 units/ml serum. Their toxic manifestations consisted of fever,
some transient leukopenia, no hypotension, but one incident of
hypertension, and transient but sometimes severe myalgia.

        Encouraging results have been seen in patients with paralytic
neurological diseases. Eight patients have been treated so far, as
summarized in Figure 11. These patients were all severely to totally
paralyzed. They all were benefited by poly(ICLC) treatment to the
point where they were able to appriximate a normal life (Engel et al.,
1978, 1981). The first seven cases involved peripheral nerve
neuropathies; the last one was a central nervous system neuropathy
resembling an acute phase of multiple sclerosis. The patient was
completely paralyzed and unresponsive to other therapy. Two days
following each weekly treatment with poly(ICLC), there was a
definite increase in muscular strength. After 9 months of therapy,

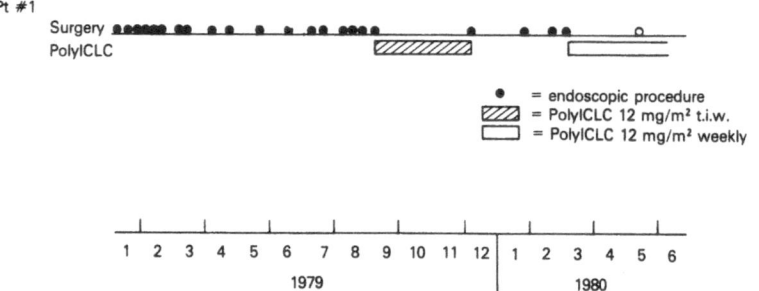

Fig. 9.   Effect of poly(ICLC) on frequency of surgical intervention in juvenile laryngopapilloma in a child.   Each dot represents one surgical intervention.

M-Component Type	Comments
kappa light chains	67% decrease in B-J excretion, plus correction of hypercalcemia and disease stabilization with first period of treatment. 44% decrease in B-J excretion when poly (ICLC) restarted 2 months later. Normocalcemic for 5 months.
IgG kappa	M-component decrease from 5.2 gm to 3.9 gm, Improved bone pain and performance status (became ambulatory)for 2-3 months.
IgG lambda	Trial stopped because of toxicity (malignant hypertension).
IgM kappa	Plasmaphoresis requirement decreased from q 14 days to q 28 days.
IgG kappa	Stable disease parameters for 2 months.
lambda light chains	Died one week after initiation of poly (ICLC).
IgG kappa	50% decrease in serum IgG M-component, plus symptomatic benefit.

Fig. 10.   Effect of poly(ICLC) on refractory multiple myeloma in man.

```
8PATIENTS WITH DYSIMMUNE NEUROLOGICAL DISEASES
REFRACTORY TO HIGH PROLONGED DOSES OF PREDNISONE
PLUS AZATHIOPRENE OR CYCLOPHOSPHAMIDE HAVE SHOWN
MODERATE TO DRAMATIC IMPROVEMENT. THESE INCLUDE
CHRONIC DYSCHWANNIAN NEUROPATHY
   "        DYSNEURONAL       "
   "     MYOPATHY (POLYMYOSITIS)
ACUTE DYSNEURONAL NEUROPATHY

POST INFECTIOUS DEMYELINATING ENCEPHALOMYELITIS,
INVOLVING CEREBELLUM,CEREBRUM, AND BRAINSTEM.
```

Fig. 11.   Types of neurologic diseases treated with poly(ICLC).

he was dismissed and was able to walk with a cane for balance.

Comparison of interferon inducers with exogenous interferon

There are some advantages offered by exogenous interferon and
some by inducers of interferon. The inducers are relatively cheap,
available in unlimited quantity, very stable and are capable of
inducing higher levels of serum interferon than are currently at-
tainable by administration of exogenous interferon. Poly(ICLC) is
an immune adjuvant; interferon generally does not appear to be so.
That can be an advantage or a disadvantage for poly(ICLC), depending
on the circumstances. On the other hand, interferon inducers require
response on the part of the host to produce interferon. If, for some
reason, the host is not able to respond, then little or no interferon
will be made. Also, while the general nature of the toxicity of
interferon and poly(ICLC) appear to be qualitatively similar, it may
be that the inducer causes more intense side reactions, possibly
because of the higher levels of interferon made. Finally, the human
race has been exposed to interferon since its beginning. The
inducers to some extent represent an unknown.

Andzhaparidze, D.G., Bektemirov, T.A., and Burgasova, M.P., 1977,
    The effect of poly(I)-poly(C) complex with poly-L-lysine on the
    course of vaccina infection monkeys, Vopr Virusol USSR, 3:339.
Baer, G.M., Shaddock, J.H., Moore, S.A., Yager, P.A., Baron, S.S.,
    and Levy, H.B., 1977, Successful prophylaxis against rabies in
    mice and rhesus monkeys: the interferon system and vaccine,
    J Infect Dis, 136:286.
Billiau, F., Buckler, C., Dianzani, F., Uhlendorf, C., and Baron, S.,
    1970, Influence of basic substances on the induction of the
    interferon mechanism. Ann NY Acad Sci, 178:657.
Burgasova, M.P., Andzhaparidze, D.G., Bektemirov, T.A., Bogomolova,
    N.N., and Yu, S.B., 1977, Influence of poly(I)-poly(C) complex
    with poly-L-lysine on experimental tick-borne encephalitis.
    Vopr Virusol USSR, 4:438.
DeClercq, E., Eckstein, F., and Merigan, T.C., 1969, Interferon in-
    duction increased through chemical modification of a synthetic
    polyribonucleotide, Science, 163:1137.
Dianzani, F., Cantagalli, P., Gagnoni, S., and Rita, G., 1968, Effect
    of DEAE-Dextran on production of interferon induced by synthetic
    double-stranded RNA in L-cell cultures, Proc Soc Exp Biol Med,
    128:708.
Engel, W.K., Cuneo, R.A., and Levy, H.B., 1978, Poly ICLC treatment
    of neuropathy, Lancet, 1:503.
Felsenfeld, G., and Huang, S., 1959, The interaction of polynucleo-
    tides with cations, Biochem Biophys Acta 34:234.
Field, A.K., Tytell, A.A., Lampson, G.P., and Hilleman, M.R., 1967,
    Inducers of interferon and host resistance. II. Multistranded
    synthetic polynucleotide complexes, Proc Natl Acad Sci USA,

58:1004.

Harrington, D.G., Hilmas, D.E., Elwell, M.R., Whitmore, R.E., and
    Stephen, E.L., 1977, Intranasal infection of monkeys with
    Japanese encephalitis virus: clinical response and treatment
    with a nuclease resistant derivative of poly (I)·poly C, Am
    J Trop Med Hyg, 26:1191.

Harrington, D.G., Crabbs, C.L., Hilmas, D.E., Brown, J.R., Higbee, G.
    A., Cole, F.E., Jr, and Levy, H.B., 1979, Adjuvant effects of
    low doses of a nuclease-resistant derivative of polyinosinic-
    polyribocytidylic acid on antibody responses of monkeys to in-
    activated Venezuelan equine encephalomyelitis virus vaccine,
    Infect Immun 24:160.

Haynes, M., Garrett, R.A., and Gratzer, W.B., 1970, Structures of
    nucleic acid-poly base complexes, Biochemistry, 9:4410.

Isaacs, A., and Lindemann, J., 1957, Virus interference. I. The
    interferon, Proc R Soc Lond [Biol.] 147:258.

Krown, S., Oetgen, H., Stewart, W., and Levy, H.B., 1980, Phase I
    trial of poly ICLC in cancer patients. In: Chirigos M (ed)
    Proceedings of symposium on biological modifiers in treatment of
    cancer, March, 1980, Raven Press, New York.

Levine, A.S., Sivalich, M., Viernick, P.H., and Levy, H.B., 1979,
    Initial clinical trials in cancer patients of polyriboinosinic-
    polyribocytidylic acid stabilized with poly-L-lysine, in carboxy-
    methylcellulose [poly(ICLC)], a highly effective interferon in-
    ducer, Cancer Res, 39:1645.

Levy, H.B., Law, L.N., and Rabson, A.S., 1969, Inhibition of tumor
    growth by polyinosinic-polycytidylic acid, Proc Natl Acad Sci
    USA, 62:357.

Levy, H.B., Baer, C., Baron, S., Buckler, C.E., Gibbs, C.J., Iadarola,
    M.J., London, W.T., and Rice, J., 1975a, A modified polyribo-
    inosinic-polyribocytidylic acid complex that induces interferon
    in primates, J Infect Dis, 132:434.

Levy, H.B., London, W., Fuccillo, D.A., Baron, S., and Rice, J.,
    1975b, Prophylactic control of simian hemorrhagic fever in
    monkeys by an interferon inducer, polyribinosinic-polyribocyti-
    dylic acid poly-L-lysine, J Infect Dis, 133:A256.

Levy, H.B., and Lvovsky, E., 1978, Topical treatment of vaccina virus
    infection with an interferon inducer in rabbits, J Infect Dis
    137:78.

Levy, H.B., Lvovsky, E., Riley, F.L., Harrington, D., Anderson, A.,
    Moe, J., Hilfenhaus, J., and Stephen, E., 1980, Immune modulating
    effects of poly(ICLC), Ann NY Acad Sci 350:33.

Levy, H.B., Lvovsky, E., Riley, F.L., and Stephen, E., 1981, Inter-
    feron induction in primates by stabilized poly I·poly C (poly
    ICLC): effect of size of polynucleotides, Infect Immun, in
    press.

Merigan, T.C., 1973, Non-viral substances which induce interferons.
    In: Finter, N. (ed), Interferon and interferon inducers, North-
    Holland/American Elsevier, New York, p. 45.

Nordlund, J.J., Wolff, S.M., and Levy, H.B., 1970, Inhibtion of
     biologic activity of poly I·poly C by human plasma, Proc Soc
     Exp Biol Med, 133:439.
O'Malley, J.A., Ho, Y.K., Chakrabarti, P., DiBerardino, L., Chandra,
     P., Orinda, D.A.O., Byrd, D.M., Bardos, T.J., and Carter, W.A.,
     1975, Antiviral activity of partially thiolated polynucleotides,
     Mol Pharmacol, 11:61.

APPLICATION OF THE CONGENER APPROACH TO THE DESIGN AND SYNTHESIS

OF PEPTIDE-CATECHOLAMINE CONJUGATES

M.S. Verlander,* K.A. Jacobson,* A.B. Reitz,*
R.P. Rosenkranz,** K.L. Melmon,** and M. Goodman*

*Department of Chemistry, U.C. San Diego, La Jolla, 92093;
and **Department of Medicine and Pharmacology, Stanford
University Medical Center, Stanford, CA 94305

INTRODUCTION

The potential for increasing the therapeutic index of drugs
through covalent attachment to natural or synthetic macromolecules
was first demonstrated almost 30 years ago.[1] Since that time, a
wide variety of both drugs and polymeric carriers has been studied
in this context.[2-4] However, these studies have failed to result
in a generalized understanding of the factors which are important
for optimization of the design of polymeric drug conjugates. This
may be attributable, in part at least, to the fact that the vast
majority of carriers studied have been high molecular weight and
also polydisperse. Characterization of the polymer-drug conjugates --
both chemical and biological -- is therefore difficult. Correlations
between the properties of the carrier and the resultant conjugate
are often impossible.

For reasons outlined previously[5] our studies in this area have
concentrated on the catecholamines **1a−c** and related β-adrenergic
drugs as model systems. Our reasons for this choice relate mainly
to the fact that the receptors for these drugs are extracellular so
that biological characterization is not complicated by transport

1a	R = H	= norepinephrine
1b	R = CH$_3$	= epinephrine
1c	R = CH(CH$_3$)$_2$	= isoproterenol

phenomena. Furthermore, a number of sensitive in vitro and in vivo
assay systems are available for these drugs. Conjugates of catecholamines and related compounds have been prepared using a variety
of different carriers, including porous glass,[6,7] Sepharose,[7] random
copolypeptides[8,9] and proteins.[8] However, none of the systems
studied to date has allowed a systematic investigation of the role
of both the chemical linkage and spacer groups between the drug and
the carrier as well as the properties of the carrier itself. In
fact, attempts to characterize completely conjugates of random copolypeptides with isoproterenol,[9] both directly and through the
synthesis of model compounds, failed to yield conclusive information
concerning the chemical nature of the biologically active species.[10]

Peptides have a number of unique advantages which make them
amongst the most versatile of potential carriers for drugs.[5] Perhaps
the most important of these, from our standpoint, is the possibility
of synthesizing completely monodisperse carrier structures, using
well-developed methodology, whose properties may be varied widely
through appropriate choice of amino acids (natural or synthetic) and
blocking groups. Conjugates derived from these carriers may therefore be purified to homogeneity and characterized completely by
conventional techniques. Thus, the chemical structure of the conjugate may be established with certainty and structure-activity
relationships developed for this novel class of drugs. Finally, the
methodology developed for the synthesis of low molecular weight,
monodisperse peptide conjugates may be translated directly to the
synthesis of conjugates of higher molecular weight, naturally-
occurring peptide carriers, such as those derived from peptide hormones and proteins, including monoclonal antibodies. The direct
targeting of small drugs through these means thus becomes a very
intriguing possibility.

The details of our systematic approach -- "the congener
approach" -- have been described in depth previously[5] and are outlined schematically in Scheme 1. In brief, drug derivatives
(congeners) which are functionalized specifically for attachment to
carrier molecules, are first prepared. Next, derivatives of the
congeners which model their attachment to carriers are studied, in
order to optimize pharmacological activity as a function of the
length of the spacer and also the chemistry of attachment of the
congeners to carriers. Finally, carriers such as monodisperse
oligopeptides are designed and synthesized and coupled to the congeners under optimized conditions to provide conjugates.

In applying this approach to the catecholamines **1a – c** we first
considered functionalization of the aromatic ring.[11,12] However,
none of our derivatives had the potency or efficacy as β-adrenergic
agonists that isoproterenol does. The most promising approach has
proven to be derivatization through the amine function in the side
chain of the molecules.[5] These studies have led to the synthesis

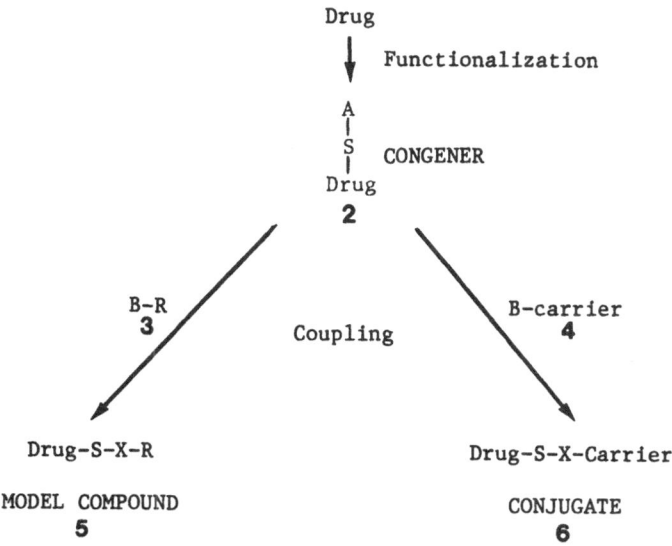

A = e.g. CO_2H, NH_2, OH, etc.; B = e.g. NH_2, CO_2H, SO_3H, NCO, etc.; X = e.g.
CO-NH, CO-O, NH-CO, NH-CO-O, NH-CO-NH, O-CO-NH, O-CO, etc.; S = e.g. $(CH_2)_n$;
R = alkyl, aryl, etc.

  Scheme 1.   Schematic Representation of the Congener Approach for
                 Drug Conjugate Design.

and evaluation of a variety of congeners of norepinephrine **1a**.[13-16]
Though the congeners themselves rarely have high potencies, on
transformation to model derivatives or conjugates,[17-18] many have
shown interesting and unique activities when tested in both _in vitro_
and _in vivo_ assay systems.

DISCUSSION

Synthesis and Characterization of Congeners and Model Derivatives

     The congener molecules we have studied **7**[13-15] formalistically
may be considered analogs of isoproterenol **1c** in which the N-
isopropyl group has been extended by a methylene chain of varying
length (the spacer), terminated by a carboxyl, amine or hydroxyl
group. The synthesis of these congeners **7** and related model derivatives **8** has been described in detail[13-15] and is outlined in
Scheme 2. Briefly, the most versatile general method for the

7 X = CO_2H, NH_2, OH

8 X = CO-NHR, NH-COR, NH-CO-OR
   NH-SO_2R, NH-CO-NHR, O-CO-NHR,
   O-COR, etc.

   R = alkyl, aryl; Y = Cl,
   CH_3CO_2, H_2PO_4.

synthesis of these compounds has proven to be the reductive amination of norepinephrine 1a with an excess of the appropriate methyl
ketone derivative 9. The reaction may be carried out using Adam's
catalyst (platinum dioxide)[19] in glacial acetic acid under atmospheric hydrogen or using sodium cyanoborohydride in methanol at
pH 5.[20] Since racemic norepinephrine was used and a second, optically active center was generated during the reductive amination
reaction, the products were actually mixtures of four possible
diastereomers. No attempt was made to resolve these. However, the
products were purified rigorously by reverse-phase HPLC in order to
eliminate possible, biologically active contaminants prior to pharmacological evaluation of the compounds.[13-16]

     The carboxyl congeners (7, X = CO_2H) and hydroxyl congeners
(7, X = OH) could be prepared directly by reductive amination of
norepinephrine with the appropriate ketoacids or ketoalcohols[13,14]
(i.e. 9, X = CO_2H, OH respectively). In the amine congener series,
this direct route was not feasible. However the ketoisocyanates 11,
prepared from the corresponding ketoacids using diphenylphosphoryl-
azide, proved to be versatile intermediates (Scheme 3).[15] The amine
congeners themselves were prepared via addition of benzyl alcohol
to the ketoisocyanates 11 to provide the benzyl urethanes 12 (R =
CH_2Ph). Reductive amination of these intermediates with norepinephrine (Scheme 2) provided the protected amine congeners which could
be deprotected using HBr/acetic acid.[15] Alternatively, the ketoisocyanates 11 could be reacted with amines, other alcohols, etc.
to provide intermediates for congener derivatives.

     Congener derivatives were synthesized which modeled the attach-
ment of the congeners to various amino acid side chains in potential

Scheme 2. Synthesis of Congeners and Model Derivatives

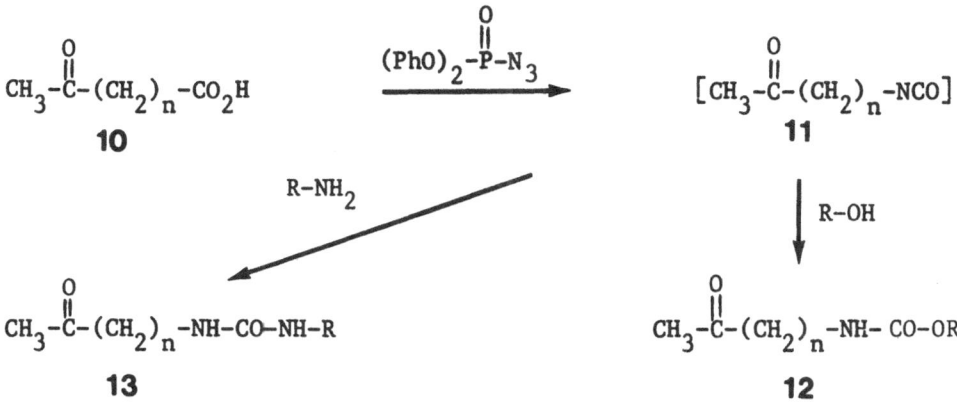

Scheme 3. Synthesis of Precursors for Amine Congeners and Derivatives

peptide conjugates. p-Toluidine, for example, was the model amine
for the side chain of p-aminophenylalanine in peptide conjugates.
Similarly, n-butylamine served as the model for the side chain of
lysine. As outlined in Scheme 2, these model derivatives were synthesized through reductive amination of preformed methyl ketone
derivatives with norepinephrine.

The congeners prepared and their in vitro β-adrenergic activities, as assessed using the response of wild type S49 cells in
production of cyclic AMP,[21] are summarized in Table 1. The length
of the spacer group (i.e., number of methylenes, n) was varied from
2 to 5 carbons. Activities are reported as relative potencies, i.e.
the ratio of the Ka (molar concentration of drug for half-maximal
activity) for isoproterenol (measured simultaneously) to that of the
test compound. As can be seen from Table 1, the congener molecules
generally had low activities, especially the carboxyl (**14-17**)[13] and
amine (**18-21**)[15] congeners which contain an additional charged group.
However, when the additional charged groups were blocked through the
formation of model derivatives, potencies were generally increased
dramatically. Conversion of the congeners to aromatic amides,ureas, or
urethanes, for example (Table 2), resulted in derivatives whose
potencies were often comparable to, or, in some instances, greater
than isoproterenol. However, in all three congener series there was
a marked dependence of relative potency on the length of the spacer
group. This was especially true for derivatives of amine congeners
(compounds **29-32**) which showed a variation of 6 orders of magnitude
in potency for spacer groups from 2 to 5 carbons. In all three
series, potency was maximal when the spacer contained four methylenes
(i.e. compounds **27, 31**, and **34**).

In view of these findings and the increased potency of several
model derivatives compared to isoproterenol, we undertook the synthesis of a series of congener derivatives in which the length of
the spacer group was kept constant (at the equivalent of 4

Table 1

In Vitro Biological Activity of Congeners of Norepinephrine

$$\underset{HO}{\overset{HO}{}} \quad \overset{OH}{\underset{}{CH}}-CH_2-\overset{\oplus}{NH_2}-\overset{CH_3}{\underset{}{CH}}-(CH_2)_n-X \quad Y^{\ominus}$$

Compound No.	X	Y	n	Relative Potency[a]
14	CO_2H	H_2PO_4	2	1.6×10^{-4}
15	"	"	3	6.8×10^{-4}
16	"	"	4	9.7×10^{-4}
17	"	"	5	8.3×10^{-4}
18	$NH_3^{\oplus}Cl^{\ominus}$	Cl	2	2.0×10^{-7}
19	"	"	3	1.1×10^{-7}
20	"	"	4	0.5×10^{-7}
21	"	"	5	0.27×10^{-7}
22	OH	"	2	1.2×10^{-2}
23	"	"	3	1.3
24	"	"	4	1.3×10^{-5}

[a] Biological activity relative to isoproterenol as measured by cyclic AMP accumulation in S49 cells[21]. Relative potency is expressed as the ratio of Ka for isoproterenol to Ka for the compound.

methylenes) but the nature of the chemical linkage and the substituent group varied. The structure of these derivatives and their in vitro biological activities are summarized in Table 3. As can be seen from these data, the potency of the compounds is remarkably sensitive to relatively minor structural variations. While reversal of the direction of the amide bond in model amides (compounds 27 vs. 31) or methylation of the amide (compound 35) had little effect, the nature and position of substituent groups in aromatic amides had a dramatic effect on potency (compounds 36 to 41). Replacement of the p-methyl substituent by an electron-donating methoxyl group (compound 36) reduced potency by more than an order of magnitude.

Table 2

In Vitro Biological Activity of Aromatic Derivatives of Congeners of
Norepinephrine

Compound No.	X	Y	n	Relative Potency[a]
25	CO-NH-⟨⟩-CH₃	H_2PO_4	2	7.1×10^{-1}
26	"	"	3	3.6×10^{-1}
27	"	"	4	5.4×10^{1}
28	"	"	5	1.1×10^{-1}
29	NH-CO-⟨⟩-CH₃	Cl	2	0.72×10^{-5}
30	"	"	3	1.93×10^{-1}
31	"	"	4	9.0
32	"	"	5	3.6×10^{-4}
33	O-CO-NH-⟨⟩-CH₃	H_2PO_4	3	9.2×10^{-2}
34	"	"	4	1.3

[a]Biological activity relative to isoproterenol as measured by cyclic
AMP accumulation in S49 cells[21]. Relative potency is expressed as
the ratio of Ka for isoproterenol to Ka for the compound.

On the other hand, derivatives containing p-n-butyl (37) or p-tri-
fluoromethyl (38) substituents showed substantially (2-3 orders of
magnitude) increased potencies compared to the p-methyl analog.
Perhaps most surprising were the results for the derivatives containing meta- and ortho-substituents, such as the meta- and ortho-
trifluoromethylanilides 39 and 40 which were 4 and 8 orders of magnitude, respectively, less potent than the para-isomer 38. ·Similarly,
the ortho-methylbenzoyl amide 41 was 6 orders of magnitude less
potent than the para-isomer 31.

Potencies were also sensitive to the type of linkage used --

Table 3

In Vitro Biological Activity of Derivatives of Congeners of
Norepinephrine

$$\text{HO}\!\!-\!\!\!\bigcirc\!\!\!-\!\!\overset{\text{OH}}{\underset{}{\text{CH}}}\!\!-\!\!\text{CH}_2\!\!-\!\!\overset{\oplus}{\underset{\text{Y}^{\ominus}}{\text{NH}_2}}\!\!-\!\!\overset{\text{CH}_3}{\underset{}{\text{CH}}}\!\!-\!\!(\text{CH}_2)_n\!\!-\!\!\text{X}$$

Compound No.	X	Y	n	Relative Potency[a]
27	CO–NH–⟨⟩–CH$_3$	H$_2$PO$_4$,	4	5.4×10^1
35	CO–N(CH$_3$)–⟨⟩–CH$_3$	"	4	7.3×10^1
36	CO–NH–⟨⟩–OCH$_3$	"	4	3.1
37	CO–NH–⟨⟩–(CH$_2$)$_3$–CH$_3$	"	4	3.2×10^3
38	CO–NH–⟨⟩–CF$_3$	"	4	1.1×10^4
39	CO–NH–⟨⟩	"	4	9.8×10^{-1}
40	CO–NH–⟨⟩ (CF$_3$, CF$_3$)	"	4	2.5×10^{-4}
31	NH–CO–⟨⟩–CH$_3$	Cl	4	9.0
41	NH–CO–⟨⟩	"	4	3.9×10^{-6}
42	NH–CO–O–CH$_2$–⟨⟩ (CH$_3$)	"	4	0.21
43	NH–CO–NH–⟨⟩–CH$_3$	"	4	2.1×10^{-3}
44	CO–NH–⟨S⟩	H$_2$PO$_4$	4	1.7×10^{-1}
45	O–CO–NH–⟨S⟩	"	3	2.9×10^1
46	NH–CO–O–⟨S⟩	Cl	3	1.0
47	CO–NH–(CH$_2$)$_3$–CH$_3$	H$_2$PO$_4$	4	2.5×10^{-1}
48	O–CO–NH–(CH$_2$)$_3$–CH$_3$	"	3	1.9×10^2

[a] Biological activity relative to isoproterenol as measured by cyclic AMP accumulation in S49 cells[21]. Relative potency is expressed as the ratio of Ka for isoproterenol to Ka for the compound.

aromatic urethanes and ureas (e.g. compounds **42** and **43** respectively) were generally less potent than their amide counterparts. Aliphatic amide derivatives (e.g. compounds **44** and **47**) were also substantially less potent than aromatic amides, although the analogous urethanes (**45** and **48** respectively), containing approximately the same length spacer groups, were 2-3 orders of magnitude more potent than the amides (**44** and **47**). Reversal of the direction of the urethane linkage (compound **45** vs. **46**) also resulted in a substantial decrease in potency.

Our results suggest that the potency of these derivatives is a complicated function of electronic, hydrophobic, steric and even conformational effects. The dramatic differences in potency between ortho- and para-substituted aromatic amide derivatives suggest the existence of a folded conformation, possibly stabilized by a hydrogen bond between the benzylic hydroxyl group and functional groups (e.g. an amide) in the side chain, in which the aromatic rings are stacked. The stability of such a conformation would certainly be sensitive to electronic (if hydrogen-bonding is important), steric and hydrophobic effects. We are currently investigating this hypothesis in greater detail. The large magnitude of these effects, however, suggests the possibility of using this approach to probe the structural requirements of the β-receptor in detail.

Despite their relative potencies, the effects of each derivative could be inhibited completely by appropriate concentrations of the β-antagonist, propranolol,[22] indicating that the observed activity occurred through a specific interaction with the β-receptor. Furthermore, competitive binding studies with [126]I-hydroxybenzylpindolol and ^{3}H-dihydroalprenolol[22] indicated that the variations in potency of these compounds relative to isoproterenol were likely to be explained on the basis of altered affinity for the β-adrenergic receptors on the S49 cell surface.

Synthesis and Characterization of Monodisperse Peptide Carriers and Conjugates

The model studies, summarized above, have indicated that congener model derivatives prepared from carboxylic acid congeners and aromatic amines are generally more active in vitro than those derived from aliphatic amines or other congeners. Since this suggests that peptide conjugates based on p-aminophenylalanine as the site of attachment of the congener are likely to be more active than those based on lysine, our initial studies have focused on the synthesis of a series of peptide carriers containing this aromatic amino acid. The carriers ranged in size from a single, blocked amino acid to pentapeptides.[17,18] The amino acids which were incorporated into the carriers, together with the blocking and solubilizing groups on the different amino acid residues, were chosen to provide a variety of properties in order to assess the importance of hydrophilic/

hydrophobic balance and charge as well as the molecular weight of the carrier.

The peptides were synthesized by conventional peptide synthetic techniques in solution and have been described in detail.[17,18] In general, although the direct attachment of congeners to the carriers is an attractive synthetic possibility, in practice, the approach which was followed routinely required the prefunctionalization of the peptide with a keto acid and reductive amination as the final step (cf. the synthesis of model derivatives, Scheme 2). This approach is summarized schematically in Scheme 4. As in the case of congeners and model amides, the conjugates were purified rigorously by semipreparative reverse-phase HPLC in order to ensure homogeneity and the absence of potential, biologically active contaminants.

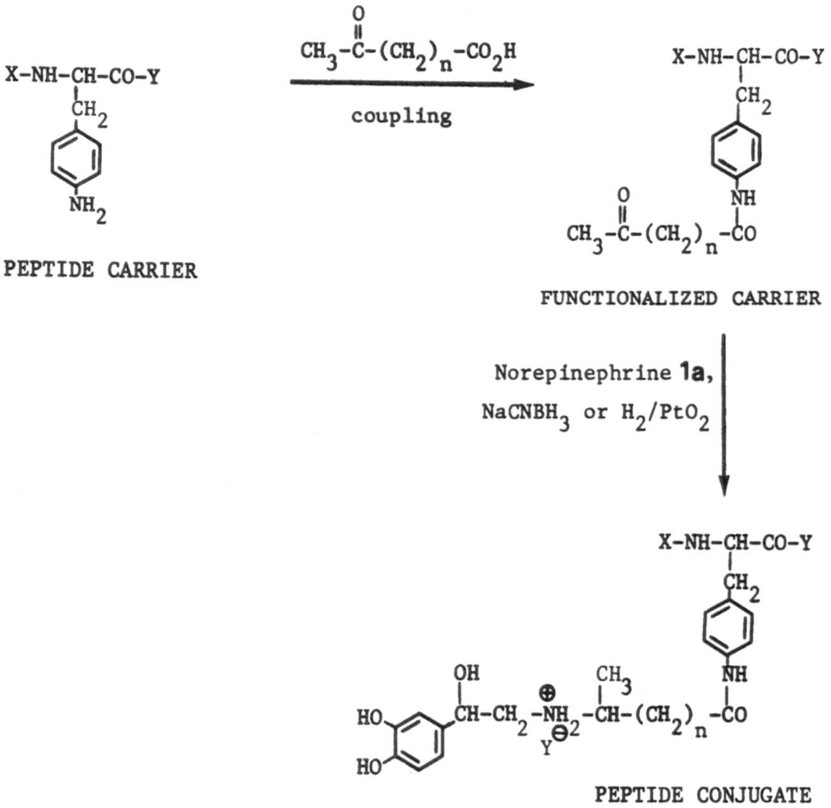

Scheme 4: General Synthetic Route to Peptide Catecholamine Conjugates

Table 4

In Vitro Biological Activities of Amino Acid and Dipeptide
Conjugates Related to Isoproterenol

$$\text{HO} \diagdown \diagup \text{CH-CH}_2\text{-}\overset{\oplus}{\underset{\ominus}{N}}\text{H}_2\text{-}\overset{CH_3}{CH}\text{-(CH}_2)_4\text{-CO-X}$$

with substituents OH, $\overset{\ominus}{H}_2PO_4$, HO

Compound No.	X^a	Relative Potency^b
49	Ac-(L)-Phe-HPA | NH	1.67×10^{-3}
50	Ac-(D)-Phe-HPA | NH	7.41×10^{-4}
51	H-Phe-Gly-NH-CH$_3$ | NH	2.41×10^{-6}
52	Ac-Phe-Gly-OH | NH	8.15×10^{-3}
53	Ac-Phe-Gly-NH-CH$_3$ | NH	2.04×10^{-2}
54	Boc-Phe-Gly-NH-CH$_3$ | NH	7.59×10^{-2}
55	Boc-Gly-Phe-NH-CH$_3$ | NH	1.26×10^{-2}

^{a}Abbreviations: Ac $= CH_3-\overset{O}{\overset{\|}{C}}-$; HPA $= NH-(CH_2)_3-OH$; Boc $= (CH_3)_3C-O-\overset{O}{\overset{\|}{C}}-$
Phe $=$ phenylalanine; Gly $=$ glycine.

^{b}Biological activity relative to the model derivative 27 as measured
by cyclic AMP accumulation in S49 cells[21].

The peptide conjugates synthesized and their in vitro biological
activities relative to the appropriate congener model amide **27** are
summarized in Tables 4 and 5. Since model studies of p-toluides had
indicated that the optimum length of the spacer group was five carbon
atoms (the secondary carbon adjacent to the nitrogen plus four
methylenes), the majority of conjugates studied to date were derived
from this congener.

Examination of the data for the amino acid and dipeptide conjugates (Table 4) reveals a number of interesting trends. First, the
two amino acid conjugates, **49** and **50**, derived from L- and D-p-amino-
phenylalanine hydroxypropylamides respectively, have essentially the
same in vitro biological activity, suggesting that the chirality of
the carrier is not a determinant of biological activity. The series
of phenylalanyl-glycyl dipeptide conjugates **51 – 55** show a marked dependence of potency on the nature of the amine and carboxyl blocking
groups. The dipeptide conjugate **51**, which lacks an amine blocking
group and therefore has a net charge of +2 at physiological pH, was
approximately six orders of magnitude less potent than compound **27**.
On the other hand, the corresponding dipeptide **52** which lacks a carboxyl blocking group and therefore has a net charge of 0 at physiological pH, was two orders of magnitude less potent than compound **27**.
This result is surprising, in view of the very low potencies found
for the carboxylic acid congeners (Table 1) and suggests that the net
charge is not the only determining factor in potency. The relative
orientation or distance between the charges may also be a key determinant of potency. The three fully-blocked dipeptide conjugates
53-55 were one or two orders of magnitude less potent than the model
amide **27,** the most potent compound being the t-butyloxycarbonyl
dipeptide N-methylamide derivative **54**. The fact that reversal of
the sequence of the peptide (compound **55**) decreased the potency of
the conjugate by approximately an order of magnitude suggests that the
orientation of hydrophobic groups relative to the pharmacophore may
be an important factor.

The sequences of the tri- and pentapeptide carriers were designed to provide information concerning the contribution of particular amino acid side chains to the solubility and also biological
activity of the conjugates derived from them. The conjugates prepared
and their in vitro biological activities are summarized in Table 5.
The series of tripeptide conjugates **56-58** allows a comparison of the
effect of three hydrophilic amino acid residues -- N-methylglutamine,
α-amino-δ-hydroxyvaleric acid (Hyv, a rare but naturally occurring
amino acid) and citrulline -- as well as the effect of the addition
of a single amino acid to the dipeptide conjugate **53**. The tripep-
tides had surprisingly low potencies, compared to the dipeptide, the
least active **56** being more than five orders of magnitude less active
than the dipeptide. The most active tripeptide conjugate **58**, which
contained citrulline, was approximately 20 times less potent than the
dipeptide, but surprisingly, its potency was decreased by three orders

Table 5

In Vitro Biological Activities of Tri- and Pentapeptide
Conjugates Related to Isoproterenol

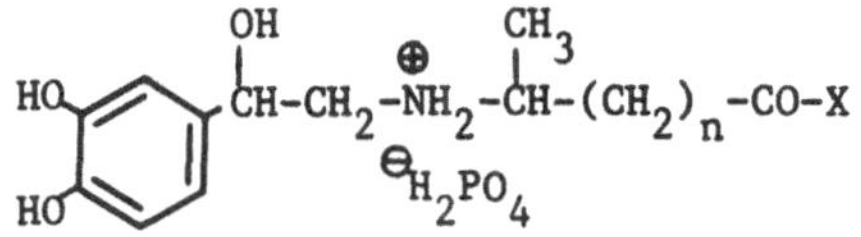

$$\text{HO} - \bigcirc - \underset{\text{OH}}{\overset{}{\text{CH}}}-\text{CH}_2-\overset{\oplus}{\text{NH}_2}-\underset{\text{CH}_3}{\overset{}{\text{CH}}}-(\text{CH}_2)_n-\text{CO-X}$$
$$\underset{\ominus_{\text{H}_2\text{PO}_4}}{}$$

Compound No.	n	X[a]	Relative Potency[b]

56	4	Ac-Glu——Phe-Gly-NH-CH$_3$
		    | |
		  NHCH$_3$ NH
			6.85×10^{-8}

57	4	Ac-Hyv-Phe-Gly-NH-CH$_3$
		       |
		       NH
			1.48×10^{-6}

58	4	Ac-Cit-Phe-Gly-NH-CH$_3$
		       |
		       NH
			1.02×10^{-3}

59	4	Boc-Cit-Phe-Gly-NH-CH$_3$
		        |
		        NH
			6.94×10^{-7}

60	3	Ac-Glu——Phe-Glu-(NH-CH$_3$)$_2$
		    | |
		  NHCH$_3$ NH
			8.33×10^{-8}

61	4	Ac-Glu——Phe-Glu-(NH-CH$_3$)$_2$
		    | |
		  NHCH$_3$ NH
			4.26×10^{-7}

62	5	Ac-Glu——Phe-Glu-(NH-CH$_3$)$_2$
		    | |
		  NHCH$_3$ NH
			1.59×10^{-7}

63	4	Boc-Phe-(Hyv)$_3$-Gly-NH-CH$_3$
		            |
		            NH
			1.46×10^{-2}

64	4	Boc--(Hyv)$_3$-Phe-Gly-NH-CH$_3$
		                |
		                NH
			1.85×10^{-2}

---

[a]Abbreviations:

Ac = CH$_3$-$\overset{\text{O}}{\overset{\|}{\text{C}}}$; Boc = (CH$_3$)$_3$C-O-$\overset{\text{O}}{\overset{\|}{\text{C}}}$; Hyv = L-α-amino-δ-hydroxyvaleric
acid; Cit = citrulline; Phe = L-phenylalanine;
Gly = glycine.

[b]Biological activity relative to the model derivative **27** as measured
by cyclic AMP accumulation in S49 cells [21].

of magnitude when converted to the more hydrophobic t-butyloxycar-
bonyl-blocked tripeptide derivative **59**. Evidently, the in vitro
biological potencies of these tripeptide conjugates is not a simple
function of their overall hydrophilic/hydrophobic balance, although
their substantially lower potencies compared with the dipeptide
conjugate **53** could simplistically be attributed to their greater
hydrophilic character. It is likely that more complex factors, which
may relate to secondary structure, play a significant role in influencing the binding of these catecholamine derivatives to the receptor
in vitro.

The conjugates **60-62** (Table 5), derived from a tripeptide,
containing p-aminophenylalanine flanked by two N-methylglutamine
residues, were synthesized in order to assess the effect of the
length of the spacer group on the biological activity of the conjugates. Although the in vitro potencies of these derivatives were
rather low (6 to 7 orders of magnitude less than compound **27**) there
was a small but significant effect of the length of the spacer group
(n = 3-5). The most active of the three conjugates, compound **61**,
contained a 5-carbon spacer group (i.e. n = 4), confirming the results
obtained for the model amide derivatives (see above).

Two isomeric pentapeptide conjugates **63** and **64**, representing
the addition of tri-(δ-hydroxy-α-aminovaleric acid)[(Hyv)$_3$] to the
most active dipeptide conjugate **54**, were also synthesized. The
position of the point of attachment (p-aminophenylalanine) of the
congener was varied in these two peptides, in order to assess the
relative effects of the hydrophobic t-butyloxycarbonyl group, the
hydrophilic tripeptide block and the C-terminal glycine-N-methylamide
residue. This variation had little effect on the biological activity
of the conjugates which both had very similar potencies (two orders
of magnitude less than compound **27**). Perhaps most surprising, in
view of the hydrophilicity of these two pentapeptides, was the fact
that these two conjugates had in vitro potencies which were not dramatically less than the considerably more hydrophobic dipeptide **54**.
This, together with the data for the tripeptide conjugates (above)
tends to confirm our hypothesis that, as the size of the peptide
carriers increases, the structure of the peptide and relative orientation of hydrophilic and hydrophobic groups in the conjugate may be
more significant, in terms of receptor interactions, than simply the
overall hydrophilic/hydrophobic balance.

The two pentapeptide conjugates **63** and **64** have additional significance. Their sequences were designed specifically for the purpose of investigating larger oligopeptides (decapeptides, pentadeca-
peptides, etc.) through appropriate fragment condensations. The C-
terminal glycine residues were included in the sequences specifically
in order to eliminate the possibility of racemization during these
syntheses. Through appropriate combinations of the two pentapeptide
carriers (head-to-tail, head-to-head, etc.) we will be able to answer

a number of important questions, concerning, for example, molecular
weight, number of drug molecules per carrier molecule, relative
position of drug molecules on the same carrier molecule, etc., without the added complication of varying the overall amino acid composition of the carriers.

In Vivo Biological Activity of Congener Model Amides and Conjugates

Preliminary in vivo data have been obtained for a number of the
congener model amides and conjugates. Potencies were assessed by
monitoring the changes in blood pressure following intravenous administration of the drugs to anesthetized rats.[23] The results,
summarized in Table 6, are expressed as relative potencies, i.e.,
the ratio of the E.D.$_{50}$ for the model amide **27** to that of the derivative. For the purpose of comparison, the in vitro relative potencies are also shown.

In general, the ranking of derivatives, suggested by the in
vitro assay, was confirmed by the in vivo data, although the range
of values in vivo was much smaller. However, a number of important
exceptions are evident. In the congener and model amide series, the
least potent (**16**) and most potent (**38**) compounds in vitro were also
the least potent and most potent of the congeners and model amides
in vivo. However, the second most potent compound in vitro, the p-n-
butylanilide **37**, was less potent than the p-toluide **27** in vivo.

In the case of the conjugates, the results were most striking in
the dipeptide and tripeptide series. The two dipeptides **54** and **55**,
which had comparable potencies in vitro, had strikingly different
potencies in vivo (Table 6). In fact, the dipeptide **54** had the
highest in vivo potency of any of the compounds tested to date. The
potencies of the three tripeptide conjugates **56-58** in vivo were
also completely reversed compared with the in vitro data. The transformation of the citrulline tripeptide **58** to its t-butyloxycarbonyl
derivative **59**, which dramatically decreased the in vitro potency,
had a similar effect on in vivo potency, leading, in fact, to the
least potent of the conjugates tested. Perhaps most surprising was
the data for the pentapeptides **63** and **64**, which were the most potent
of the peptide conjugates in vitro but amongst the least potent of
the peptide conjugates in vivo.

During the course of evaluating the peptide conjugates in vivo
significant increases in duration of action were noted, being most
pronounced for the pentapeptide conjugates **63** and **64**. These effects
are not reflected in the relative potency values, but will obviously
be an important factor in assessing the value of this approach for
the production of clinically useful drugs.

As noted above, the data provided by in vitro studies clearly
relate to effects which appear to be solely due to interactions with

Table 6

Relative In Vitro and In Vivo Potencies of Selected Congener
Model Amides and Conjugates Related to Isoproterenol

$$\underset{HO}{\overset{HO}{\diagdown}}\;\; \overset{OH}{\underset{|}{C}}H-CH_2-\overset{\oplus}{NH_2}-\overset{CH_3}{\underset{|}{C}}H-(CH_2)_4-COR \quad \overset{\ominus}{H_2PO_4}$$

Compound No.	R	Relative Potency	
		In Vitro[a]	In Vivo[b]
16	OH	1.80×10^{-5}	0.10
27	NH—⟨ ⟩—CH_3	1.0	1.0
37	NH—⟨ ⟩—$(CH_2)_3$-CH_3	5.93×10^{1}	0.69
38	NH—⟨ ⟩—CF_3	2.04×10^{2}	2.07
54	Boc-Phe-Gly-NH-CH_3 　　| 　　NH	7.6×10^{-2}	2.31
55	Boc-Gly-Phe-NH-CH_3 　　| 　　NH	1.26×10^{-2}	0.67
56	Ac-Glu—Phe-Gly-NH-CH_3 　　|　　| 　NHCH_3 NH	6.85×10^{-8}	1.09
57	Ac-Hyv-Phe-Gly-NH-CH_3 　　| 　　NH	1.48×10^{-6}	0.91
58	Ac-Cit-Phe-Gly-NH-CH_3 　　| 　　NH	1.02×10^{-3}	0.49
59	Boc-Cit-Phe-Gly-NH-CH_3 　　| 　　NH	6.94×10^{-7}	0.14
63	Boc-Phe-(Hyv)_3-Gly-NH-CH_3 　　| 　　NH	1.46×10^{-2}	0.36
64	Boc-(Hyv)_3-Phe-Gly-NH-CH_3 　　| 　　NH	1.85×10^{-2}	0.47

[a]Biological activity relative to model derivative 27 as measured by
cyclic AMP accumulation in S49 cells [21].

[b]Biological activity relative to model derivative 27 as measured by
changes of blood pressure in anesthesized rats[23].

the β-receptor, together with effects such as variations in distribution, metabolism, etc. Therefore, discrepancies between in vitro
and in vivo data clearly must be attributable to changes in the
pharmacokinetics, including metabolism of the drug, or its pharma-
codynamics as key determinants, in addition to the pure β-receptor
effects noted in vitro. These anomalies will allow us to define the
limits of structural variations in the congeners, model amides and
conjugates and therefore assist in the design of drugs which include
or exclude these structural variations, depending on their clinical
significance.

CONCLUSIONS

Our studies have shown that dramatic alterations in the potency
of catecholamine derivatives can be effected through structural modifications at a point which is far-removed from the previously assumed
biologically active portion of the molecule. These effects have been
demonstrated both for low molecular weight, model derivatives and
also for a series of small, monodisperse peptide conjugates of varying structure and molecular weight. Thus, high molecular weight
carriers may not be required for effective carrier-drug conjugates.
Since the changes in in vitro potency can be directly related to
differences in affinity for the β-receptor, our results suggest
exciting potential for both a clearer understanding of the mechanism
of binding of these drugs to the β-receptor and also a novel structure-activity approach for the design of new and useful drugs.
Furthermore, preliminary in vivo data has suggested that higher
molecular weight conjugates have significantly increased durations
of effect, suggesting a route for increasing the therapeutic index
and also possibly the selectivity of these drugs.

We are currently extending our studies to the synthesis of
related congeners of other drugs such as β-antagonists and histamine,
as well as conjugates derived from other carriers such as proteins.
Our in vitro studies are also being expanded to include biodisposi-
tion and metabolic studies of selected congener model amides and
conjugates, in order to address some of the questions raised above.
However, our preliminary studies suggest that some interesting and
therapeutically useful new drugs may result from these studies.

ACKNOWLEDGEMENT

The authors wish to thank Hoffmann-La Roche, Inc., and the
Burroughs-Wellcome Foundation for grants-in-aid which allowed us to
carry out the earlier part of this research and the National Institutes of Health (HL 26340) for subsequent financial support. We also
wish to thank Moon Ja-Choo and Lynn Anderson for their excellent
technical assistance with the pharmacological testing.

REFERENCES

1. H. Jatzkewitz, A peptamine (glycyl-L-leucylmescaline) bound to a
    colloidal blood plasma substitute (polyvinylpyrrolidone) as a
    new type of depot for a biologically active primary amine
    (mescaline), Z. Naturforsch. 10:27 (1955).
2. H. Ringsdorf, Structure and properties of pharmacologically
    active polymers, J. Polymer Sci. Symp. 51:135 (1975).
3. L.G. Donaruma, Synthetic biologically active polymers, Prog.
    Polymer Sci. 4:1 (1975).
4. H.-G. Batz, Polymeric drugs, Adv. Polymer Sci. 23:25 (1977).
5. M.S. Verlander, K.A. Jacobson, R.P. Rosenkranz, K.L. Melmon and
    M. Goodman, Some novel approaches to the design and synthesis
    of peptide-catecholamine conjugates, Biopolymers (1983), in
    press.
6. J.C. Venter, J.E. Dixon, P.R. Maroko and N.O. Kaplan, Biologically
    active catecholamines covalently bound to glass beads, Proc.
    Natl. Acad. Sci. U.S.A. 69:1141 (1972).
7. M.S. Yong and J.B. Richardson, Stability and biological activity
    of catecholamines and 5-hydroxytryptamine immobilized to
    sepharose and glass beads, Can. J. Physiol. Pharmacol. 53:616
    (1975).
8. K.L. Melmon, Y. Weinstein, B.M. Shearer, H.R. Bourne and S.
    Bauminger, Separation of specific antibody-forming mouse cells
    by their adherence to insolubilized endogenous hormones, J.
    Clin. Invest. 53:22 (1974).
9. M.S. Verlander, J.C. Venter, M. Goodman, N.O. Kaplan and B. Saks,
    Biological activity of catecholamines covalently linked to
    synthetic polymers: proof of the immobilized drug theory,
    Proc. Natl. Acad. Sci. U.S.A. 73:1009 (1976).
10. M. Goodman, M.S. Verlander, K.L. Melmon, K.A. Jacobson, A.B.Reitz,
    J.P. Taulane, M.A. Avery and N.O. Kaplan, Characterization of
    catecholamine-polypeptide conjugates, Eur. Polymer J. (1983),
    in press.
11. M.A. Avery, M.S. Verlander and M. Goodman, Synthesis of 6-amino-
    isoproterenol, J. Org. Chem. 45:2750 (1980); ibid. 46:5459
    (1981).
12. A.B. Reitz, M.A. Avery, M.S. Verlander and M. Goodman, Synthesis
    of ring alkylated isoproterenol derivatives, J. Org Chem.
    46:4859 (1981).
13. K.A. Jacobson, D. Marr-Leisy, R.P. Rosenkranz, M.S. Verlander,
    K.L. Melmon and M. Goodman, Conjugates of catecholamines.
    I. N-Alkyl functionalized carboxylic acid congeners and
    amides related to isoproterenol, J. Med. Chem. (1983), in
    press.
14. A.B. Reitz, M.A. Avery, R.P. Rosenkranz, M.S. Verlander, K.L.
    Melmon, and M. Goodman, Conjugates of catecholamines. VI.
    Synthesis and β-adrenergic activity of N-hydroxyalkyl cate-
    cholamine derivatives, J. Med. Chem. submitted.

15. A.B. Reitz, E. Sonveaux, R.P. Rosenkranz, M.S. Verlander, K.L. Melmon and M. Goodman, Conjugates of catecholamines. V. Synthesis and β-adrenergic activity of novel N-aminoalkyl derivatives of norepinephrine, J. Med. Chem. submitted.

16. R.P. Rosenkranz, K.A. Jacobson, M.S. Verlander, A. Welton, L. Klevans, M. Goodman and K.L. Melmon, Conjugates of catecholamines. II. In vitro and in vivo pharmacological activity of N-alkyl functionalized carboxylic acid congeners and amides related to isoproterenol, Mol. Pharm. submitted.

17. K.A. Jacobson, R.P. Rosenkranz, M.S. Verlander, K.L. Melmon and M. Goodman, Conjugates of catecholamines. III. Synthesis and characterization of monodisperse oligopeptide conjugates related to isoproterenol, Int. J. Peptide Protein Res. (1983), in press.

18. R.P. Rosenkranz, K.A. Jacobson, M.S. Verlander, A. Welton, L. Klevans, M. Goodman and K.L. Melmon, Conjugates of catecholamines. IV. In vitro and in vivo pharmacological activity of monodisperse oligopeptide conjugates. J. Pharmacol. Exp. Ther. submitted.

19. W.S. Emerson, The preparation of amines by reductive amination, Org. Reactions 4:174 (1948).

20. R.F. Borch, M.D. Bernstein and H.D. Durst, The cyanohydridoborate anion as a selective reducing agent, J. Amer. Chem. Soc. 93:2897 (1971).

21. (a) P. Coffino, H.R. Bourne, P.A. Insel, K.L. Melmon, G. Johnson and J. Vigne, Studies of cyclic AMP action using mutant tissue culture cells, In Vitro 14:140 (1978); (b) A.G. Gilman, A protein binding assay for adenosine 3',5'-cyclic monophosphate, Proc. Natl. Acad. Sci. U.S.A. 67:305 (1970).

22. P.A. Insel and L.M. Stoolman, Radioligand binding to beta-adrenergic receptors of intact cultured S-49 cells, Mol. Pharm. 14:549 (1978).

23. R.P. Rosenkranz, K.A. Jacobson, M. Goodman, M.S. Verlander and K.L. Melmon, In vitro and in vivo betamimetic activity of congeners of isoproterenol, Proc. W. Pharm. Soc. 25:19 (1982).

NEW POLYMERIC AND OLIGOMERIC MATRICES AS DRUG CARRIERS

Paolo Ferruti *, Gian Carlo Scapini *, Luisa Rusconi **, and Maria Cristina Tanzi **

*Istituto di Chimica degli Intermedi dell' Università, Via Risorgimento 4, 40136 Bologna, Italy
** Dipartimento di Chimica Industriale e Ingegneria Chimica del Politecnico, Piazza L.da Vinci 32 - 20133 Milano, Italy

INTRODUCTION

The use of synthetic polymers in the pharmacological field is at present one of the newest and most exciting areas of polymer chemistry. In particular, the synthesis and the bioactivity of macromolecular drugs are receiving increasing attention by several groups[1-38].

Two main routes are actually followed to prepare pharmacologically active polymers: i) the preparation of macromolecular drugs in which the activity is due to the whole macromolecule, which is supposed to be not chemically altered in the body fluids before exerting its activity; ii) the preparation of polymeric derivatives of drugs, in which the activity is related to the presence of moieties whose structure is strictly related to that of non macromolecular substances of well established activity. In some instances the activity does not seem to be related to a release of active fragments in the body fluids. However the most interesting results can be obtained in the case of macromolecular drugs acting because they are able to release non-macromolecular active fragments once introduced in the body fluids.

The main results which in principle may be achieved by this technique are: the maintenance on time of suitable blood levels of drugs, resulting in a longer duration of activity, and a minimization of side effects; a preferential localization at the level of given target cells or groups of cells, which may be obtained in some cases by the incorporation of purposely tailored segments in the matrix; and, in the case of some oligomeric derivatives, an improved bioavailability due to vehiculation across physiological barriers, resulting for instance in a better absorption through the gastrointestinal walls (see below).

The chemical problems involved in the preparation of macromolecular drugs of the second category are usually related to the selection of a suitable matrix, and the selection of a suitable bond between the drug and the matrix.

The matrix may be purposely tailored to impart the appropriate solubility to the final product (in most cases water solubility), and/or to be selectively localized in the body. This is usually obtained by adding to the active units suitable hydrophylic units which are by themselves pharmacologically inert, and/or preparing block copolymers in which "active" blocks are coupled with "directing" blocks [5].

An important point concerning high molecular weight matrices is their biodegradability in the body fluids. An undegradable backbone might be acceptable only in polymer-drug adducts to be taken orally, since high polymers usually are not absorbed through the gastrointestinal tract, if previously freed from low molecular weight fractions. On the contrary, an undegradable backbone is not acceptable in polymer-drug adducts to be administered parenterally, in order to avoid any long-term permanence in the body of macromolecular residues.

These problems, of course, are not encountered with low molecular weight matrices.

Broadly speaking, the preparation of oligomeric or polymeric derivatives of drugs may be achieved by two main routes: i) the preparation of a polymerizable derivative of the drug, and its polymerization; ii) the preparation of oligomeric or polymeric matrices carrying chemical functions able to react selectively with some groups present in the drug molecule. Even if both routes can be employed, the latter is more convenient, in principle, because

a single type of matrix may be used to prepare derivatives of a number of drugs. Furthermore, in many cases, the drug moieties contain
chemical functions which may interfere with the polymerization processes. As far as the type of bond between the drug and the matrix
is concerned, ester bonds or iminic bonds are usually preferred because, as a rule, they are easily cleaved in the body fluids. Amidic
bonds may be also considered, but in most cases they are not easily
cleaved in physiological environments. At any rate, every covalent
bond which may undergo degradation by restoring the original drug
molecule, may be convenient.

HIGH MOLECULAR WEIGHT DRUG BINDING MATRICES

Our work on multifunctional drug-binding matrices was firstly
concerned with the synthesis of polymeric imidazolides and benzotriazolides. To this purpose, a series of high molecular weight polymers
with a polyvinylic backbone have been synthesized [39-41]. These polymers are listed in Table 1.

Other activated derivatives of polymeric acids have been synthesized by Ringsdorf and his group [42].

All the polymers listed in Table 1 are able to enter into exchange reactions with alcohols and amines, giving ester or amidic
bonds. The order of reactivity is:
succinimides < benzotriazolides < imidazolides .

On the whole, the polymeric benzotriazolides offer the best
combination of ease of preparation, stability, and chemical reactivity. In reacting with compounds containing several reactive
functions, they also show a fairly good selectivity, reacting first
with primary or secondary amino groups, then with primary or phenol-
ic hydroxyl groups, and finally with secondary hydroxyl groups. As
expected, bulky substituents around the reactive functions tend to
slow down the exchange reaction.

For all these reasons, the polymeric benzotriazolides appear
to provide an excellent route to prepare polymeric derivatives of
many drugs, as well as a number of multifunctional polymers.

As pointed above, polymeric matrices with a polyvinylic backbone are not expected to be easily degraded by themselves in biological environments. On the other hand, activated derivatives of
polymeric acids with a main backbone degradable to safe metabolites

Table 1.   Activated matrices of polyvinylic structure

polymer	structure of the repeating unit
poly(N-acryloxysuccinimide)	
poly(N-methacryloxysuccinimide)	
poly(N-acryloxylimidazole)	
poly(1-acryloylbenzotriazole)	

polymer	structure of the repeating unit
poly(1-(N-methacryloyl)-ε-aminocaproyl-benzotriazole)	
poly(1-(4-methacryloxy)benzoylbenzo-triazole)	

would offer, in most cases, the best opportunities to prepare polymer-drug adducts suitable, at least in principle, for all ways of
administration. To this purpose, we thought it interesting to prepare activated amides of the succinic half-esters of natural polysaccharides, namely starch and dextrane, and to study their ability
to enter into exchange reactions with alcohols and amines [43].   The
succinic half-esters of starch and dextrane were prepared according
to the following scheme:

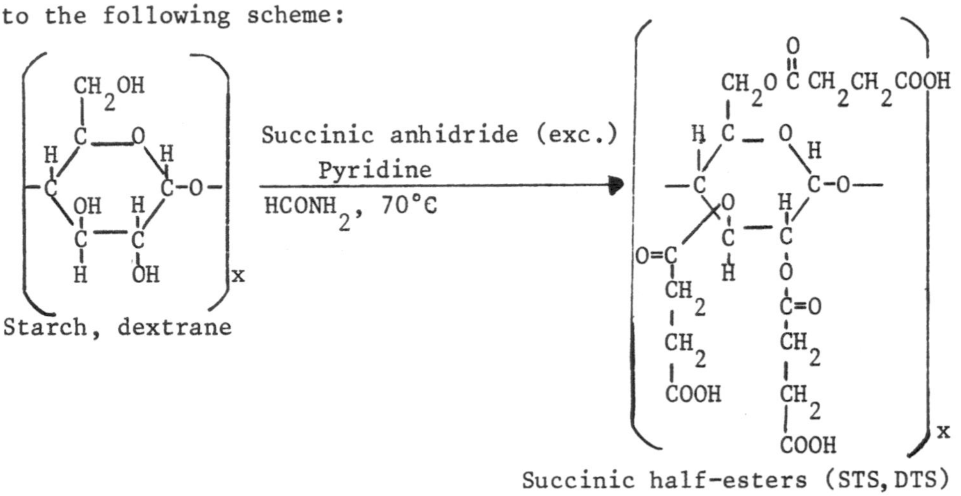

Starch, dextrane

Succinic half-esters (STS, DTS)

Their imidazolides were prepared by reacting the half-esters
either with carbonyldiimidazole (CDI) or with imidazole in the
presence of dicyclohexylcarbodiimide (DCCI) in dimethylsulphoxide
solution.   Usually they were not isolated, since they were very
sensitive to the atmospheric moisture, but were directly reacted
with the appropriate alcohol or amine.   The yields of the exchange
reactions varied from 60% (by the carbodiimide method) to about
100% (by the carbonyldiimidazole method). These figures apply to
primary alcohols, and aliphatic or cycloaliphatic amines.

Secondary alcohols usually give reaction yields of about 30%.

Scheme

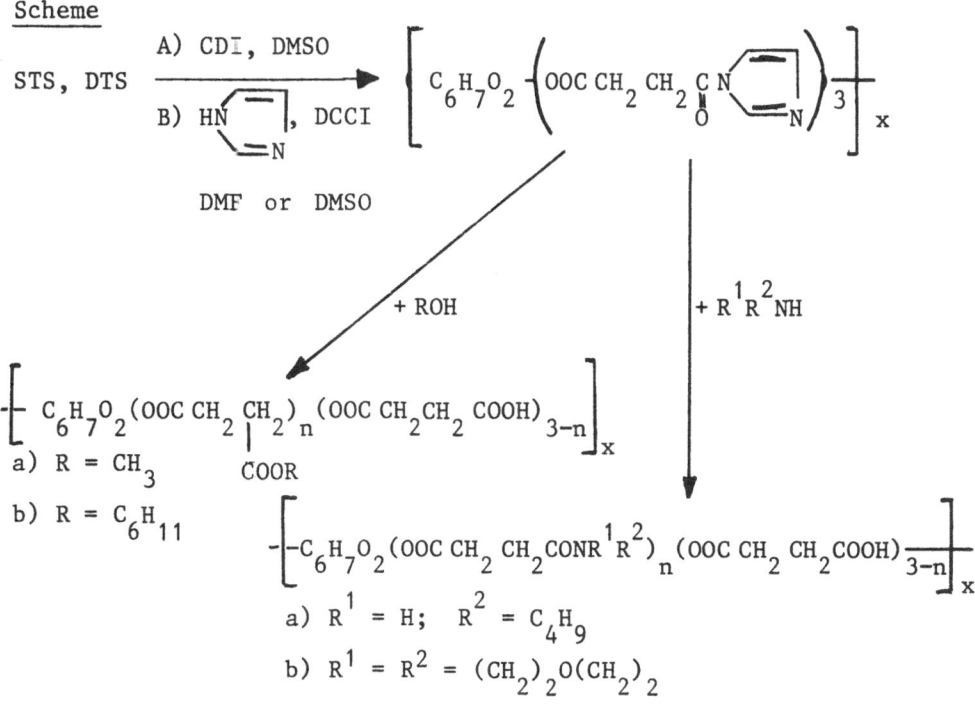

$$STS, DTS \xrightarrow[\substack{B) HN \\ DMF \ or \ DMSO}]{A) CDI, DMSO} \left[C_6H_7O_2 \left(OOC\ CH_2\ CH_2\ \underset{O}{C}\ N \underset{N}{\bigcirc} \right)_3 \right]_x$$

+ ROH | + R^1R^2NH

$$\left[C_6H_7O_2 (OOC\ CH_2\ \underset{\underset{COOR}{|}}{CH_2})_n (OOC\ CH_2CH_2\ COOH)_{3-n} \right]_x$$

a) $R = CH_3$

b) $R = C_6H_{11}$

$$\left[-C_6H_7O_2 (OOC\ CH_2\ CH_2 CONR^1R^2)_n (OOC\ CH_2\ CH_2 COOH)_{3-n} \right]_x$$

a) $R^1 = H$; $R^2 = C_4H_9$

b) $R^1 = R^2 = (CH_2)_2O(CH_2)_2$

By this technique, polymeric adducts of model multifunctional compounds, such as 4-aminobenzoic acid, or L-dopa, are easily pre-pared.

The benzotriazolide of the succinic half-ester of starch was prepared by reaction with 1-H-benzotriazole in the presence of ci-cyclohexylcarbodiimide. The reaction yield was about 66%, i.e., only about two out of three carboxyl groups were transformed into benzotriazolide groups:

$$STS + \underset{}{\bigcirc} \overset{H}{\underset{N}{\overset{N}{\bigcirc}}} N + DCCI$$

DMF, 0-15°C

$$\left[C_6H_7O_2 (OOC\ CH_2\ CH_2 COOH) \left(OOC\ CH_2\ CH_2 \underset{O}{\overset{C}{\|}} N \underset{N=N}{\bigcirc} \right)_2 \right]_x$$

On the whole, we found that STS benzotriazolide has about the same reactivity towards alcohols and amines as other previously described polymeric benzotriazolides [40]. When caused to react with methanol or morpholine under proper conditions, STS benzotriazolide gave products identical to those obtained from the corresponding imidazolides.

OLIGOMERIC MATRICES

As we shall see later, the use of oligomeric instead of high molecular weight matrices to prepare derivatives of drugs may still lead to products with a considerably prolonged pharmacological activity. Furthermore, in the case of oral administration, the oligomeric matrix is often able to vehiculate the active principles across the gastrointestinal walls, thus facilitating absorption and increasing bioavailability.

α-Hydro-ω-hydroxy-poly(oxyethylene)s (poly(ethyleneglycol)s) appear to be particularly convenient as oligomeric matrices, since they are available as fractions with well defined molecular weights, are highly hydrophilic, and have a good biocompatibility. They have been used as such as carriers for some carboxyl-bearing drugs (see below).

In order to obtain oligomeric matrices able to react with drugs bearing hydroxyl- or amino groups, succinic half esters of poly(ethyleneglycol)s were prepared, and transformed into the corresponding benzotriazole and imidazole derivatives. These proved to be able to enter into exchange reactions with hydroxylated or aminated compounds [37], in a way similar to that of high molecular weight matrices bearing similar functions (see above)

$$H-(OCH_2CH_2)_n-OH + \cdots \longrightarrow \left[\begin{array}{c} O \\ \| \\ C-(OCH_2CH_2)_n-OC=O \\ | \qquad\qquad\qquad\quad | \\ CH_2 \qquad\qquad\qquad CH_2 \\ | \qquad\qquad\qquad\qquad | \\ CH_2 \qquad\qquad\qquad CH_2 \\ | \qquad\qquad\qquad\qquad | \\ COOH \qquad\qquad\quad COOH \end{array} \right] \longrightarrow$$

(A)

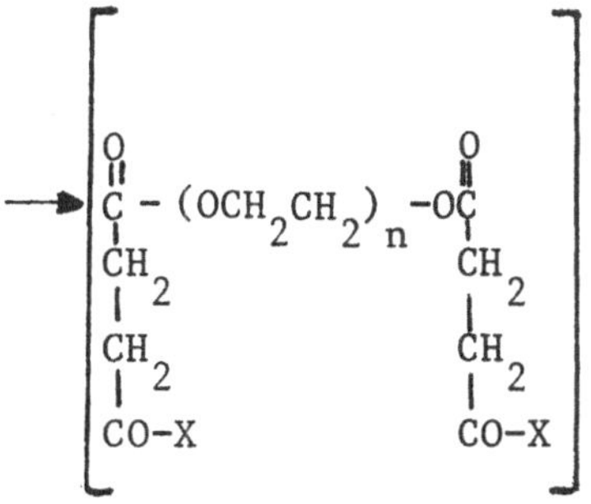

$$\left[\begin{array}{c} \overset{O}{\overset{\|}{C}} - (OCH_2CH_2)_n -O\overset{O}{\overset{\|}{C}} \\ \underset{|}{CH_2} \qquad \underset{|}{CH_2} \\ \underset{|}{CH_2} \qquad \underset{|}{CH_2} \\ CO-X \qquad CO-X \end{array} \right]$$

(B)

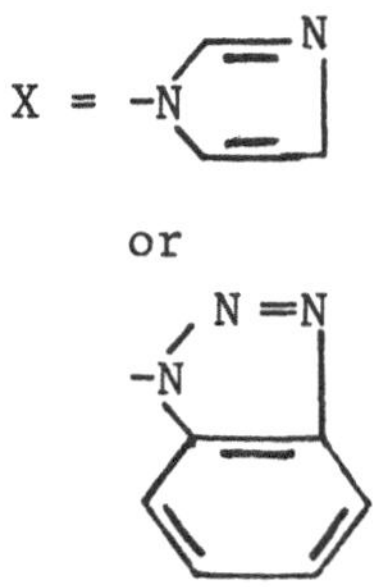

$$X = -N \quad \text{or} \quad -N$$

R-OH

(B) + $\xrightarrow{\;-\,2\,H\,X\;}$ Products

R_1R_2-NH

The same technique has been applied to poly(propyleneglycol)s, in order to obtain less hydrophilic, more lipophilic oligomeric drug-binding matrices [38].

PHARMACOLOGICAL RESULTS

Among the macromolecular drugs belonging to the first of the two main categories defined in Introduction, we have studied anti-silicotigen polymers [44, 45] and antimetastatic polymers [46]. Among the macromolecular drugs of the second category, very good results in experimental animals have been obtained with polymeric derivatives of nicotinic acid [8]. Polymeric derivatives of prostaglandin $F_2\alpha$ have also been studied [7]. We think it interesting to report here some recent results obtained with oligomeric derivatives of 4-isobutylphenyl-2-propionic acid (Ibuprofen), a well known anti-inflammatory agent, and with an oligomeric derivative, and a high molecular weight derivative, of $3\alpha,7\beta$-dihydroxy-6β-cholan-24-oic acid (ursodeoxycholic acid), a cholagogic agent.

(a) Poly(ethyleneglycol)derivatives of Ibuprofen

The synthesis of poly(ethyleneglycol)derivatives of Ibuprofen with $\overline{M}_n$ ranging from 400 to 2400, and containing either one (Scheme A), or two (Scheme B) drug residues per molecule, has been perform-ed [47, 48] starting from the corresponding Ibuprofen imidazolide:

Scheme A

Scheme B

The results of the pharmacological evaluation on test animals (rats) of some of the above derivatives are summarized in Figures 1, 2, and 3.

It may be observed that in all cases a more sustained activity in the time can be achieved. In the lower-molecular-weight compounds, also the initial activity increases with respect to the free drug. This means that the bioavailability is considerably increased, what is probably due to a vehiculating effect of the matrix across the gastrointestinal walls, leading to better absorption. More recently, these results have been confirmed with similar derivatives of (+)-6-methoxy-α-methyl-naphtaleneacetic acid (Naproxen).

The ulcerogenic potency of all these derivatives was somewhat lower than that of the free drugs. Their oral LD_{50} in mice were always higher than 2000 mg/Kg.

(b) Oligomeric and polymeric derivatives of ursodeoxycholic acid

A derivative of ursodeoxycholic acid with poly(ethyleneglycol) of $\overline{M}_n$ 200 was prepared in a way similar to that previously reported in the case of Ibuprofen derivatives. This derivative was esterified only at one end, and, consequently, had a primary hydroxyl group at the opposite end:

a) $U-COOH +$ $\longrightarrow U-CON$ $+ CO_2 + Imid$

b) $U-CON$ $+ HO(CH_2CH_2O)_4CH_2CH_2OH$ (exc.)

$$CHCl_3 \quad \Big| \quad 60°C$$

$U-COO(CH_2CH_2O)_4CH_2CH_2OH + Imid$

$U =$

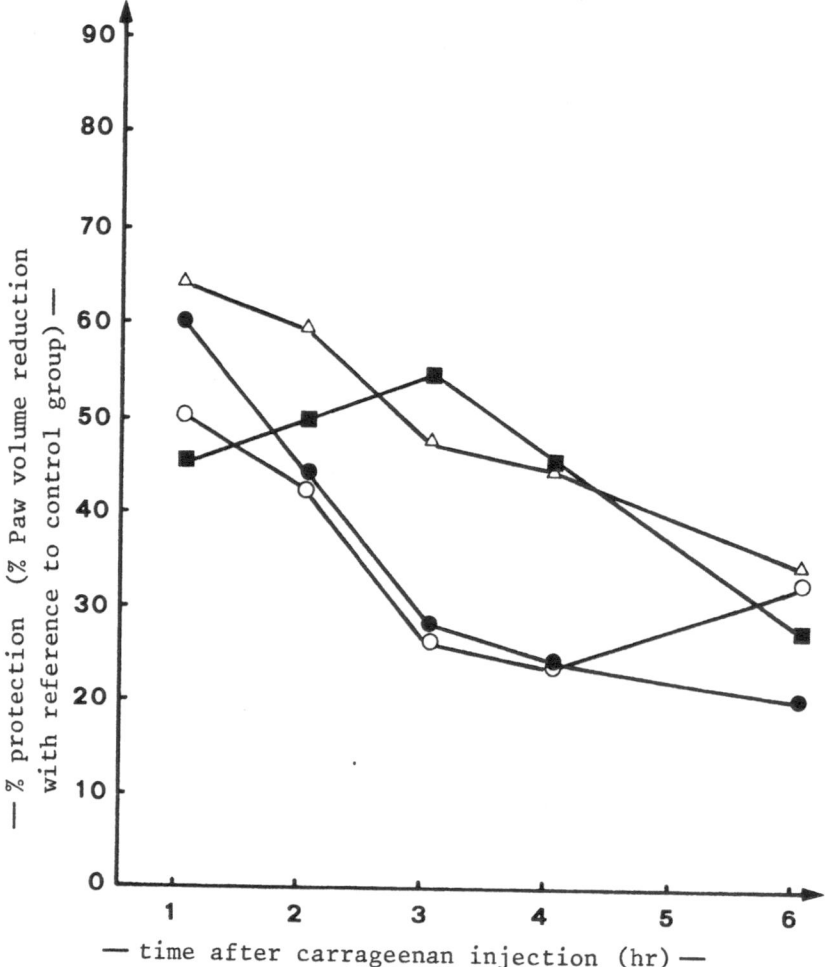

Fig. 1.   ●——● Ibuprofen

△——△ IBU-COO(CH$_2$CH$_2$O)$_3$CH$_2$CH$_2$OH   (IBU-1)

□——□ IBU-COO(CH$_2$CH$_2$O)$_{20.3}$CO-IBU   (IBU-2)

○——○ IBU-COO(CH$_2$CH$_2$O)$_{45}$CO-IBU   (IBU-3)

(IBU = $\underline{i}$.C$_3$H$_7$CH$_2$—⟨○⟩—CH—)
                                    |
                                   CH$_3$

All drugs were administered orally at the abscissa times
before subplantar injection of carrageenan, in a dose corresponding to 100 mg/kg Ibuprofen, to male and female
Wistar rats (12 for every compound and every measurement).
Paw volumes were measured 4 hr after carrageenan injection.

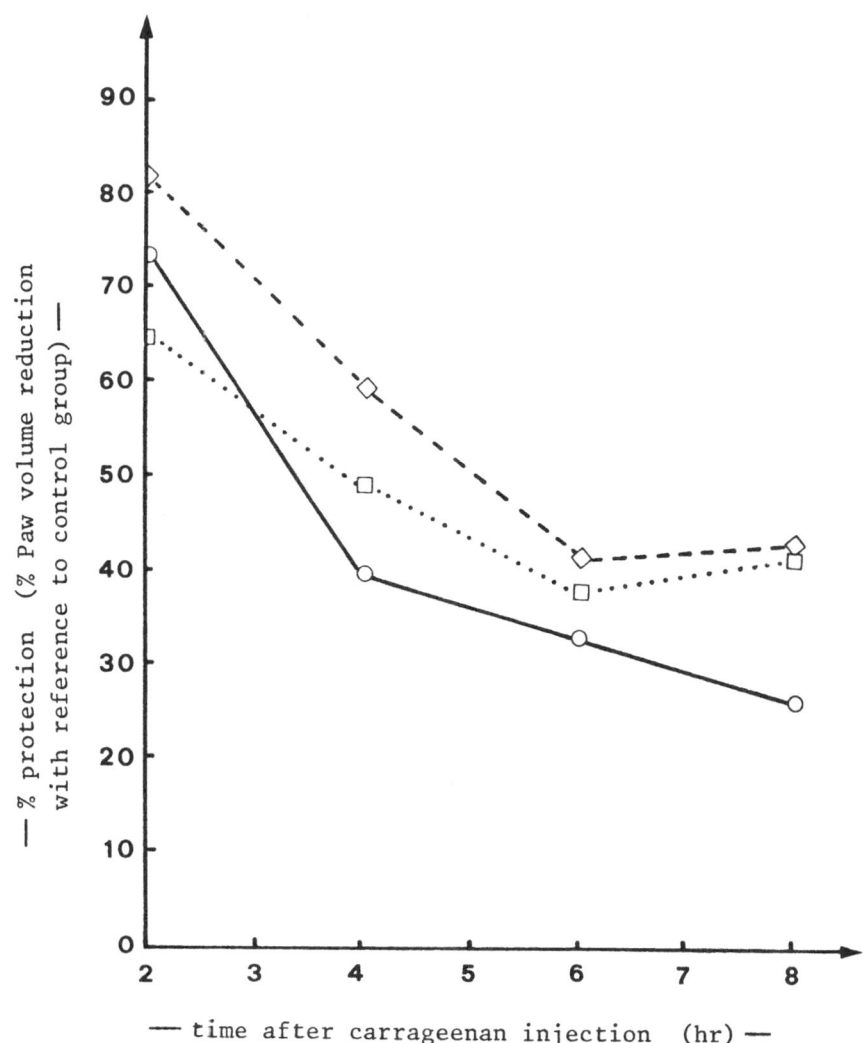

Fig. 2. O—O Ibuprofen (50 mg/kg)
         ◇—◇ IBU-1 (dose corresponding to 50 mg/kg Ibuprofen)
         □—□ IBU-1 (dose corresponding to 30 mg/kg Ibuprofen)

Drugs were administered orally 1 hr before carrageenan
injection in 28 male Wistar rats, random assigned to
four groups. Paw volumes were measured at the abscissa
times, after carrageenan injection.

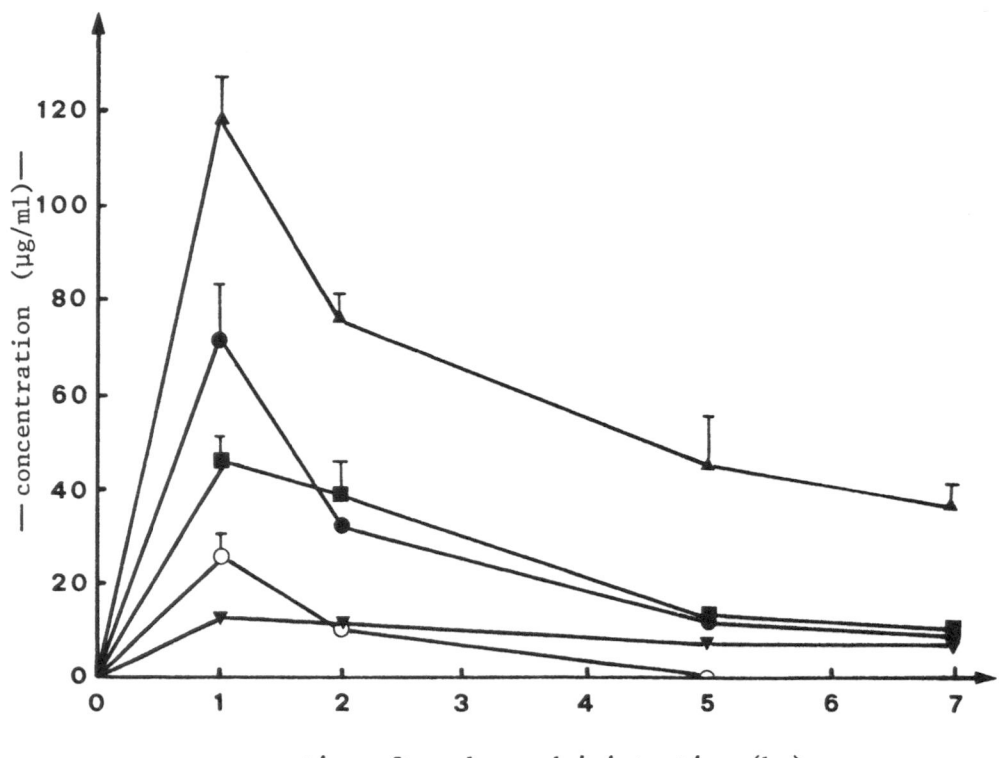

— time after drug administration (hr) —

Fig. 3. Ibuprofen plasma levels in male Wistar rats, after oral ad-
ministration of Ibuprofen (O 19.2 mg/kg; ● 58.8 mg/kg)
IBU-1 (▲, dose corresponding to 58.8 mg/kg Ibuprofen),
IBU-2 (■, dose corresponding to 38.8 mg/kg Ibuprofen),
IBU-3 (▼, dose corresponding to 19.2 mg/kg Ibuprofen).
Plotted values are the mean ± SE of N = 5.

This oligomeric derivative was further reacted through the free end with a copolymer of N-acryloylmorpholine and N-acryloylbenzotriazole, thus obtaining a high molecular weight adduct of the same acid:

$$RCOOH \longrightarrow RCON \longrightarrow RCO(OCH_2CH_2)_n OH \qquad (I)$$

(I) +

Since the benzotriazolides show selectivity towards primary hydroxy-groups (see above), no crosslinking reactions were observed.

Both compounds have been fed to fasting human volunteers at a dose corresponding to 300 mg of free drug, and the plasma levels of ursodeoxycholic acid determined over a 12 hrs period, and compared with those obtained with an equivalent dose of the free drug.

It has been found that the polymeric derivative gives a pharmacologically significant plasma level of drug $(1,8 \mu mol/ml)$, which is practically constant over the whole observation period, and probably more. The oligomeric derivative gives a peak of $10 \mu mol/ml$ after 3 hrs, then the plasma levels steadly decrease, but are still pharmacologically significant after 10 hrs. The free drug gives a peak of $3,5 \mu mol/ml$ after 2hrs, then drops to undetectable levels after 4 hrs.

The results are reported in Figure 4.

Thus, both derivatives show a considerable pharmacological interest, and in the case of the oligomeric one, the results obtained with Ibuprofen have been duplicated. This further confirms that binding of drugs to oligomeric and polymeric matrices is a general technique which allows to modulate both the initial activity, and the duration of the activity with time of many drugs.

It may be further observed that using poly(ethyleneglycol)s as oligomeric matrices, an increased adsorption after oral administration is almost generally observed, provided the molecular weight of the matrix is sufficiently low. This may allow to admi-

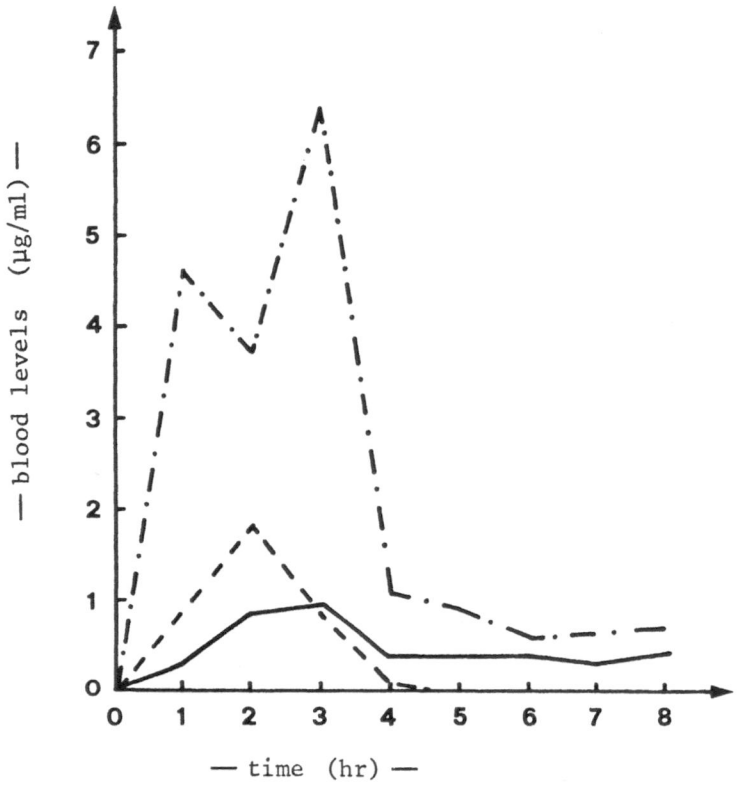

Fig. 4. Blood levels of ursodeoxycholic acid after administration
of oligomeric and polymeric derivatives, in comparison with
the free acid. All derivatives were administered orally in
a single dose corresponding to 300 mg of free drug, to
fasting human volunteers.
----- free acid;   -·-·- PEG oligomeric derivative;
——— polymeric derivative.

nister orally a number of drugs which in their free state are poorly absorbed by the gastrointestinal tract.

REFERENCES

1. B.Z. Weiner, M. Tahan, A. Zilkha, J. Med. Chem. 15:410 (1972)

2. H.G. Batz, H. Ringsdorf, H.Ritter, Makromol. Chem. 175:2229 (1974)

3. F. Scrollini, L. Molteni, Eur. J. Med. Chem. 9:621 (1974)

4. A. Havron, B.Z. Weiner, A. Zilkha, J.Med.Chem. 17:770 (1974)

5. H. Ringsdorf, J. Polym. Sci., Polym.Symp. 51:135 (1975)

6. H. Ringsdorf, H. Ritter, H. Rolly, Makromol. Chem. 177:741 (1976)

7. P. Ferruti, R. Paoletti, L. Puglisi, in "Advances in Prostaglandin and Thromboxane Research", Vol. 1, Raven Press, New York, 1976, p. 231

8. L. Puglisi, V. Caruso, R. Paoletti, P. Ferruti, M.C. Tanzi, Pharmacol. Res. Commun. 8:379 (1976)

9. P. Ferruti, Farmaco, Ed. Sci. 32:220 (1977)

10. H.G. Batz in "Advances in Polymer Science", Springer Verlag, Berlin 1977, Vol. 23, p. 25

11. M. Przybylski, E. Fell, H. Ringsdorf, D.S. Zaharko, Makromol. Chem. 179:1719 (1978)

12. C. Pinazzi, J.C. Rabadeux, A. Pleurdeau, Eur.Polym.J. 14:205 (1978)

13. C. Pinazzi, J.C. Rabadeux, A. Plaurdeau, P. Nivière, J.P. Paubel, J.P. Benoit, Makromol. Chem. 179:1699 (1978)

14. V. Hoffmann, H. Rinsgdorf, E. Schaumlöffel, Makromol. Chem. 180:595 (1979)

15. C. Pinazzi, J.P. Benoit, J.C. Rabadeux, A. Pleurdeau, Eur. Polym. J. 15:1069 (1979)

16. E.P. Goldberg, in Polymeric Drugs, eds. G.Donaruma and O.Vogl, Academic Press, New York, 1978, pp. 239-261

17. E.P. Goldberg, H. Iwata, R.N. Terry, W.E. Longo, M. Levy, T.A.
     Lindheimer and J.L. Cantrell, in Affinity Chromatography and
     Related Techniques, eds. T.C.J. Gribnau, J. Visser and R.J.
     F. Nivard, Elsevier, Amsterdam, 1982, pp.375-386

18. L.G. Donaruma, J.R. Dombroski, J.Med.Chem. 14:460 (1971)

19. L.G. Donaruma, J. Razzano, J. Med. Chem. 14:244 (1971)

20. L.G. Donaruma, O. Vogl, Eds., Polymeric Drugs, Academic Press,
     New York, 1978

21. R.M. Ottenbrite, E. Goodelle, A. Munson, Polymer, 18:461
     (1976)

22. L.G. Donaruma, R.M. Ottenbrite, O. Vogl, "Anionic   Polymeric
     Drugs", John Wiley and Sons, New York, 1980

23. C.M. Samour, Chemtech. 8:494 (1978)

24. J. Pitha, J. Zjawiosci, R.J. Lefkowitz, M.G. Caron, Proc.Natl.
     Acad. Sci. USA, 77(4):2219 (1980)

25. E. Hurwitz, M. Wilchok, J. Pitha, J. Appl.Biochem. 2:25 (1980)

26. J. Pitha, J. Zjawioscy, R.J. Lefkowitz, M.G. Caron, Makromol.
     Chem. 182:1945 (1981)

27. J. Pitha, S. Zawadzki, B.A. Hughes, Makromol. Chem. 183:781
     (1982)

28. M. Vert, R.W. Lenz, A.C.S. Polym. Prept. 20(1):608 (1979)

29. E. Schacht, L. Ruys, E. Goethals, P. Gyselinck, R. Van Severen
     P. Braeckman, J. Pharm. Belg., 36(2):113 (1981)

30. J. Kopecek, P. Rejmanova, V. Chytry, Makromol.Chem. 182:799
     (1981)

31. R. Duncan, P. Rejmanova, J. Kopecek, J.B. Lloyd, Biochim.
     Biophys. Acta 678:143 (1981)

32. P. Rejmanova, B. Obereigner, J. Kopecek, Makromol.Chem.
     182:1899 (1981)

33. R. Duncan, P. Rejmanova, J. Kopecek, J.B.Lloyd, Biochim.
     Biophys. Acta 678:143 (1981)

34. B.Z. Weiner, A. Zilkha, G. Porah, Y. Grundfeld, Eur.J.Med.
     Chem.-Chim. Ther. 11:525 (1976)

35. C. Pinazzi, J.C. Rabadeux, A. Pleurdeau, J. Polym. Sci., Polym. Chem. Ed. 15:1319 (1977)

36. G. Bauduin, D. Bondon, Y. Pietrasanta, B. Pucci, F. Vial-Reveillon, Eur. J. Med. Chem.-Chim. Ther. 14:119 (1979)

37. P. Ferruti, M.C. Tanzi, L. Rusconi, R. Cecchi, Makromol.Chem. 182:2183 (1981)

38. L. Rusconi, M.C. Tanzi, C. Zambelli, P. Ferruti, Polymer 23:1689 (1982)

39. P. Ferruti, F. Vaccaroni, J. Polym. Sci., Polym. Chem. Ed. 13:2859 (1975)

40. P. Ferruti, F. Vaccaroni, M.C. Tanzi, J. Polym. Sci., Polym. Chem. Ed. 16:1435 (1978)

41. P. Ferruti, M.C. Tanzi, F. Vaccaroni, J. Polym. Sci., Polym. Chem. Ed., 17:277 (1979)

42. H.G. Batz, G. Franzmann, H. Ringsdorf, Makromol. Chem. 172:27 (1973)

43. P. Ferruti, M.C. Tanzi, F. Vaccaroni, Makromol. Chem. 180:375 (1979)

44. G. Natta, E.G. Vigliani, F. Danusso, B. Pernis, P. Ferruti, M.A. Marchisio, Atti Accad. Naz. Lincei 40:11 (1966)

45. P. Ferruti, M.A. Marchisio, Med. Lavoro 57:481 (1966)

46. P. Ferruti, F. Danusso, G. Franchi, N. Polentarutti, S. Garattini, J. Med. Chem. 16:496 (1973)

47. R. Cecchi, L. Rusconi, M.C. Tanzi, F. Danusso, P. Ferruti, J. Med. Chem. 24:622 (1981)

48. P. Ferruti, F. Danusso, M.C. Tanzi, G. Quadro, Italian Patent Application 19878A/80 and 19879A/80 (to Medea Researches s.r.l., Via Pisacane 34/A, Milan, Italy)

DEVELOPMENT OF N-(2-HYDROXYPROPYL)METHACRYLAMIDE COPOLYMERS AS

CARRIERS OF THERAPEUTIC AGENTS

Ruth Duncan, Jindřich Kopeček* and John B. Lloyd

Department of Biological Sciences, University of Keele
Keele, Staffs., U.K. and *Institute of Macromolecular
Chemistry, Czechoslovak Academy of Sciences, Prague
Czechoslovakia

INTRODUCTION

Most pharmacological agents are low molecular weight compounds which readily penetrate many cell types and hence, unless they specifically interfere with a unique biochemical pathway, produce unwanted effects as well as therapeutic benefit. Attachment of drugs to soluble macromolecular carriers restricts their uptake by cells to the mechanism of pinocytosis and thus provides the opportunity to:-

i) target drug specifically to the cell type where its action is required.

ii) deliver drug intracellularly over a prolonged period of time at concentrations chosen to maximize therapeutic efficiency.

Although much attention has been paid to the use of insoluble synthetic polymers (in the form of gels or networks) as depots for controlled drug release,[1,2] relatively little attempt has been made to evaluate the potential of soluble synthetic polymers as a targetable drug delivery system.[3] Most studies on macromolecular carriers have confined themselves to using naturally occurring macromolecules, e.g. dextran,[4] albumin,[5] DNA[6] and antibodies,[7] as the carrier vehicle. We feel that synthetic polymers offer several theoretical advantages over these systems, because they can be tailor-made to meet the individual requirements of any particular drug delivery system under consideration.[8]

Macromolecules, unless they are highly lipid-soluble, can only

penetrate cells by the mechanism of pinocytosis.[9]   During
pinocytosis the plasma membrane invaginates and then pinches off
forming a vesicle that contains macromolecules that were either
adherent to the membrane or simply dissolved in the extracellular
fluid.   Pinocytic vesicles then migrate into the cell fusing with
each other and also with vesicles of intracellular origin called
lysosomes; the latter contain some seventy different hydrolytic
enzymes[10] which between them are capable of digesting all naturally
occurring macromolecules they would normally expect to meet.   The
monomeric constituents thus liberated are able to traverse the
lysosome membrane and enter the cytoplasm for reutilization or loss
from the cell.   In theory it should be possible to devise a
synthetic polymeric carrier system with the following properties:-

   i) a suitable molecular weight to facilitate efficient
       capture by the desired target cells

  ii) targeting residues known to enhance membrane adsorption
       and thus pinocytic capture by target cells

 iii) polymer-drug linkages that are stable during transport to
       the target cells but degradable by lysosomal enzymes once
       the carrier complex is internalized

  iv) the polymer itself should be biologically inert to avoid
       provoking an immune reaction or inducing of polymer-
       related toxic effects.

HYDROXYPROPYLMETHACRYLAMIDE COPOLYMERS AS POTENTIAL DRUG CARRIERS

Preparation of N-(2-hydroxypropyl)methacrylamide copolymers

     Some years ago a new hydrophilic polymer, poly N-(2-hydroxy-
propyl) methacrylamide (HPMA) (see Fig. 1), was synthesized in
Czechoslovakia with the aim of preparing a polymer having the
properties needed for a blood plasma expander.[11,12,13]   This
polymer proved suitable for this purpose[14,15] and so is also
potentially useful as a carrier of therapeutic agents.[16,17,18]
Although it is possible to bind therapeutic agents to the

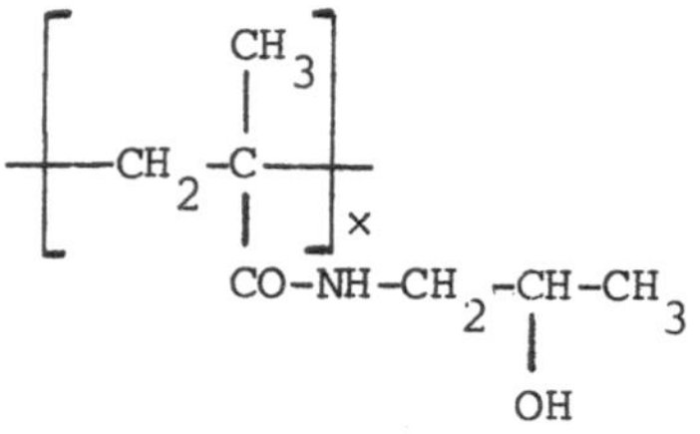

$$\left[-CH_2-\underset{\underset{CO-NH-CH_2-\underset{\underset{OH}{|}}{CH}-CH_3}{|}}{\overset{\overset{CH_3}{|}}{C}} - \right]_x$$

Fig. 1

homopolymer via the secondary alcohol groups we have concentrated on the preparation of copolymers of HPMA with oligopeptidyl side-chains terminating in the drug analogue.   These are prepared by copolymerizing HPMA and p-nitrophenyl esters of methacryloylated aminoacids or oligopeptides.[19,20,21]   The p-nitrophenyl ester group was chosen as the reactive group because active esters are frequently used in the synthesis of peptides[22] and in polymer chemistry.[23,24,25]

. As an example the preparation of a copolymer of HPMA with the p-nitrophenyl ester of N-methacryloylglyglycine is given below.

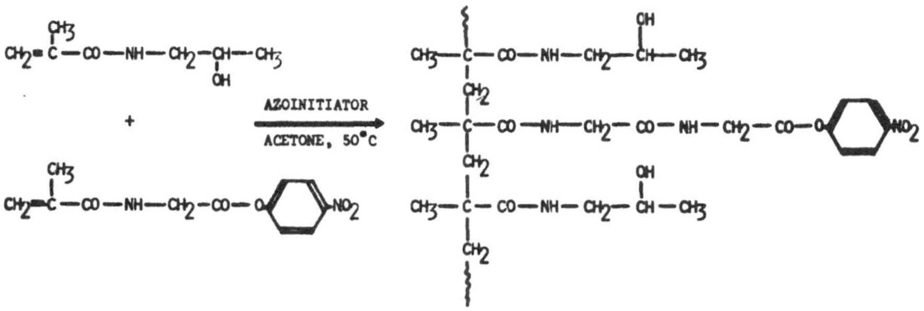

Fig. 2

The p-nitrophenoxy group reacts readily with compounds that contain an aliphatic amino group, with the formation of an amide bond.   The reaction may be easily detected spectrophotometrically in the UV region.[26]

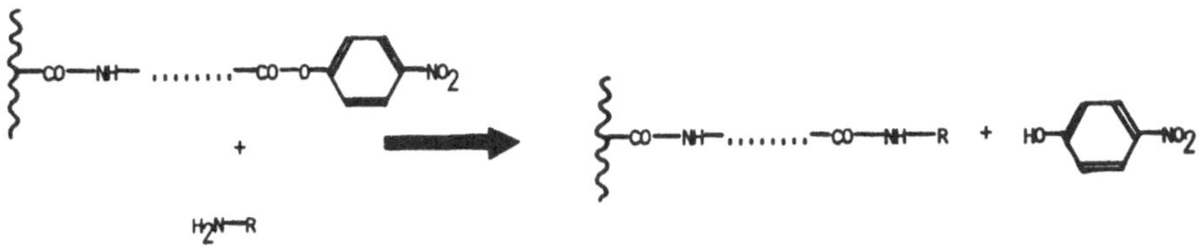

Fig. 3

Reaction between amines and the reactive HPMA copolymers forms the basis for attachment of oligopeptide sequences, targeting residues or therapeutic agents to the polymer precursor (Fig. 3). Reaction of aliphatic amines (tertiary butylamine and diisopropyl-amine) in dimethylsulphoxide at 25°C with reactive copolymers con-taining different side-chains has been used to characterize copolymer reactivity.[21]   It has been shown that copolymers of HPMA with p-nitrophenyl esters of α-(acylamino) acids react faster with amines than copolymers with p-nitrophenyl esters of β-(acyl-amino) acids or, ε-(acylamino) acids.[21]

Drugs or model drugs (p-nitroaniline, tyrosinamide) can be bound to the polymeric carrier by an oligopeptide spacer, the amino acid sequence the latter being chosen to suit known enzyme specificities.   Similarly copolymers can also be prepared containing residues that can specifically (e.g. galactosamine) or non-specifically (e.g. tyrosine) enhance their rates of pinocytic capture.   An advantage of the system under discussion is that several compounds containing an amino group can be sequentially bound to the polymer precursor, e.g. it is possible to bind consecutively a drug, a targeting moiety and even a second drug.

The above-mentioned techniques have been used to prepare HPMA copolymers which have been used to:

i) follow the cleavage of oligopeptidyl side-chains by enzymes in vitro.

ii) measure their rate of pinocytic capture by rat visceral yolk sacs cultured in vitro (a model system for monitoring pinocytosis and intracellular degradation)

iii) investigate the ability of sugar residues to target copolymers in vivo

iv) evaluate the possibility of drug delivery.

Degradation of HPMA copolymers by lysosomal enzymes in vitro

Previously it has been shown that HPMA copolymers bearing oligopeptidyl side-chains terminating in p-nitroaniline can be cleaved at the terminal bond by chymotrypsin.[27]   Hydrolysis was found to be dependent on the presence in the sequence of amino acids that are known to interact with the active site of the enzyme; also the rate of cleavage was shown to increase with increasing length of the peptidyl side-chain.   In context of the polymeric carrier system, the oligopeptidyl side-chains of HPMA are viewed as drug attachment/release sites, and therefore we wished to investigate their susceptibility to hydrolysis by lysosomal enzymes. In an initial investigation[28] we showed that rat liver lysosomal enzymes (prepared as tritosomes) were able to liberate p-nitroaniline from four oligopeptidyl side-chains during incubation at 37°C in potassium phosphate buffer (0.2M containing 0.2% Triton X-100) at pH 5.5.   Eighteen other side-chains were resistant to degradation under these conditions.   Subsequently we have repeated these experiments but with the addition of reduced glutathione (5mM) and EDTA (3mM) to the buffer.   These additives are essential for the expression of lysosomal thiol-proteinase activity, and p-nitroaniline was liberated from many more of the side-chains under these conditions (Table 1).   Of the oligopeptidyl sequences

Table 1.   Degradation of oligopeptidyl side-chains of HPMA copolymers
           by rat liver lysosomal enzymes

Code No.	Side Chain	Amount of polymer (g) that contains 1 mole active groups	Rate of NAp release (%/H)	
			No addition	+GSH
1	-Gly-Ala-Phe-NAp	6591	0	ND
2	-Gly-Ileu-Phe-NAp	8610	0	1.38
3	-Gly-Val-Phe-NAp	8065	0	3.71
4	-Gly-Gly-Phe-NAp	5550	0	0.44
5	-Gly-Gly-Phe-Phe-NAp	6043	0.52	1.81
6	-Gly-Ala-Tyr-NAp	6448	0	ND
7	-Gly-Ileu-Tyr-NAp	8770	0	5.96
8	-Gly-Val-Tyr-NAp	8293	0	ND
9	-Gly-Gly-Tyr-NAp	5683	0	ND
10	-Gly-Gly-Phe-Tyr-NAp	6125	1.01	ND
11	-Acap-Phe-NAp	5627,7447	1.75	0.6
12	-Acap-Leu-NAp	5193	0	0
13	-Gly-Phe-NAp	5712	0	0
14	-Gly-Leu-Phe-NAp	4773	0.74	2.59
15	-Gly-Phe-Phe-NAp	9253	0	2.96
16	-Gly-D-Phe-Phe-NAp	7538	0	3.62
17	-Ala-Gly-Val-Phe-NAp	8014	0	12.56
18	-Gly-Gly-Val-Phe-NAp	8503	0	3.06
19	-Gly-Phe-Tyr-NAp	8707	0	2.44
20	-Gly-β-Ala-Tyr-NAp	5717	0	0
21	-Gly-Leu-NAp	4420	0	0

studied, only four proved completely resistant to cleavage by
lysosomal thiol-proteinases. The side-chain -Ala-Gly-Val-Phe-NAp
was the most rapidly cleaved.   In some cases the liberation of
p-nitroaniline was not linear with time throughout the course of the
5h incubation period and the rates of cleavage shown in Table 1
represent the maximum linear rate.

     When it became apparent that the lysosomal thiol-proteinases
were particularly important in the degradation of HPMA copolymer
side-chains, a new series of side-chains was designed to meet known
specificities of particular lysosomal enzymes.[29]   Two side-chains
were synthesized (-Gly-Phe-Leu-Gly-Phe-NAp and -Gly-Gly-Phe-Leu-Gly-
Phe-NAp) which contain a pentapeptide sequence previously shown[30]
to be susceptible to the lysosomal proteinase cathepsin D, a thiol-
independent enzyme.   Five further sequences were prepared containing hydrophobic amino acids in the P_2 and P_3 positions in
relation to the terminal bond (according to the terminology of

Schechter and Berger[31]).   These side-chains were designed for
cleavage by the lysosomal thiol-proteinase cathepsin L, a rat liver
lysosomal enzyme[32] known to hydrolyse oxidized B-chain of insulin,
such that hydrophobic residues are present in the P_2 and P_3
positions.[33]

    All seven sequences used in these experiments were readily
hydrolysed by rat liver lysosomal enzymes in the presence of GSH
(5mM) and EDTA (3mM) Table 2.   No liberation of p-nitroaniline
occurred in their absence.   Degradation was effectively inhibited
by leupeptin, a thiol-proteinase inhibitor, and Sephadex G-15
chromatography of incubation mixtures showed that the major
hydrolysis product was always p-nitroaniline, indicating that
cleavage usually occurred at the terminal amide bond.   In some
instances peptidyl nitroanilide fragments were also liberated.
Side-chains 27 and 28 were not cleaved by cathepsin D at any point
along their length.   Although p-nitroaniline was released steadily
from 22-26 immediately on adding enzyme, its liberation from 27 and
28 took place only after a lag phase of 50-100 min.   This probably
reflects a multistep enzyme cleavage of these side-chains before
eventual p-nitroaniline release.

    The requirement for thiol in the incubation buffer and the

Table 2. Degradation of oligopeptidyl side chains of HPMA copolymers
              by rat liver lysosomal enzymes

Code No.	Side chain	Amount of polymer (g) that contains 1 mole active groups	Rate of NAp release (%/h)	NAp released after 5h %
22	-Gly-Phe-Phe-Ala-NAp	8310	8.31	22.26
23	-Gly-Phe-Phe-Leu-NAp	8804	16.20	26.50
24	-Ala-Val-Ala-NAp	13204	7.65	23.62
25	-Gly-Phe-Phe-Gly-NAp	8010	7.29	21.94
26	-Gly-Phe-Leu-Gly-NAp	10450	20.37	45.18
27	-Gly-Phe-Leu-Gly-Phe-NAp	11351	NL	> 52.00
28	-Gly-Gly-Phe-Leu-Gly-Phe-NAp	9114	NL	> 40.00

inhibition by leupeptin confirm the importance of thiol-proteinase activity in side-chain degradation, but do not pinpoint which particular enzyme or enzymes are responsible.   There are three main thiol-proteinases in rat liver, cathepsins B, H and L. Although cathepsin B readily hydrolyses small synthetic oligopeptide substrates containing arginine or a diarginyl sequence[34] it also shows some specificity towards bonds adjacent to hydrophobic amino acids in oxidized insulin B-chain.[35]   The specificity of cathepsin H is still not well understood.

Incubation of HPMA copolymers with purified lysosomal enzymes should help identify the enzyme(s) most active in side-chain degradation.   Recently it has been found that incubation of cathepsin B with HPMA copolymers bearing oligopeptidyl side-chains terminating in p-nitroaniline (incubated in phosphate buffer (0.1M) pH 6.0 containing cysteamine (25mM) and EDTA (1mM) causes p-nitroaniline liberation from many sequences.[36]   The side-chain most susceptible to cleavage by tritosomes (26) is certainly very effectively hydrolysed by cathepsin B.[36]

Whatever the precise enzymic mechanism, we have certainly demonstrated that it is possible to incorporate into HPMA copolymers oligopeptidyl side-chains that are very effectively cleaved by lysosomal enzymes to liberate drug analogue.   These side-chains should provide a useful drug attachment/release point.   The next task is to demonstrate that sequences can be chosen that are susceptible to lysosomal hydrolysis but resistant to attack in plasma. Any carrier that parts from its load during transit has little value unless there is no synergistic benefit of the carrier on drug action.

Capture of HPMA copolymers by mammalian cells in culture

The rat visceral yolk sac is a tissue composed of three cellular layers that surrounds the developing rat embryo during gestation[37] and serves to provide nutrients for the growing fetus,[38] protect it from harmful substances and is possibly involved in the transfer of passive immunity.[39]   The outer layer of cells are of the columnar epithelial type and these cells are actively engaged in pinocytic capture of material from the uterine cavity.   At Keele over the past decade we have developed an organ culture system which has been used to study the physiological functions of the yolk sac and also, by using the tissue as a model system, to identify some of the important factors governing the pinocytic process[40,41] and to investigate the kinetics of uptake and degradation of macromolecules.[42,43]

HPMA copolymers whose side-chains contained tyrosine residues were radiolabelled with [^{125}I]iodide and their uptake by yolk sacs

and subsequent intracellular fate investigated.[44]   Copolymers
bearing four different oligopeptidyl side-chains were used, and all
were captured by the yolk sac at the same rate (Table 3), this rate
being similar to value reported previously for the rate of uptake
of ^{125}I-labelled poly(vinylpyrrolidone),[42] a substrate known to be
internalized entirely in the fluid-phase (i.e. with no binding to
the yolk sac surface).   The finding that HPMA copolymers have
little natural affinity for plasma membranes was encouraging, as
this maximized the potential to incorporate additional residues
chosen to target the polymer specifically.

Although all the copolymers studied were captured by the tissue
at the same rate, their side-chains were hydrolysed to differing
extents by lysosomal enzymes.   The most susceptible sequence was
-Gly-Gly-Tyr-NAp, some 60% of the [125I]iodotyrosine associated
with this copolymer being released back into the culture medium
during a 5h incubation.   Inclusion of leupeptin (40 µg/ml) in the
incubation medium completely inhibited the degradation of this side-
chain, again suggesting the importance of lysosomal thiol-
proteinases in the process.   The differential rates of liberation
of [125I]iodotyrosine from copolymer side-chains observed in these
experiments serves to illustrate the possibility of manipulating
side-chain composition to mediate the desired rate of release of a
pharmacon.

Targeting of HPMA copolymers

Efficiency of pinocytic capture can be greatly enhanced if substrates adsorb to the internalizing plasma membrane.[45]   Physio-
logically this phenomenon is utilized by cells to ensure that they
capture efficiently those macromolecules they require.   Some macro-

Table 3.   Pinocytic uptake and intracellular degradation of HPMA
copolymers by rat visceral yolk sacs cultured in vitro

Code Nos.	Side chain	Content of side chain in polymer (%mol)	Rate of pinocytic uptake (µl/mg protein/h)	Side chain degradation after 7h incubation (%)
29	-Gly-Ileu-Tyr-NAp	1.71	3.36	25.10
30	-Gly-Gly-Tyr-NAp	2.68	2.91	57.71
31	-Gly-Phe-Tyr-NAp	1.73	3.71	37.53
32	-Gly-β-Ala-Tyr-NAp	2.67	3.21	5.28

molecules have properties, such as an overall positive charge[40] or increased hydrophobicity,[46] that lead to a non-specific attraction for cell membranes and thus high rates of capture by many cell types.   Other contain specific residues that are complementary to and interact with others that form an integral part of membrane receptors unique to specific cell types.   The role of carbohydrate residues in specific pinocytic uptake systems is now well documented.[47]   Hepatocytes have membrane receptors that recognize galactose residues[48] and this receptor-mediated uptake system is important in the clearance of denatured serum glycoproteins from the circulation.   Fibroblasts efficiently capture exogenous lysosomal enzymes by their surface adsorption via mannose-6-phosphate,[49] whereas reticulendothelial cells have been shown to recognize the mannose, glucose or N-acetylglucosamine components of a variety of different substrates.[50]   Non-carbohydrate-dependent recognition systems also exist, one example being the highly specific capture of LDL by fibroblasts[51] which appears to involve ionic interaction between positive charges of the apoprotein B component of LDL and specific negative charges in the LDL receptor.

As mentioned above, HPMA copolymers containing approximately 2% oligopeptidyl side-chains have no affinity for the plasma membrane and enter cells by fluid-phase pinocytosis.   Increasing the hydrophobicity of macromolecules appears to enhance, non-specifically, their rate of capture by rat visceral yolk sacs[52] and we have shown that incorporation of 20% tyramine residues into the synthetic polymer poly-α-β-N(2-hydroxyethyl) -D,-L- aspartamide greatly enhances its rate of uptake by this tissue.[53]   More recently we have found that incorporation of tyrosine residues (approximately 18%) into HPMA oligopeptidyl side-chains produces the same results[54] (Table 4). The existence of a correlation between the percentage tyrosine in the molecule and the rate of pinocytic uptake indicates that synthetic polymers can be designed in such a way as to control their non-specific affinity for membranes.

In order to test the possibility of targeting polymers to specific organs within the body, we prepared HPMA copolymers containing a small percentage of the side-chains -Gly-Gly-Tyr-NH_2 (the tyrosine being present to permit radiolabelling and so allow the fate of the polymer to be followed) and also containing approximately 2% mol of side-chains -Gly-Gly-R, where R was either 1-aminopropanol-2, glucosamine, galactosamine or mannosamine (Table 5).   Radio-labelled polymers were injected intravenously and their blood clearance and subsequent body distribution followed.   The pattern of blood clearance of copolymers showed that the galactosamine-bearing copolymer[47] was cleared much more rapidly than the others, only 10-15% of the radioactivity associated with this polymer remaining in circulation after 60 min (Fig. 4).

Table 4. Tyrosinamide containing copolymers

Code No.	Side chain	Side chain content %mol	Mean uptake after 5h (μl/mg protein)
33	$-Fly-Gly-Tyr-NH_2$	0.9	13.11
34	$-Gly-Gly-Tyr-NH_2$	3.7	14.33
35	$-Gly-Gly-Tyr-NH_2$	9.0	25.72
36	$-Gly-Gly-Tyr-NH_2$	10.9	27.20
37	$-Gly-Gly-Tyr-NH_2$	15.4	57.03
38	$-Gly-Gly-Tyr-NH_2$	18.9	181.18
39	$-Tyr-NH_2$	0.6	12.22
40	$-Tyr-NH_2$	3.8	-
41	$-Tyr-NH_2$	7.3	22.66
42	$-Tyr-NH_2$	8.5	-
43	$-Tyr-NH_2$	11.5	30.66
44	$-Tyr-NH_2$	15.4	145.57

Examination of the tissue distribution of radioactivity after 60 min indicated that the rapid loss of plasma radioactivity associated with the galactosamine-containing copolymer could be accounted for by its rapid capture by the liver (Fig. 5). Comparison of the percentage of all four polymers resident in the

Table 5. HPMA copolymers containing carbohydrate residues

Code No.	Side chain	Content of side chain in polymer (%mol)
45	$-Gly-Gly-Tyr-NH_2$	0.4
	$-Gly-Gly-NH-CH_2-CH-CH_3$ with OH	$\sim$ 2.0
46	$-Gly-Gly-Tyr-NH_2$	0.4
	$-Gly-Gly-Glucosamine$	$\sim$ 2.0
47	$-Gly-Gly-Tyr-NH_2$	0.4
	$-Gly-Gly-Galacotsamine$	$\sim$ 2.0
48	$-Gly-Gly-Tyr-NH_2$	0.4
	$-Gly-Gly-Mannosamine$	$\sim$ 2.0

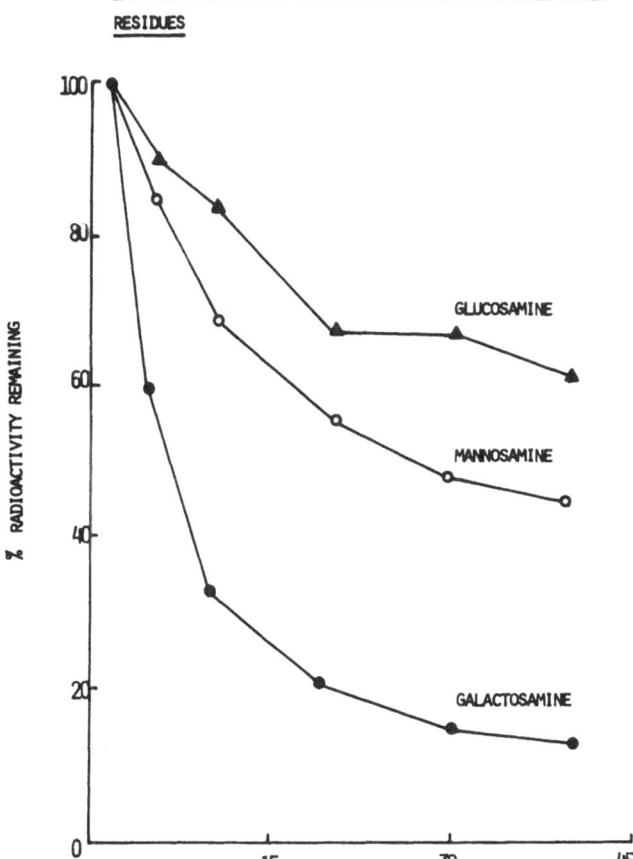

BLOOD CLEARANCE OF HPMA COPOLYMERS BEARING CARBOHYDRATE RESIDUES

Fig. 4

liver after 60 min illustrated the efficiency of just 2-4 galacto-samine residues per HPMA copolymer molecule in polymer targeting.

The tissue distribution of radioactivity 5h after adminis-tration of copolymers (Fig. 6) shows that most radioactivity is by then detectable in the urine/faeces. This observation results from the fact that the side-chain –Gly-Gly-Tyr-NH_2 is digestible by lysosomal enzymes. If we assign [^{125}I] iodotyrosine the role of drug analogue, we can conclude that in addition to the demonstration of successful targeting to liver, we have shown that drug is sub-sequently liberated intracellularly and ultimately excreted.

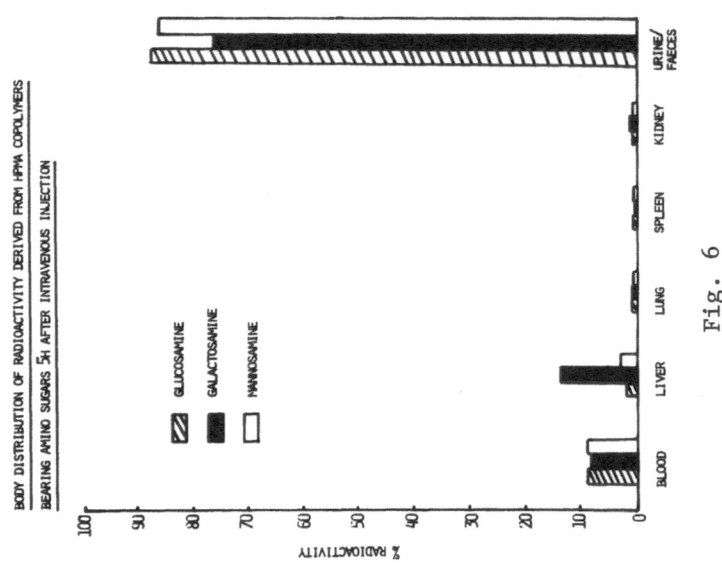

Fig. 6

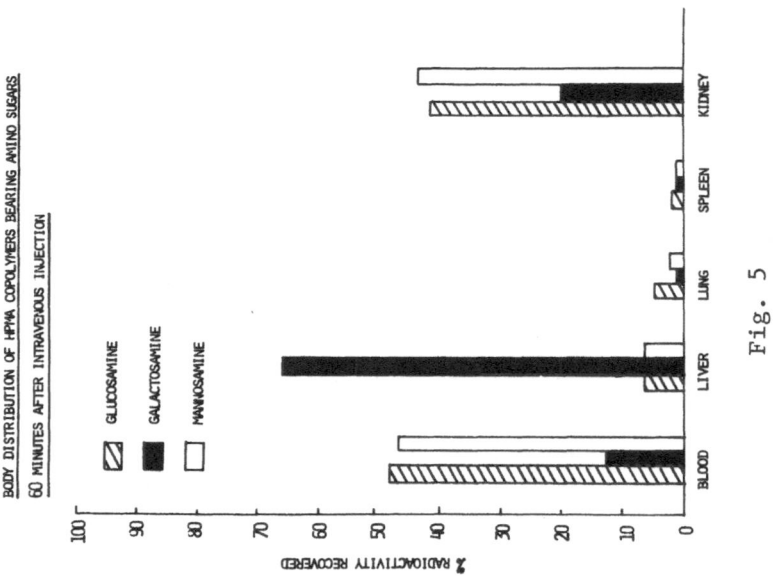

Fig. 5

CONCLUSIONS

So far we have been reporting results obtained with copolymers bound to drug analogues chosen in preference to 'real' drugs because they are less expensive and easier to monitor in living systems. Using the model systems described we have demonstrated that HPMA copolymers can adequately perform the basic requirements of a drug carrier system. We are now beginning to evaluate the system with authentic pharmacological agents attached to the copolymers.

In conclusion, we believe that synthetic polymers have great potential as carriers of chemotherapeutic agents. Their molecular weight range gives them access to a much wider variety of cell types than the larger vesicular vehicles such as liposomes of nano-capsules which tend to be accumulated exclusively by cells of the reticuloendothelial system. The chemical plasticity of synthetic polymers allows the synthesis of multicomponent systems in which the individual elements retain some chance of expressing their biological activity.

This work is supported by the Cancer Research Campaign and the British Council under their Academic Links with Eastern Europe Scheme.

REFERENCES

1. L.G. Donaruma and O. Vogl, "Polymeric Drugs", Academic Press, New York (1978).
2. R. Baker, "Controlled Release of Bioactive Materials", Academic Press, New York (1980).
3. H. Ringsdorf, Structure and properties of pharmacologically active polymers, J. Polymer Sci. Polymer Symp. 51:135 (1975)
4. L. Molteni, Dextrans as drug carriers, in: "Drug Carriers in Biology and Medicine", G. Gregoriadis, ed., Academic Press, London (1979).
5. B.C.F. Chu and J.M. Whiteley, The interaction of carrier-bound methotrexate with L1210 cells, Mol. Pharmacol. 17:382 (1980).
6. A. Trouet, R. Baurain, D. Deprez-De Campeneere, D. Layton and M. Masquelier, DNA, liposomes, and proteins as carriers for antitumoral drugs, in: "Recent Results in Cancer Research Vol.75", G. Mathé and F.M. Muggia, eds., Springer-Verlag, Berlin (1980).
7. G.J. O'Neill, The use of antibodies as drug carriers, in: "Drug Carriers in Biology and Medicine", G. Gregoriadis, ed., Academic Press, London (1979).
8. J. Kopeček, Biodegradation of polymers for biomedical use, in: "IUPAC Macromolecules", H. Benoit and P. Rempp, eds., Pergamon Press, Oxford (1982).

9. M.K. Pratten, R. Duncan and J.B. Lloyd, Adsorptive and passive pinocytic uptake, in: "Coated Vesicles", C.D. Ockleford and A. Whyte, eds., Cambridge University Press, Cambridge (1980).

10. A.J. Barrett and M.F. Heath, Lysosomal enzymes, in: "Lysosomes, a Laboratory Handbook", J.T. Dingle, ed., Elsevier/North Holland Biomedical Press, Amsterdam (1977).

11. J. Kopeček and H. Bažilová, Poly[N-(2-hydroxypropyl)methacrylamide].I. Radical polymerisation and copolymerisation, Eur. Polymer J. 9:7 (1973).

12. M. Bohdanecký, H. Bažilová and J. Kopeček, Poly[N-(2-hydroxypropyl)methacrylamide].2. Hydrodynamic properties of diluted polymer solutions, Eur. Polymer J. 10:405 (1974).

13. J. Strohalm and J. Kopeček, Poly[N-(2-hydroxypropyl)methacrylamide].IV. Heterogenous polymerization, Angew Makromol. Chem. 70:109 (1978).

14. J. Kopeček, L. Šprincl and D. Lím, New types of synthetic infusion solutions. 1. Investigation of the effect of solutions of some hydrophilic polymers on blood, J. Biomed. Mater. Res. 7:179 (1973).

15. L. Šprincl, J. Exner, O. Štěrba and J. Kopeček, New types of synthetic infusion solutions. III. Elimination and retention of Poly[N-(2-hydroxypropyl)methacrylamide] in a test organism, J. Biomed. Mater. Res. 10:953 (1976).

16. J. Kopeček, Soluble biomedical polymers, Polymers in Medicine 7:191 (1977).

17. J. Kopeček and P. Rejmanová, Enzymatically degradable bonds in synthetic polymers, in:"Controlled Drug Delivery", S.D. Bruck, ed., CRC Press, Florida (in press).

18. J. Kopeček, Biodegradation of polymers for biomedical use, Pure Appl. Chem. (in press).

19. J. Kopeček, K. Ulbrich, J. Vacik, J. Strohalm, V. Chytrý, J. Drobník and J. Kalal, Czech. Pat. 173846 (7.10.1976, priority 23.4.1974); U.S. Pat. 4, 062,831 (13.12.1977).

20. J. Drobník, J. Kopeček, J. Labský, P. Rejmanová, J. Exner, V. Saudek and J. Kalal, Enzymatic cleavage of side chains of synthetic water-soluble polymers, Makromol. Chem. 177:2833 (1976).

21. P. Rejmanová, J. Labský and J. Kopeček, Aminolyses of monomeric and polymeric 4-nitrophenyl esters of N-methacryloylamino acids, Makromol. Chem. 178:2159 (1977).

22. H.G. Garg, Amino acid active esters and their potentialities for peptide synthesis, J. Scient. Ind. Res. 29:236 (1970).

23. H.G. Batz, G. Franzmann and H. Ringsdorf, Pharmakologisch aktive polymere. 5. Modellreaktionen zur Umsetzung von Pharmaka und Enzymen mit monomeren und polymeren reaktiven Estern, Makromol. Chem. 172:27 (1973).

24. P. Ferrutti, A. Betteli and A. Feré, High polymers of acrylic
     and methacrylic esters of N-hydroxysuccinimide as
     polyacrylamide and polymethacrylamide precursors,
     Polymer 13:462 (1972).
25. C.P. Su and H. Morawetz, Reactivity of polymer substituents.
     Aminolysis of p-nitrophenyl ester residues attached to
     various polymer backbones, J. Polymer Sci. 15:185 (1977).
26. P. Rejmanová, PhD Thesis, Institute of Macromolecular
     Chemistry, Prague, Czechoslovakia (1981).
27. J. Kopeček, P. Rejmanová and V. Chytrý, Polymers containing
     enzymatically degradable bonds. I. Chymotrypsin catalyzed
     hydrolysis of p-nitroanilides of phenylalanine and
     tyrosine attached to the side-chains of copolymers of
     N-(2-hydroxypropyl)methacrylamide, Makromol. Chem.
     182:799 (1981).
28. R. Duncan, J.B. Lloyd and J. Kopeček, Degradation of side
     chains of N-(2-hydroxypropyl)methacrylamide copolymers by
     lysosomal enzymes, Biochem. Biophys. Res. Commun. 94:284
     (1980).
29. R. Duncan, H.C. Cable, P. Rejmanová, J. Kopeček and J.B. Lloyd,
     Design of synthetic copolymer side-chains to facilitate
     their degradation in lysosomes, Cell Biol. Int. Reports
     5 (Suppl.A) (1981)
30. H. Keilová, On the specificity and inhibition of cathepsins
     D and B, in: "Tissue Proteinases", A.J. Barrett, ed.,
     North-Holland Publishing Company, Amsterdam (1971).
31. I. Schechter and A. Berger, On the size of the active site in
     proteases. I. Papain, Biochem. Biophys. Res. Commun.
     27:157 (1967).
32. H. Kirschke, J.Langer, B. Weideranders, S.Ansorge and
     P. Bohley, Cathepsin L, a new proteinase from rat-liver
     lysosomes, Eur. J. Biochem. 74:293 (1977).
33. H.-J. Kärgel, R. Dettmer, G. Etzold, H. Kirschke, P. Bohley
     and J. Langer, Action of cathepsin L on the oxidized
     B-chain cf bovine insulin, FEBS Letts. 114:257 (1980).
34. A.J. Barrett, Anal. Biochem. 47:280 (1972).
35. K. Otto, Cathepsins B1 and B2, in: "Tissue Proteinases",
     A.J. Barrett, ed., North Holland Publishing Company,
     Amsterdam (1971).
36. P. Rejmanová, J. Pohl, M. Baudyš and J. Kopeček, unpublished
     results.
37. W. Seibel, An ultrastructural comparison of the uptake and
     transport of horseradish peroxidase by rat visceral yolk-
     sac placenta during mid- and late gestation, Am. J. Anat.
     140:213 (1974).
38. S.J. Freeman, PhD Thesis, University of Keele, Keele, U.K.
     (1982).
39. G.E. Ibbotson, PhD Thesis, University of Keele, Keele, U.K.
     (1978).

40. R. Duncan and J.B. Lloyd, Pinocytosis in the rat visceral
     yolk sac.   Effects of temperature, metabolic inhibitors
     and some other modifiers, Biochim. Biophys. Acta 544:647
     (1978).
41. R. Duncan, M.K. Pratten and J.B. Lloyd, Mechanism of poly-
     cation stimulation of pinocytosis, Biochim Biophys.
     Acta 587:463 (1979).
42. K.E. Williams, E.M. Kidston, F. Beck and J.B. Lloyd,
     Quantitative studies on pinocytosis. I. Kinetics of up-
     take of [^{125}I]polyvinylpyrrolidone by rat yolk sac cultured
     in vitro, J. Cell Biol. 64:113 (1975).
43. K.E. Williams, E.M. Kidston, F. Beck and J.B. Lloyd,
     Quantitative studies on pinocytosis. II. Kinetics of
     protein uptake and digestion by rat yolk sac cultured
     in vitro, J. Cell Biol. 64, 123 (1975).
44. R. Duncan, P. Rejmanová, J. Kopeček and J.B. Lloyd, Pinocytic
     uptake and intracellular degradation of N-(2-hydroxypropyl)-
     methacrylamide copolymers, Biochim. Biophys. Acta 678:143
     (1981).
45. P.S. Jacques, Endocytosis, in: "Lysosomes in Biology and
     Pathology Vol.2", J.T. Dingle and H.B. Fell, eds.,
     North-Holland Publishing Company, Amsterdam (1969).
46. T. Kooistra and K.E. Williams, Adsorptive pinocytosis of
     ^{125}I-labelled lactate dehydrogenase isoenzymes H4 and M4
     by rat yolk sacs incubated in vitro, Biochem. J. 198:587
     (1981).
47. E.F. Neufeld and G. Ashwell, Carbohydrate recognition systems
     for receptor mediated pinocytosis, in: "The Biochemistry
     of Glycoproteins and Proteoglycans", W.J. Lennarz, ed.,
     Plenum Press, New York (1980).
48. G. Ashwell and A.G. Morrell, The role of surface carbohydrates
     in the hepatic recognition and transport of circulating
     glycoproteins, in: Advances in Enzymology Vol.41",
     A. Meister, ed., Wiley and Sons, New York (1974).
49. H.D. Fischer, M. Natowicz, W.S. Sly and R.K. Bretthauer,
     Fibroblast receptor for lysosomal enzymes mediates
     pinocytosis of multivalent phosphomannan fragment,
     J. Cell Biol. 84:77 (1980).
50. P. Stahl, J.S. Rodman, M.J. Miller and P.H. Schlesinger,
     Evidence for receptor mediated binding of glycoprotein,
     glycoconjugates and lysosomal glycosidases by alveolar
     macrophages, Proc. Natl. Acad. Sci. USA 75:1399 (1978).
51. J.L. Goldstein and M.S. Brown, The low-density lipoprotein
     pathway and its relation to atherosclerosis, Ann. Rev.
     Biochem. 46:897 (1977).

52. A.T. Moore, K.E. Williams and J.B. Lloyd, The effect of chemical
     treatments of albumin and orosomucoid on rate of clearance
     from the rat bloodstream and rate of pinocytic capture by
     rat yolk sac cultured in vitro, Biochem. J. 164:607 (1977).
53. R. Duncan, D. Starling, F. Rypáček, J. Drobník and J.B. Lloyd,
     Pinocytosis of poly-α,B-N(2-hydroxyethyl) - D,L-aspartamide
     and a tyramine derivative by rat visceral yolk sacs cultured
     in vitro, Biochem. Biophys. Acta 717:248 (1982).
54. R. Duncan, P. Rejmanová, J. Kopeček and J.B. Lloyd, unpublished
     results.

BIOLOGICALLY ACTIVE COMPOUNDS IMMOBILIZED ON CELLULOSE DERIVATIVES

Cristofor Simionescu and Severian Dumitriu

Department of Organic Chemistry and Macromolecular
Chemistry, Polytechnic Institute of Jassy
6600 Jassy, Romania

INTRODUCTION

The enzymes, the principal biocatalysts of the living matter, are present in the organisms in most cases either bound to the cell membrane or included in the cytoplasmatic network. In order to re-produce in vitro some cell metabolic processess as close to reality as possible a wide variety of enzymes, vitamins, hormones, nucleic acids and organite were immobilized on different natural and synthe-tic polymers[1-8] Subsequently, the immobilized enzymes were widely applied industrially in continuous biosyntheses with high yields and economizing raw materials[8-15]

Recently the immobilization procedure was extended to drugs[16-20] with the aim of retardation and increase in their action specifici-ty. In this connection, antibiotics[21-24] insulin[25,26] and cytostatic agents[27-31] were immobilized on low molecular weight carriers which are hydrophilic or soluble in water, either bio- or nonbiodegradable.

These two major research directions have in common the perfor-ming of reactions under mild conditions so that the chemical struc-ture of the coupled compound is not affected. Besides, the formation of compounds compatible with biological systems is followed and the carriers so chosen as to assure a microsystem suitable for the good action of the coupled agent and not to form toxins hardly removable by their biodegradation.

IMMOBILIZATION OF ENZYMES

Synthesis of New Cellulose Derivatives as Carriers

A series of reactions on cellulose and carboxymethylcellulose was performed for introducing reactive groups capable of binding enzymes and drugs under mild conditions (Figure 1).

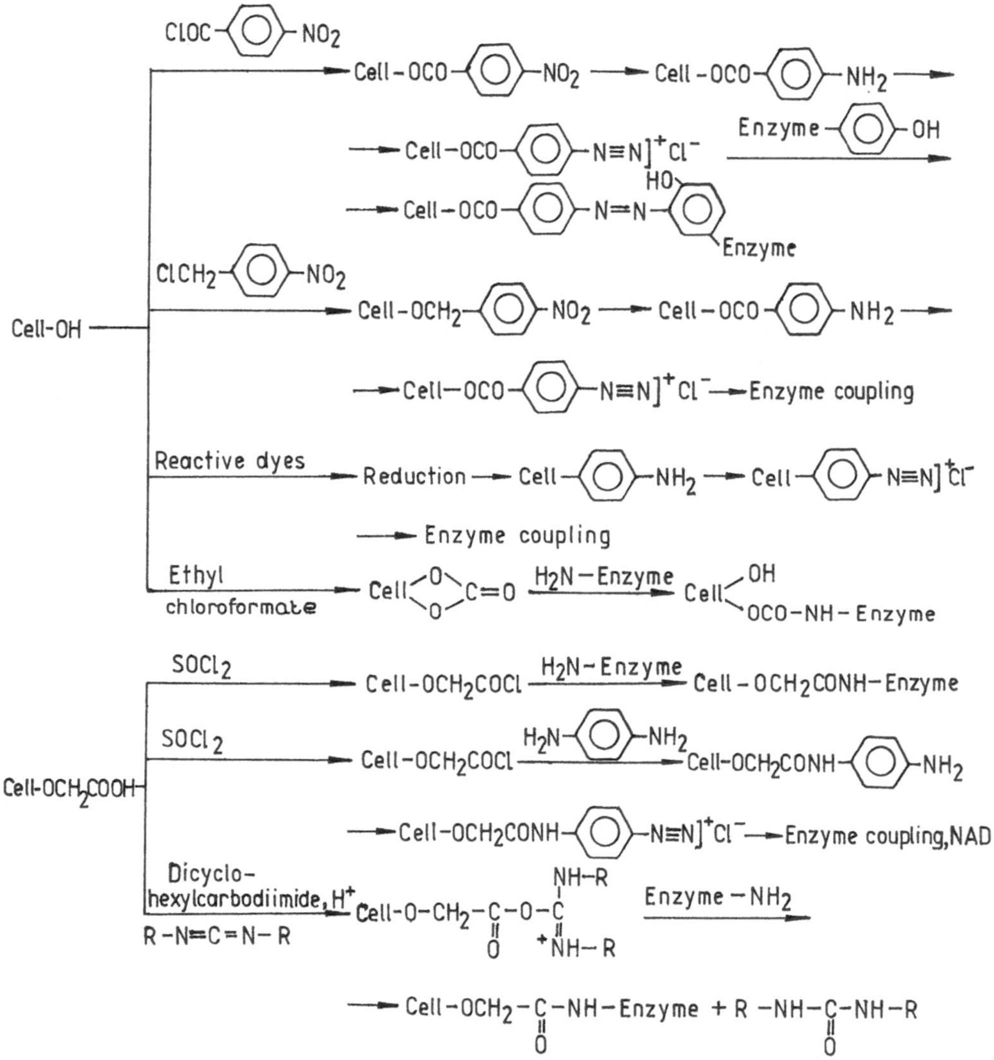

Fig. 1. Reaction schemes relevant to functionalization of cellulose and carboxymethylcellulose.

Immobilization of Oxidoreductases and Their Cofactors

Among the oxidoreductases the alcoholdehydrogenase (ADH) was studied. It was immobilized on the acid chloride of carboxymethylcellulose (ClCMC) and on the diazonium salts of the 4-aminobenzoylcellulose. Different operative conditions are required for the two carriers (Figure 2). The couplings on the diazonium salts of 4-aminobenzoylcellulose were performed with longer reaction times (210 min), an activity of only 32.2 UI/g being attained although the cellulose derivative has a high substitution degree (DS 1.76).

The cofactor of this enzyme, NAD, was immobilized distinctly and simultaneously with ADH on ClCMC having different substitution degree (DS 0.74; 1.70; 2.02). The samples obtained by simultaneous binding show oxido-reductive action toward ethyl alcohol only if either NAD or free ADH is added. The coupled NAD maintains 90% of the activity of the free form.

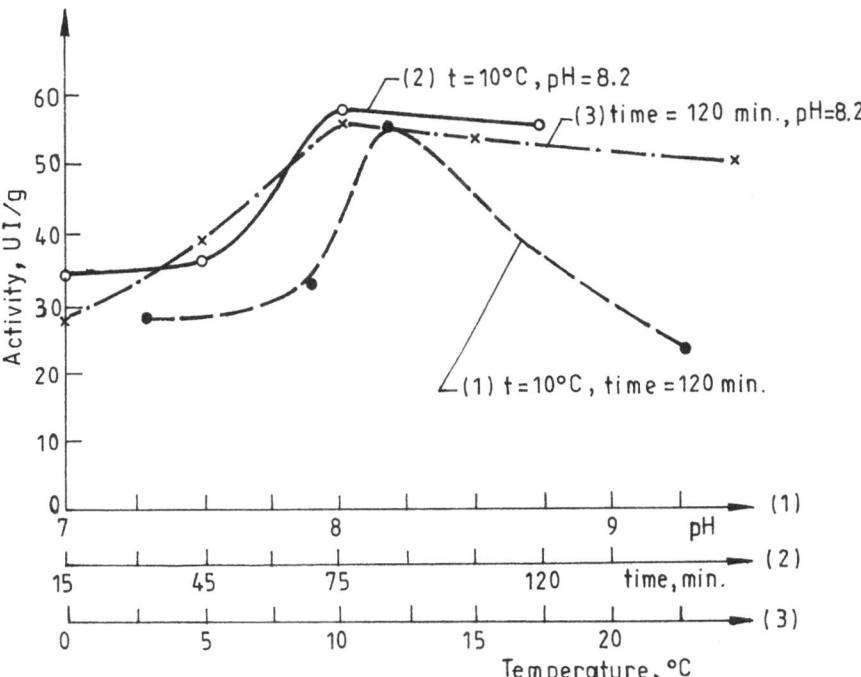

Fig. 2. Activity of alcoholdehydrogenase immobilized on ClCMC as a function of coupling pH, time and temperature (DS 0.9).

Immobilization of Hydrolases

Immobilization of Urease. The urease (E.C. 3.5.1.5) was immobilized
on three different carriers: 4-aminobenzylcellulose, 4-aminobenzoyl-
cellulose, poly(vinyl 4-aminobenzoate).
The relative activity (r $\% = m_1/m_2 \cdot 100$; m_1 = weight of coupled urease;
m_2 = weight of initial urease) of the samples depends on time, pH and
temperature (Figures 3-5).

The carrier nature is clearly evidenced to affect directly the
activity of the immobilized urease, the maximum being directly pro-
portional to the matrix hydrophilicity | 4-aminobenzylcellulose (DS
0.6) = 7.5%; 4-aminobenzoylcellulose (DS 1.2) =1.2%; poly(vinyl-
4-aminobenzoate) = 0.7%| .

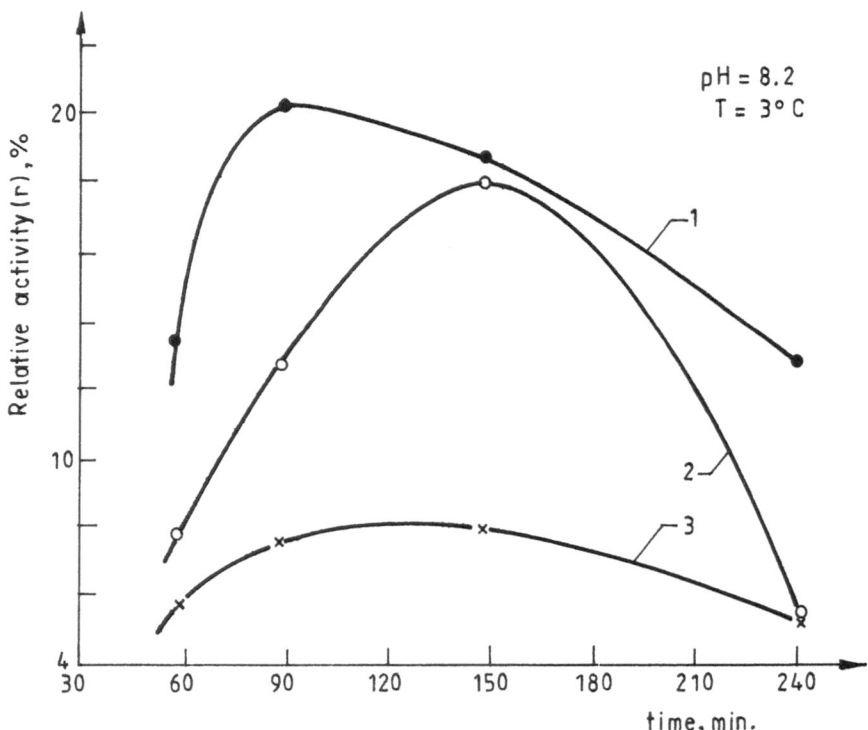

Fig. 3. Variation of the relative activity (r) of the immobilized
        urease as a function of coupling time: (1) 4-aminobenzyl-
        cellulose (DS 0.6); (2) 4-aminobenzoylcellulose (DS 1.2);
        (3) poly(vinyl-4-aminobenzoate).

Immobilization of α-Amylase. The immobilization of α-amylase (E.C. 3.2.1.1.) was performed on ClCMC at pH values between 7.5 – 9.   The coupled enzyme shows activity maximum at 38°C (Figure 6). This maximum diminishes with increasing temperature to 68°C in a lower   extent (55%) as compared with the free enzyme.[32]

By following the pH influence on the hydrolytic efficiency, a relative stabilization was found between pH = 5.8 – 8.2 (Figure 7), that is a much wider range than that observed for the free α-amylase activated by Ca ions.[32] Among the possible explanations of this behaviour, that referring to the possible stabilization of the active center and proteic chain by the free ionizable carboxy groups remaining on the CMC macromolecular chain, deserves mention.

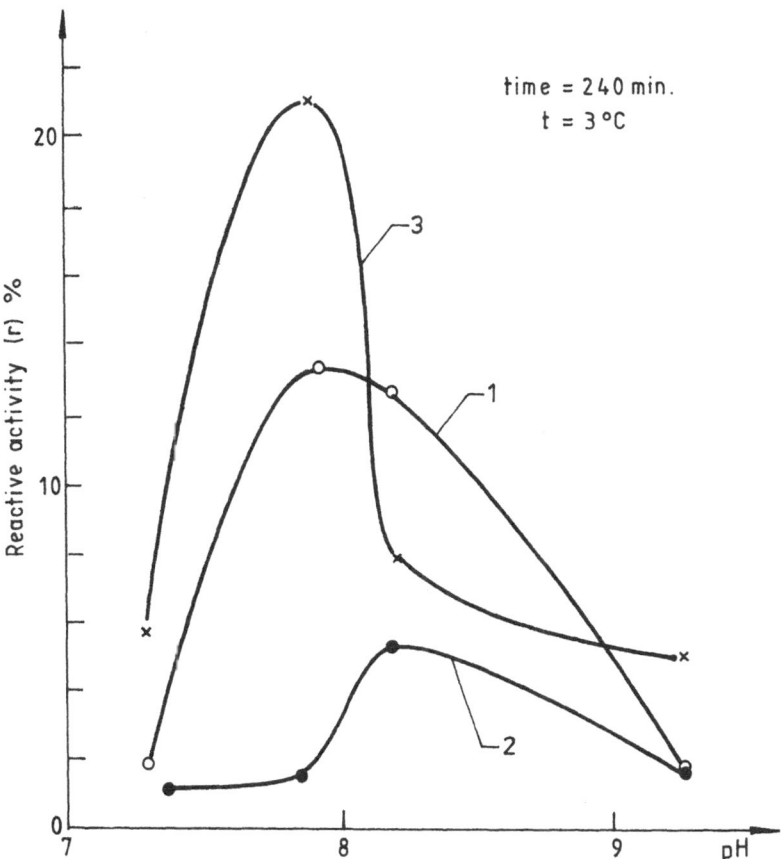

Fig. 4. Variation of the relative activity (r) of the immobilized urease as a function of the coupling pH: (1)   4-aminonenzyl-cellulose (DS 0.6); (2)   4-aminobenzoylcellulose (DS 1.2); (3)   poly(vinyl-4-aminobenzoate).

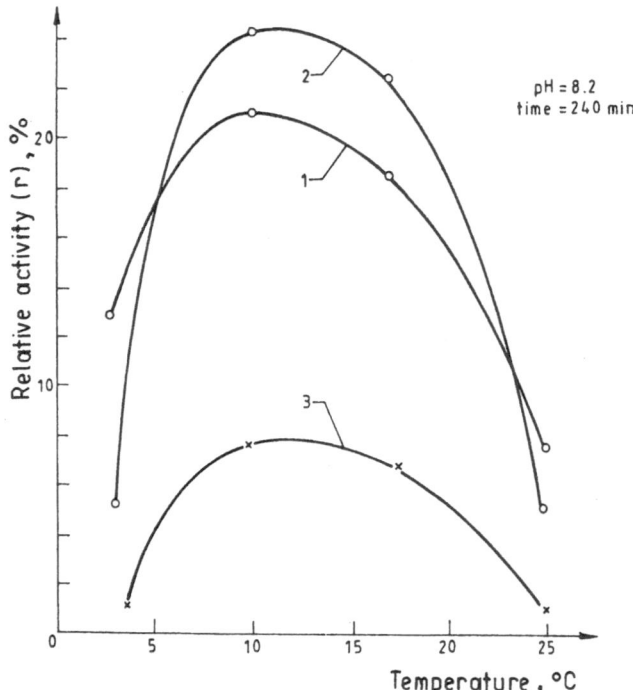

Fig. 5. Variation of the relative activity (r) of the immobilized
        urease as a function of the coupling temperature: (1) 4-ami-
        nobenzylcellulose (DS 0.6); (2) 4-aminobenzoylcellulose
        (DS 1.2); (3) poly(vinyl-4-aminobenzoate).

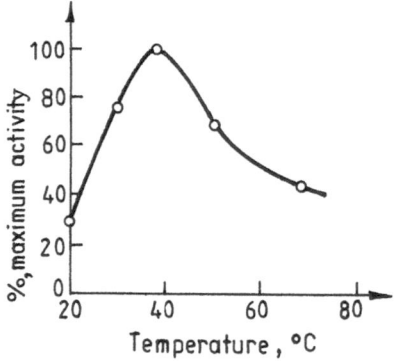

Fig. 6. Temperature profile for immobilized α-amylase at pH = 5.8 and
        t = 30 min.

The profile of starch hydrolysis (Figure 8) by the immobilized α-amylase indicates a significant increase in the conversion degree up to 30 min. In contrast with the free enzyme, the activity also increases after the limiting incubation period, hence the inhibition by the reaction products is not noticed.

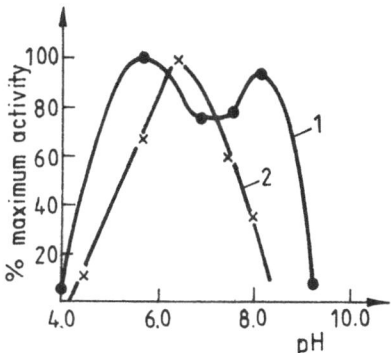

Fig. 7. pH profiles for: (1) coupled, and (2) soluble[32] α-amylase in the presence of Cl^- ions. t = 30 min; T = 38°C. Activity was plotted as percent of maximum activity at pH = 5.8 for the immobilized enzyme, and pH = 6.5 for the soluble one.

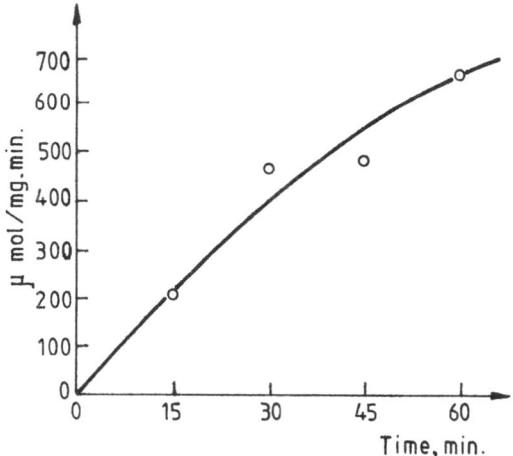

Fig. 8. Starch hydrolysis by immobilized α-amylase at pH = 5.8 and T = 38°C.

Immobilization of Amyloglucosidase. The coupling of amyloglucosidase (E.C. 3.2.1.) on C1CMC was performed at different enzyme/carrier ratios (Table 1).

The activity of immobilized amyloglucosidase decreases with increasing pH (Figure 9) at room temperature, probably due to the destruction of the secondary structure, hence of the active centers[33] A noticeable activity maintained up to 15 hours was found at pH = 8 and at the temperature of 5°C. In this case, significant activity losses were not observed even after reaction times of 20 hours.

Table 1.   Activity of Amyloglucosidase Coupled on C1CMC
            at Different Enzyme/Carrier Ratios.
            Time = 4 hr; pH = 8; T = 25°C; v = 40 ml.

Enzyme/ Carrier (g/g)	1.25/1	1/1	1/1.33	1/2	1/4
Activity (mg/g)	29.9	57.7	54.4	34.7	37.3

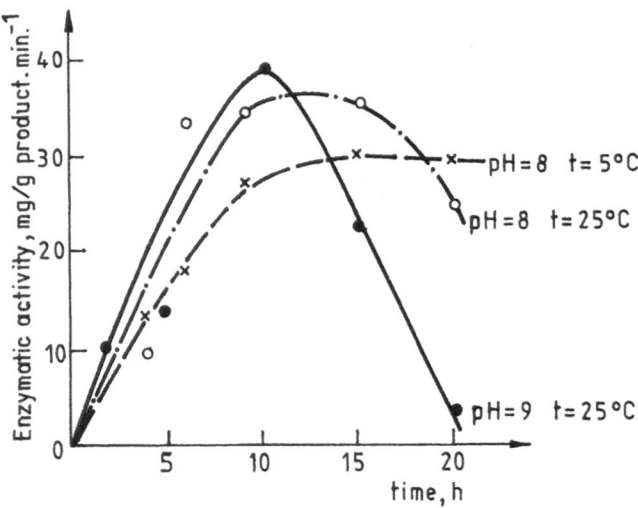

Fig. 9. Effect of pH and temperature on the activity of amyloglucosidase coupled on C1CMC.

Immobilization of β-Amylase. The immobilization of β-amylase (E.C. 3.2.1.2.) was studied on a wide variety of carriers: trans-2,3-cellulose carbonate, ClCMC (DS 2.02), N-(4-aminophenyl)CMC amide (DS 0.74; %N = 1.07), BIOZAN R, and acid chloride of BIOZAN R.

The activity of the coupling products depends on pH (Figure 10), time (Figure 11) and temperature (Figure 12).

The optimum coupling pH is seen to shift toward basic values, the maximum being shown by the polymers with diazonium salts. Under these conditions the electrostatic interactions between the carrier and the protein, hence the enzyme reaction efficiency, increase.

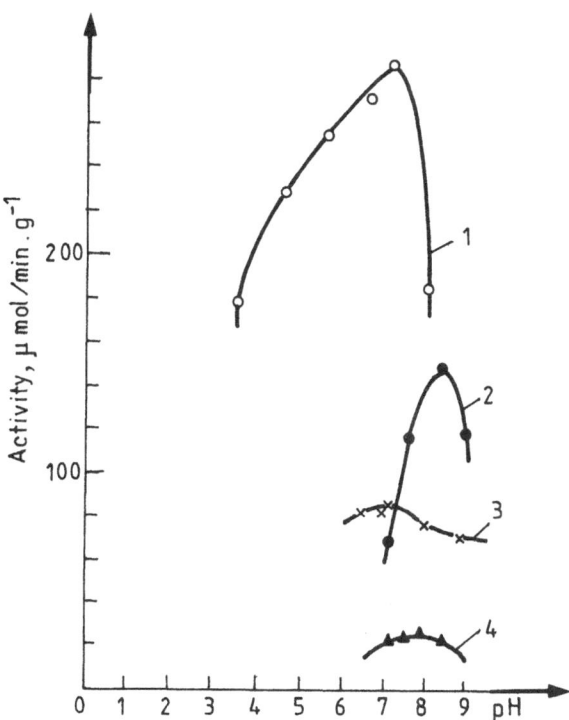

Fig. 10. Variation of the β-amylasic activity units of the samples obtained from: (1) trans-2,3-cellulose carbonate; (2) CMC acid chloride (DS 2.02); (3) diazotized N-(4-aminophenyl)CMC amide; (4) acid chloride of BIOZAN R (%Cl = 7.76); as a function of pH. Time = 6 hr; temperature = 5°C.

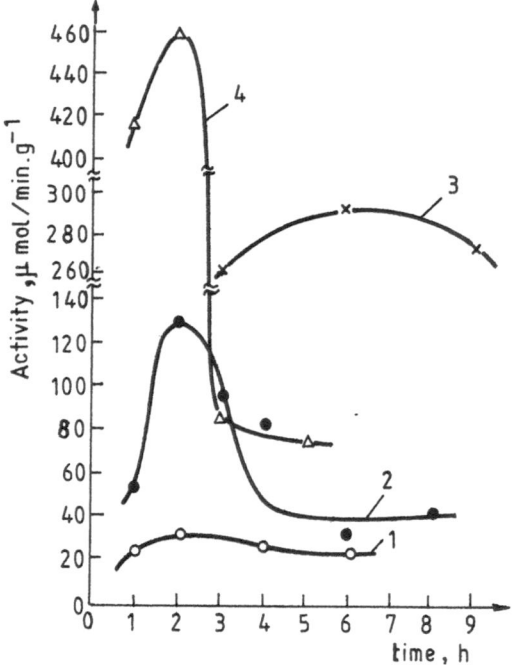

Fig. 11. Variation of the β-amylasic activity units of the samples
obtained from: (1) acid chloride of BIOZAN R; (2) CMC acid
chloride (DS 2.02); (3) trans-2,3-cellulose carbonate; (4)
diazotized N-(4-aminophenyl) CMC amide; as a function of
time. Temperature = 5°C.

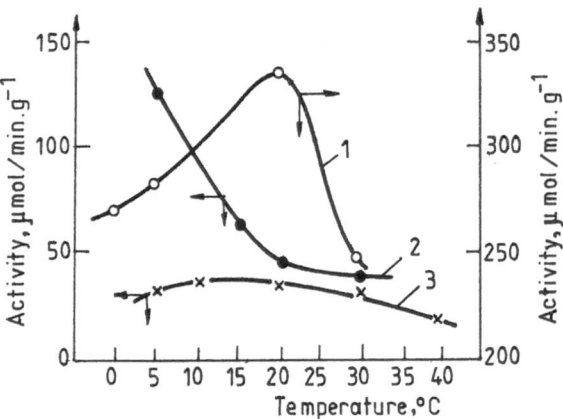

Fig. 12. Variation of the β-amylasic activity units of the samples
obtained from: (1) trans-2,3-cellulose carbonate; (2) ClCMC
(DS 2.02); (3) BIOZAN R; as a function of temperature at
optimum time and pH values.

The reaction time decreases with increasing reactivity of the functional groups and increasing intramolecular electrostatic forces. For istance, for coupling on the diazonium salts of N-(4-aminophenyl) CMC amide a time of two hours, assuring a maximum of 448 UI/g polymer, is sufficient (Figure 11).

The experimental conditions for a maximum β-amylase activity are given in Table 2.

The BIOZAN R is evidenced to be a suitable carrier for the immobilized β-amylase, since it ensures the formation of a microsystem as due to carboxylic groups, Ca, Na, K ions, and water molecules, which is apt for the catalytic activity of β-amylase.

By comparing the resistance to the severe pH values of the β-amylase coupled on N-(4-aminophenyl) CMC amide and on trans-2,3- -cellulose carbonate, the latter derivative is seen to confer a greater resistance. Within the 1.1 - 8.9 pH range, the enzymatic activity decreases by 30% (Figure 13, curve 2) in comparison with 75% for the former derivative (Figure 13, curve 3) and 95% for the free enzyme (Figure 13, curve 1).

Table 2. Optimum pH, Time and Temperature Values for a Maximum
β-Amylase Activity on Different Carriers.

Carrier	Reactive Group	Optimum pH	Optimum Time (hr)	Optimum Temperat. (°C)	Maximum Activity (UI/g)
BIOZAN R	COCl	8.0	2	10	33.46
ClCMC (DS 2.02)	COCl	8.4	2	5	125.65
N-(4-amino-phenyl) CMC amide	N_2^+ Cl^-	7.2	2	5	448.08
Cellulose Carbonate	Trans-2,3-car-bonate	7.0	6	20	341.88
BIOZAN R	COOH	—	12	4	3000.00

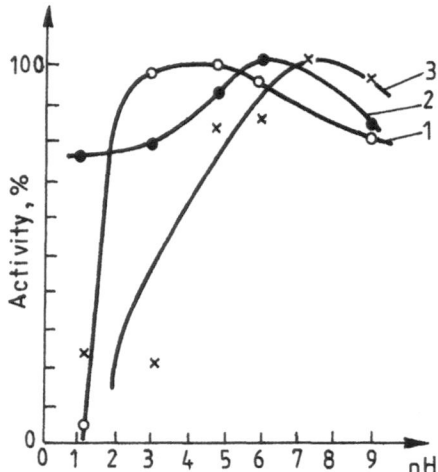

Fig. 13. Resistance of β-amylase at severe pH conditions : (1) free;
        (2) coupled on <u>trans</u>-2,3-cellulose carbonate; (3) coupled
        on N-(4-aminophenyl) CMC amide.

<u>Kinetics of the Reactions Catalyzed by Immobilized Enzymes</u>. The
kinetics of the reactions catalyzed by the immobilized enzymes is
influenced by a series of factors inexistent for the free enzymes,
such as conformational and steric transformations of the proteic
chain induced by the carrier, the diffusion of the substrate and
of the reaction products to and from the active centers, the composition of the microsystem created by the carrier. The diffusion
effects are reflected by the values of the Michaelis constants (K_M).
The K_M of an immobilized enzyme is generally higher,[34] identical[35]
and only in few cases lower than that of the free enzyme. Higher K_M
values are generally due to diffusion effects.

    The Lineweaver – Burk plots (Figures 14 – 16) permitted the cal-
culations of K_M values for the enzymes immobilized on different
carriers.

    The Lineweaver – Burk plot of the amyloglucosidase immobilized
on cellulose carbonate is practically linear, while for the enzyme
coupled on ClCMC at different pH values (Figure 14, curves 2 and 3)
the linearity is maintained only for increased concentrations of the
substrate with increased K_M value. In the former case the plot linearity attests that the amyloglucosidase molecules are located mainly

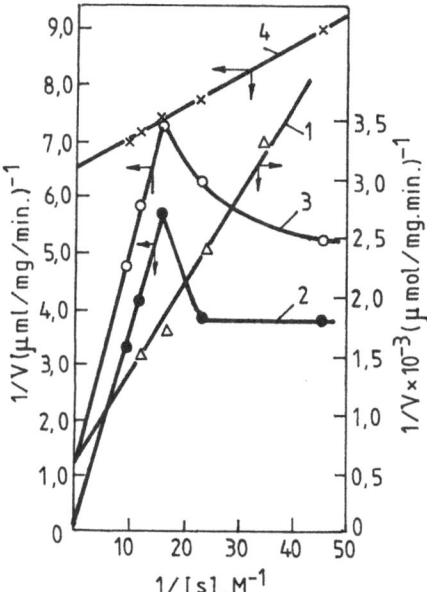

Fig. 14. Lineweaver – Burk plots for α–amylase at pH = 5.8 (1); amyloglucosidase immobilized on ClCMC, at pH = 8 (2) and at pH = 9
(3); amyloglucosidase immobilized on cellulose carbonate
at pH = 8 (4).

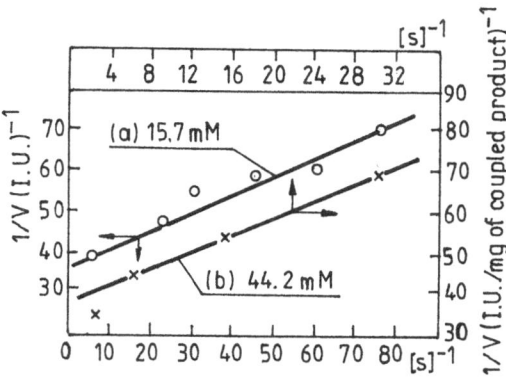

Fig. 15. Lineweaver – Burk plots for urease immobilized on ClCMC (a)
and for ADH coupled on 4-aminobenzoylcellulose (b).

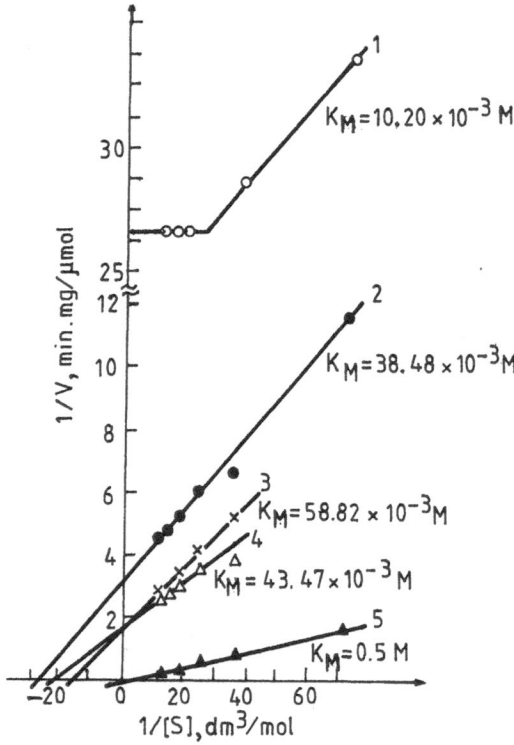

Fig. 16. Lineweaver – Burk plots for β-amylase immobilized on: (1)
acid chloride of BIOZAN R; (2) N-(4-aminophenyl) CMC amide;
(3) ClCMC; (4) trans-2,3-cellulose carbonate; (5) BIOZAN R.

on the external surface of the cellulose carbonate particles. The
diffusion of both substrate and reaction products to and from the
surface is simplified in comparison with the diffusion taking place
when the enzymes are bound within the carrier micropores and the substrate diffusion is prevented by the reaction products in the enzyme
environment. Similar results were reported by Lilly and Sharp[36] on
the chimotripsin immobilized on CMC.

An unexpectedly low K_M value (0.13 mM) (Figure 14, curve 1) was
found for α-amylase bound to ClCMC. In this case one may suppose
that in the acid medium (pH = 5.8) the substrate molecules are charged
positively by protonation, while the matrix retains a negative charge
from the unconverted carboxyl groups. The electrostatic attraction
will lead to higher substrate concentrations around each particle
in comparison with solution mass and, consequently, to a decrease in
the K_M value. The diffusion of the substrate molecules must also

be taken into account for reaction catalyzed by the urease fixed on ClCMC since a value of $K_M = 15.7$ mM was found.

The Michaelis – Menten relation is also obeyed in the processes involving the immobilized ADH ($K_M = 44.2$ mM). The diffusion effect does not explain in a satisfactory way the increase of K_M, so that conformational modifications of the active center must be considered to affect the enzyme kinetic behaviour. It is known that the ADH molecule consists of 4 protomers bound through a Zn atom to 4 mol-ecules of coenzyme.[37] In the coupling processes the carrier – enzyme chemical bond is formed at one protomer only, and the quaternary structure is destroyed. In fact, at lower coupling pH values (pH = 8) the ADH activity is strongly diminished probably due to the structure splitting in non-bound protomers, the ADH being active as a complete structural composition only.

Marked differences can also be observed in the kinetic behaviour of β-amylase coupled on the five carriers having different micro- and macrostructures (Figure 16). The results obtained with enzymes immobilized on different carriers are summarized in Table 3.

Table 3. K_M Values for Enzymes Immobilized on Cellulose
          Derivatives.

Enzyme	Carrier	Reactive Group	K_M (mM)
ADH	ClCMC	COCl	44.2
Urease	4-aminobenzoyl-Cellulose	$N_2^+ Cl^-$	15.7
α-Amylase	ClCMC	COCl	0.13
	BIOZAN COCl	COCl	10.20
	ClCMC	COCl	28.48
β-Amylase	N-(4-aminophenyl) CMC Amide	$N_2^+ Cl^-$	58.82
	Cellulose Carbonate	trans-2,3- carbonate	43.47
	BIOZAN R (DCI)	COOH	500.00

IMMOBILIZATION OF DRUGS

The immobilization of drugs was performed with the aim of re-
tarding drug uptake and increasing the action specificity to a certain
region of the body.

The carriers applied in this field must fulfill the following
principal requirements:
- to be non-toxic and not to produce toxic substances by biodegradation
- to couple and to release easily the drugs
- to be soluble in water to allow oral and injection administration

By taking these considerations into account, the immobilization
of some antibiotics on cellulose grafted with poly(acrylic acid),
acid chloride of CMC, BIOZAN R was studied and ionic and covalent
bonds achieved.

Immobilization Through Ionic Bonds

The carrier - antibiotic bonds of this type are necessary when
drug acts intracellularly and does not suffer chemical modifications.
The basic antibiotics (streptomycin, kanamycin, neomycin) were immobilized by ionic exchange on medicinal bandages grafted with poly-
(acrylic acid), oxycellulose and BIOZAN R.

A comparative study of the possibilities of coupling on the
copolymers grafted with poly(acrylic acid) and oxycellulose revealed
an increased efficiency of the latter (Table 4). The structure of
the grafted copolymer is responsible for this behaviour, in that it

Table 4.   Amount of Streptomycin Bound to Oxycellulose
           and Grafted Cellulose.

Sample	Ion-Exchange Capacity (mVal/Kg)	Coupling Efficiency (%)	Bound Streptomycin (μg/g Cell)
Irradiated Cotton (Oxycellulose)	1004	48	38800
Cotton Grafted with Acrylic Acid	1040	14	24500

is much more compact and hinders the penetration of the bulky antibiotic molecules to the reaction site.
Some grafting conditions and the amount of bound streptomycin are
given in Table 5.

  The particular structure of BIOZAN R, as well as its properties
such as the biological resistance to medium variations (pH, organic
solvents), recommend this polymer for antibiotic couplings by ion-
-exchange reactions. The conditions for the streptomycin fixation
are given in Figures 17 and 18.

Table 5. Amount of Streptomycin Bound to Medical Bandages
          as a Function of Grafting Conditions (Monom. Conc. 50%).

Grafting Conditions		Ion-Exchange	Bound	Coupling
Temp. (°C)	Time (hr)	Capacity (mVal/Kg)	Streptomycin (µg/g Cell)	Efficiency (%)
80	3	450	2520	13.1
90	3	960	22500	14.8
80	4	1760	140000	80.0
90	4	2020	194000	88.0

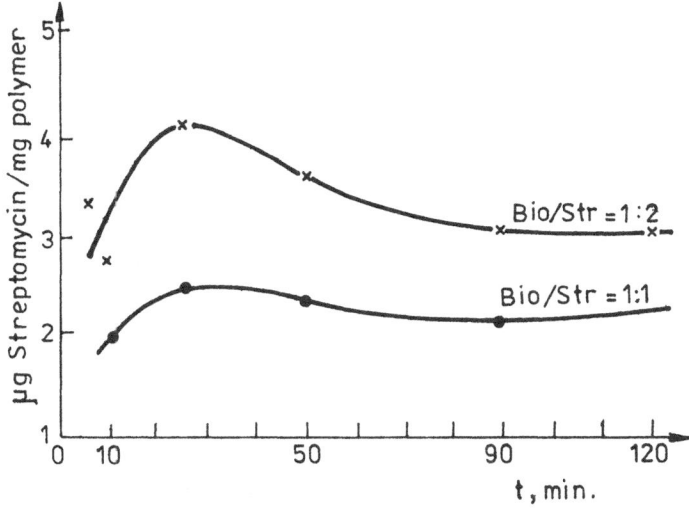

Fig. 17. Variation on time of the amount of streptomycin fixed as
         a function of the BIOZAN R/streptomycin ratio (Bio/Str)

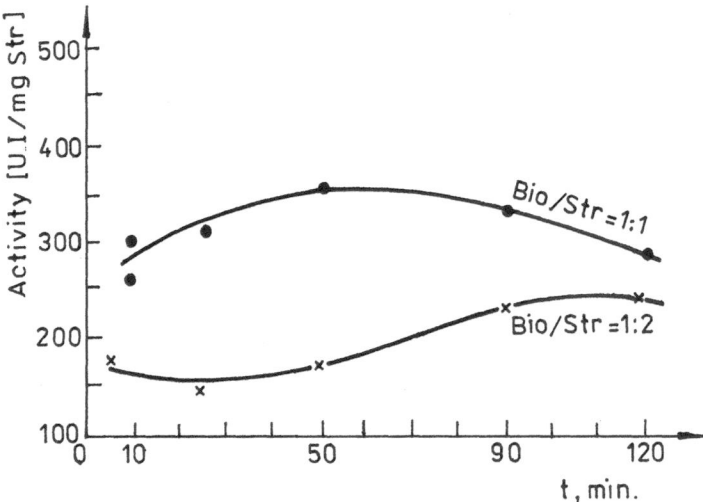

Fig. 18. Variation on time of the antimicrobial activity of the
         coupling product as a function of the BIOZAN R/streptomycin
         ratio (Bio/Str).

The fact deserves mention that the antimicrobial activity of
the samples is lower at Bio/Str ratio of 1/2 than at Bio/Str of 1/1,
where an increasing amount of coupled streptomycin is noticed.

The kanamycin and neomycin couplings were less efficient due
to their lower basicity and stability.

Immobilization Through Covalent Bonds

The covalent coupling of antibiotics (tetracyclin, 6-AP acid)
was performed on aminoaromatic derivatives of cellulose and BIOZAN R.
Tetracyclin was immobilized on medicinal bandages partially modified
with diazotized aminoaromatic groups.

The microbiological tests were carried out with gram positive
and gram negative germs and made evident their antimicrobial properties (Table 6).

The coupling of 6-AP acid on ClCMC, cellulose carbonate and
BIOZAN R led to the preparation of macromolecular semisynthetic
penicillins.

Some microbiological characteristics of the synthesized penicillins depending on the content of the bound 6-AP are given in
Table 7.

Table 6. Antibiotic Activity of 4-Aminobenzoyl Cellulose Coupled to Tetracyclin.

Sample	Coupling Temp. (°C)	Test Microorganism			
		Bacillus Subtilis C.M.I. (µg/ml)	Bacillus Cereus C.M.I. (µg/ml)	Staph. Citeus C.M.I. (µg/ml)	Bacillus Anth. C.M.I. (µg/ml)
4-Aminobenzoyl Cellulose (powder)	25	15	15	1	125
4-Aminobenzoyl Cellulose (powder)	70	23	6	5	80
4-Aminobenzoyl Cellulose (tissue)	70	10	3	0.5	40

Table 7. Antibiotic Activity of Semisynthetic Penicillins with Macromolecular Structure.

Penicillin Content		C.M.I. (µg/ml)	
%	UI/mg	Staphylococcus	Sarcina Lutea
4.3	112	55	—
5.2	140	60	—
1.3	38	250	—
—	84	100 – 200	3 – 6

It is known that by means of carbodiimides acylations of amines can be carried out under mild conditions (temperature 0 – 5°C) that are necessary for chemical modification of biologically active compounds. By using BIOZAN R and dicyclohexylcarbodiimide (DCI) the 6-AP acid was coupled according to the following reactions:

Bio-COOH + $C_6H_{11}-N=C=N-C_6H_{11}$ + $H^{\oplus}$ $\longrightarrow$ $\underset{\displaystyle \overset{O}{\|}}{Bio-C-O-\underset{\overset{\displaystyle NH^{\oplus}-C_6H_{11}}{|}}{\overset{\displaystyle NH-C_6H_{11}}{|}}C}$ $\longrightarrow$

H_2N-6AP

$\longrightarrow$ $Bio-\overset{O}{\overset{\|}{C}}-NH-6AP$ + $H^{\oplus}$ + $C_6H_{11}-NH-\overset{O}{\overset{\|}{C}}-NH-C_6H_{11}$

    The influence of the DCI/Bio ratio on the amount of 6-AP bound is reported in Figure 19.

    By the pharmacodynamic tests carried out on rats by intraperitoneal injection of Bio-6AP (0.5 ml) the lack of toxicity was found. The maintaining of a limiting concentration attack by injection once in 24 hours was also established.

IMMOBILIZATION OF VITAMIN B_{12} ON BIOZAN R

    Vitamin B_{12} was immobilized on BIOZAN R using DCI. Along with the chemical bonding an ion-exchange process takes also place. The yields of vitamin bound through covalent and ionic bonds are dependent on the DCI/Bio ratio (Figure 20), vitamin concentration (Figure 21) and reaction time.

CONCLUSIONS

    The immobilization of biologically active compounds - enzymes, antibiotics, vitamins - was performed on several modified natural polymers - 4-aminobenzoylcellulose, cellulose carbonate, acid chloride of CMC and BIOZAN R.

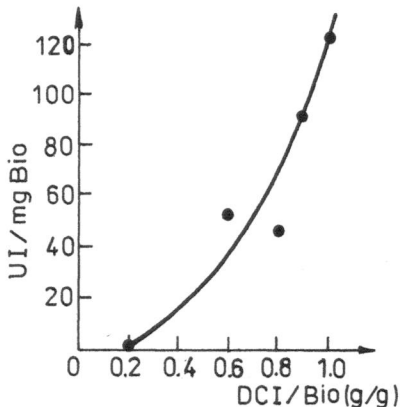

Fig. 19. Variation of the amount of 6-AP bound (UI) to BIOZAN R as
a function of DCI/Bio (g/g) ratio. Temperature 5°C.

The activity of the immobilized compound either increases or

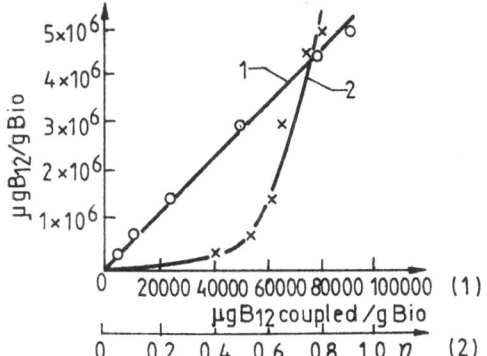

Fig. 20. Variation of the amount of vitamin B_{12} bound to BIOZAN R
as a function of the µg B_{12}/g Bio ratio. Temperature = 10°C;
time = 6 hr.

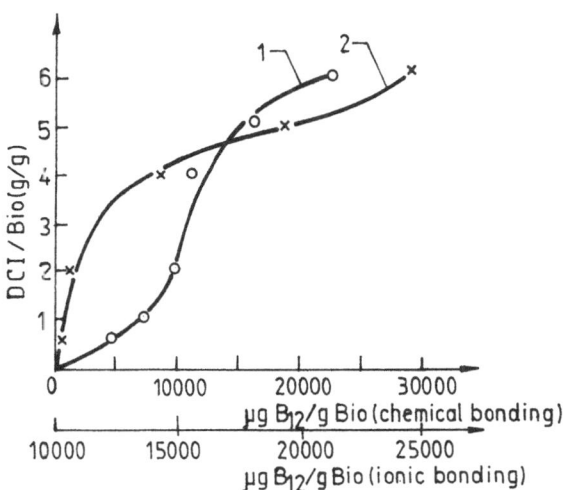

Fig. 21. Variation of the amount of vitamin B_{12} bound to BIOZAN R
as a function of the DCI/Bio (g/g) ratio.   (1) Chemical
bonding;   (2) Ionic bonding.

decreases depending on the chemical structure and physical properties (hydrophobicity, porosity) of the carrier. In the case of the enzymes the carrier hydrophilicity was found to be essential for maintaining the biocatalytical activity.

The drugs require the formation of labile bonds, preferably ionic, by immobilization. The supports used - acid chloride of CMC, BIOZAN R - fulfill the conditions required by the biodegradation, non-toxicity, solubility in water.

REFERENCES

1. R. Axen, J. Porath, and S. Ernback, Nature 214:1302 (1967)
2. H. H. Weetall and L. S. Hersh, Biochim. Biophys. Acta 185:464 (1969).
3. G. Manecke and R. Korenzecher, Makromol.Chem. 178:729 (1977).
4. S. D. Bruck, Polymer 16:409 (1975).
5. B. K. Kusserow, Trans. Am. Chem. Soc. Art. Int. Organs 17:1 (1971).
6. C. R. Lowe and P. D. G. Dean, FEBS Letters 14:313 (1971).
7. T. Hattori and C. Furusaka, J. Biochem. 48:831 (1960).
8. Cr. Simionescu, S. Dumitriu, V. Bulacovschi, and V. I. Popa J. Polym. Sci., Polym. Symp. 66:171 (1979).
9. W. R. Vieth and K. Venkatasubramanian, Chem. Tech. 3:677 (1975).
10. T. Sato, T. Mori, T. Tosa, I. Chibata, M. Furui, K. Yama-shita, and S. Sumi, Biotechnol. Bioeng. 17:1797 (1975).
11. I. Chibata and T. Tosa, "Applied Biochemistry and Bioengi-neering", Academic Press, New York (1976).
12. K. Mosbach and P. O. Larsson, Biotechnol. Bioeng. 12:19 (1970).
13. H. H. Weetall and G. Baum, Abstr. Papers, Am. Chem., No 158, Biol., 153 (1969).
14. R. Koelsch, Enzymologia 42:257 (1972).
15. E. Lagerlöf, L. Nathorst-Westfelt, B. Ekström, and B. Sjöberg, in "Methods in Enzymology", K. Mosbach, ed., Academic Press, New Yrok (1976).
16. J. Koch-Weser and E. M. Sellers, New England J. Med. 294:311,526 (1976).
17. L. G. Donaruma, Progr. Polymer Sci. 4:1 (1975).
18. W. B. Jakoby, in "Methods of Enzymatic Analysis", H. U. Bergmeyer, ed., Academic Press, New York (1974).
19. Y. Miura, S. Aoyagi, F. Ikeda, and K. Miyamoto, Biochimie 62:595 (1980).

20. J. Kalal, J. Drobnik, J. Kopecek, and J. Exner, in "Polymeric Drugs", L. G. Donaruma and O. Vogl, eds., Academic Press, New York (1978).

21. S. Dumitriu, R. Butnaru, and Cr. Simionescu, Cell. Chem. Technol. 7:553 (1973).

22. Cr. Simionescu, S. Dumitriu, and R. Butnaru, Cell. Chem. Technol. 7:641 (1973).

23. V. A. Snezhko, L. N. Samoilova, K. P. Khomyakov, A. I. Valakhanovic, R. V. Zaretskaya, A. D. Virnik, G. Ya. Rosenberg, and Z. A. Rogovin, Antibiotiki 17:48 (1972).

24. J. E. Shaw, W. Bayne, and L. Schmitt, Clin. Pharmacol. Therap. 19:115 (1976).

25. P. Cuatrecasas, Proc. Nat. Acad. Sci. USA 63:450 (1969).

26. A. Zaffaroni and P. Bonsen, in "Polymeric Drugs", L. G. Donaruma and O. Vogl, eds., Academic Press, New York (1978).

27. E. Hurowitz, R. Marov, R. Arnon, and M. Sela, Cancer Biochem. Biophys. 1:197 (1976).

28. W. H. Cole, in "Chemotherapy of Cancer", Lea & Febiger (1970).

29. F. Karush and C. L. Hornick, Int. Arch. Allergy 45:130 (1973).

30. F. L. Moolten and S. R. Cooperband, Science 169:68 (1970).

31. T. Ghose, Br. Med. J. 3:495 (1972).

32. E. H. Fischer and E. A. Stein, "The Enzymes", 2nd Ed., vol. 4, Academic Press, New York (1960).

33. G. O. Hustad, T. Richardson, and N. F. Olson, J. Dairy Sci. 56:1118 (1973).

34. E. Brown and A. Racois, Bull. Soc. Chim. Fr. 743 (1974).

35. W. E. Hornby, M. Lilly, and E. M. Crook, Biochem. J. 107:669 (1968).

36. M. Lilly and A. K. Sharp, Chem. Eng. (London) 215:CE12 (1968).

37. J. R. Kagi and B. L. Vallee, J. Biol. Chem. 235:3188 (1960).

PREPARATION OF ENZYME POLYMERS FOR THERAPEUTICAL USE:

PRELIMINARY "IN VIVO" STUDY

P. Marincol, M.-H. Remy and D. Domurado

Laboratoire de Technologie Enzymatique
E.R.A. N° 338 du CNRS
Université de Compiègne, B.P. 233, FRANCE

INTRODUCTION

Originating in the work of DeDuve's group (1), the concept
of drug targeting is intended to improve the selectivity of transfer
of a drug to target cells (2). The drug targeting consists of the
association of a therapeutic effect and of a tissular specificity
resting on the molecular association of a drug and a vector in
a complex. The drug distribution in the body, being in principle
governed by the vector, can be modified as a function of the chosen
vector.

This approach has been tested to improve the effectiveness
of antiprotozoal (3) and antitumoral (4) drugs but gave rise also
to a new class of anticancer agents : the immunotoxins (5, 6).

Enzyme replacement is the logical approach for the treatment
of lysosomal storage diseases (7) but use of an "address label"
for the injected molecule is required (8). Antibodies (5,6),
hormones (9), lectins (10), glycoconjugates (11), virus coats (12),
bacterial toxins (13) might be used as vectors. Liposomes injected
in the blood stream are taken up rapidly by cells in the reticuloendothelial system (3).

Previous work in Poznansky's group and our own showed that
the formation of enzyme-albumin polymers solves several problems
linked to the injection of enzymic preparations (14) : the
immunologically neutral homologous albumin used to coat the enzyme,
hides its antigenic sites and protects it against the proteases ;
the cross-linking increases thermal resistance ; the high molecular
weight of the polymer prevents the renal elimination.

The in vivo study of the drug distribution as a function of
the vector is the next step in the direction of a future therapeutic
application. It depends on the type of marker (radioactive,

fluorescent, enzymatic...) the visualisation of which must be
precise with an intact tissular morphology, be representative of
the drug distribution, give information about the drug integrity.

The results will rely on the sampling method (individual organ
or whole-body study) and on the techniques used (paraffin embedding,
cryomicrotomy, light or electron microscopy, autoradiography...)

In order to satisfy these prerequisites, we chose an enzymic
marker (horse radish peroxidase -HRP- Sigma type VI) because it is
a model of a therapeutic enzyme, because the enzyme visualisation
implies its integrity.

Since carbohydrates are able to protect enzymes from thermal
denaturation and proteolysis (15), since they are good candidates
as vectors (16), we also studied a carbohydrate-enzyme conjugate.

The preparation method of albumin-coated HRP, the preparation
and behaviour of the carbohydrate-enzyme derivative, the histological results will successively be exposed.

I - ALBUMIN-COATED HRP PREPARATION

In order to coat completely the peroxidase with albumin, the
knowledge of the enzyme structure is needed to choose the best
cross-linking method. Welinder (17) published the amino-acid
sequence. Six residues in 308 are lysine. About 56 carbohydrate
residues are grouped in 8 chains. The total molecular weight is
44,000.

Two methods could be used to coat the HRP with albumin.
Avrameas crosslinks the amino group with glutaraldehyde (18).
Nakane derivatizes the carbohydrate chains (19). The goal of
these two methods was the production of an immunological marker.
For a good revelation of the antigens, the smallest complex is the
best. We were looking for a complete covering of the enzyme.

1) Glutaraldehyde synthesis

Two polymerization methods were used :
* a 2-step method in which albumin is added to the mixture HRP-
  glutaraldehyde after a preactivation period.
* a 3-step method. Here, the glutaraldehyde is eliminated from
  the activated HRP before the addition of albumin.
The 2-step method. Lysine activation by glutaraldehyde.
The reaction mixture is composed of 40 mg of HRP in 2.2 ml of
phosphate buffer 20 mM pH 6,86 ; 1.7 ml of 125 mM glutaraldehyde
in the same buffer. The reaction takes place at 4°C for 24 hours.
Copolymerization. 1.1 ml of albumin (20 % in the same buffer) are
added to the reaction mixture. After 6 hours at 4°C, glycine is
added until a final concentration of 10 g/l. The polymers were

separated on a G 100 Sephadex chromatography column. Four peaks
were observed. The first one corresponds to a high molecular weight
and has very little enzyme activity (0.4 to 0.7 % of the native HRP
in the reaction mixture). It represents copolymers of albumin
with a few HRP molecules (MW around 1,000,000). The second peak
contains conjugates HRP-albumin (MW 110,000). The third one is
composed of HRP dimers (MW 90,000). The molecular weights were
checked by 5 % polyacrylamide gel electrophoresis in presence of
SDS (5 % PAGE-SDS).

Small molecules formed from glycine and glutaraldehyde are
to be found in the last peak.

The 3-step method. The activation step was the same as above
except that 80 mg of HRP were used. The excess glutaraldehyde is
eliminated by dialysis (48 hours at 4°C with the same buffer).
220 mg of albumin were added. The mixture was allowed to react for
20 hours at 4°C. Glycine was added to stop the reaction.

The gel chromatography gave only one activity peak corres-
ponding to a molecular weight in the range 70,000 - 150,000.
The 5 % PAGE-SDS allowed us to conclude the presence of the dimer
HRP-albumin.

Despite the different crosslinking conditions which eliminate
the albumin copolymers, we had only one albumin molecule reacted
with each HRP molecule. This prompted us to use Nakane's method(19).

2) Polymerization after carbohydrate oxidation

The carbohydrates are first oxidized by sodium periodate.
In 1.2 ml, 8 mg of HRP are reacted in presence of the oxidant at
a final concentration of 100 mM during 40 minutes at the laboratory temperature. The oxidant is removed by dialysis 24 hours
at 4°C. Albumin (2.6 ml of a 36 % solution in carbonate buffer)
is then added. The mixture is allowed to react 3 hours at 4°C.
Sodium borohydride is then added to reduce the Schiff bases formed.
A large albumin : HRP ratio (80 to 1) was used in order to avoid
too high molecular weight polymers.

The results of the 6 B Sepharose gel chromatography are shown
in figure 1.

The HRP is present in the fractions 50 to 55 (enzyme activity
and absorbance at 403 nm) which are also the elution fractions of
ferritin. The albumin-HRP molar ratio calculated from the absorbances
at 280 nm and 403 nm is around 6 for all the fractions of this
peak. Nothing can be concluded from the 5 % PAGE-SDS because of
the too short migration path (less than 2 mm). The second peak
(fraction 72) represents the unreacted albumin.

We can conclude that this protocol produces albumin-coated
HRP. Six albumin molecules can completely shield the enzyme from
immunoglobulins or proteases. The calculated molecular weight of
this complex (446,000) is in agreement with the observations.
This preparation was used for the in vivo studies.

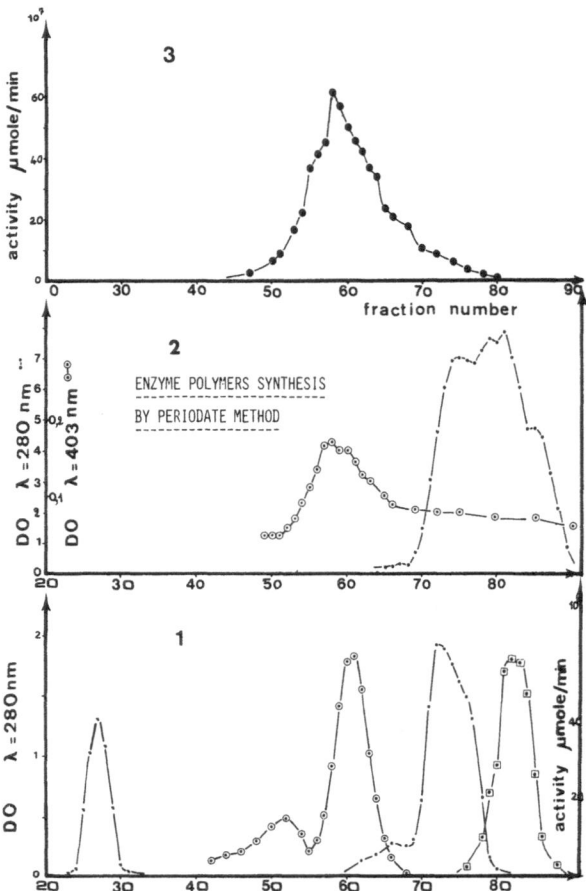

<u>Figure 1</u>

Study and purification of albumin-coated horseradish
peroxidase on a Sepharose 6 B chromatography column.
1.1 Calibration.
    From left to right, the four peaks are :
    - blue dextran (MW 2,000,000)
    - ferritin (MW 440,000)
    - albumin (MW 67,000)
    - HRP (MW 44,000)
1.2 Absorbance at 403 nm and 280 nm of the chromatographed
    polymer fractions.
1.3 Peroxidase activity of the chromatographed polymer
    fractions.

II - PREPARATION OF DISACCHARIDE-ENZYME CONJUGATES

The synthesis of neoglycoproteins has been performed by attachment of lactose residues to the initially carbohydrate-free enzyme. The reductive alkylation of the amino group of lysine residues is done using sodium cyanoborohydride (20).

Bovine pancreatic α-chymotrypsin (100 mg), lactose (100 mg) and sodium cyanoborohydride (100 mg) were dissolved in 5 ml of borate buffer (100 mM, pH 9) and incubated at 4°C for 200 hours. 33 % of activity was recovered in the neoglycochymotrypsin. The activity yield was 70 % for the control experiment (sodium cyano-borohydride omitted). The cannibalistic denaturation played an important role in the activity loss. The reaction mixture was purified on a Sephadex G 25 column. The fractions were assayed for protein by Folin's method, for total carbohydrate by the phenol-sulphuric acid procedure and for enzyme activity with an artificial substrate (GPNA).

The first protein peak coincided with the enzyme activity and sugar concentration. The second peak represents enzyme fragments generated by autoproteolysis. When a solution of enzyme and free sugar is percolated through the column, no sugar could be detected in the enzyme peak.

The active fractions were pooled and characterized. On average, 10 amino groups were modified on each molecule of enzyme. This is an important modification since only 14 lysines are present in α-chymotrypsin.

The glycosylation can further be demonstrated by affinity chromatography on a galactose-specific immobilized lectin. The test was done on a PNA-Sepharose column with native and modified chymotrypsins. The chymotrypsin-galactose peak was shifted compared to the native enzyme. All the enzyme molecules were modified : it was possible to follow the modification of the physico-chemical properties of the derivatized enzyme without further purification.

The resistance of the neoglycoenzyme to thermal denaturation was tested at 50°C. For the first study, the cannibalistic denaturation was eliminated by working at an acidic pH. The half-life of chymotrypsin was 180 minutes. On the other hand, the neo-glycoenzyme activity plateaued rapidly at 80% of the initial activity. Free galactose at the same concentration did not protect the enzyme activity. Using the optimum conditions of autolysis (Tris HCl 50 mM pH 8, $CaCl_2$), the chymotrypsin inactivation was extremely rapid (half-life : 9 minutes) compared to the neoglyco-enzyme (71 minutes). The denaturation process kinetics depended on the enzyme concentration. In this experiment, the chymotrypsin and chymotrypsin-galactose was equal to 2.8×10^{-4} M but the initial activity of chymotrypsin was two times higher. The protection effect toward proteolysis was not due to the lysine modification since this endopeptidase is specific for aromatic amino acids.

III - HISTOENZYMATIC LOCALIZATION

The use of an enzymatic marker made it preferable to work with cryomicrotomy. This technique prevents the use of high temperatures (from the enzyme point of view) paraffin baths and of organic solvents both deleterious for the enzyme activity.

The study of drug distribution would be best made by whole body imaging as can be done by cryomicrotomy with the help of a Scotch tape and autoradiography (21).

Using the staining method of Graham and Karnovsky (22), we first tried to study the whole-body distribution of HRP. Unfortunately, in spite of the use of many different types of Scotch tapes to ensure a good adhesion of the slice to the adhesive, a good behaviour of the tape during staining and mounting, we were unable to get histological preparations good enough to be observed at magnifications greater than x 100. However, these trials showed us the native myeloperoxidase in the bone marrow, in the macro-phages present in the lungs and in the axis of the intestinal villi. We also observed urinary excretion of intravenously injected HRP. Haemoglobin can give the brown derivative of DAB depending upon the state of the red blood cells.

The rest of the study was performed on organ samples. By that way, we observed the presence of HRP in the lumen of the tubules and its likely reabsorption (22). No filtration was demonstrated after intravenous injection of albumin-coated HRP. A detailed study of the distribution of this polymer is under way.

CONCLUSION

We demonstrated that proteic and/or carbohydrate coating can render an enzyme less susceptible to thermal denaturation, enzymatic proteolysis and immunological attack. Carbohydrates or proteins can also be fixed in order to target the enzyme to the right tissue(s) or cell(s). Little is still known about the in vivo behaviour of such molecular associations but the promises of enzyme technology for therapeutic applications (lysosomal storage diseases for instance) are so great as to stimulate such studies in the near future.

REFERENCES

1- DeDuve C., Debarsy T., Poole B., Trouet A., Tulkens P., VanHoof F., Biochem. Pharmacol., 23 : 2495-531 (1974)
2- Trouet A., Eur. J. Cancer, 14 : 105-11 (1978)
3- Alving C.R., Steck E.A., Trends Biochem. Sci. 4 : N175-7 (1979)
4- Mathe G., Tran B.L., Bernard J., C.R. Acad. Sci. 246 : 1626 (1958)

5- Blythman H.E., Casellas P., Gros O., Gros P., Jansen F.K., Paolucci F., Pau B., Vidal H., Nature 290: 145-6 (1981)
6- Olsnes S., Pihl A., Pharmacol. Ther. 15 : 355-81 (1981)
7- Holcenberg J.S., Ann. Rev. Biochem. 51 : 795-812 (1982)
8- Desnick R.J., Thorpe S.R., Fiddler M.B., Physiol. Rev. 56 : 57-99 (1976)
9- Cawley D.B., Herschman H.R., Gilliland D.G., Collier R.J., Cell, 22 : 563-70 (1980)
10- Harper C.G., Gonatas J.O., Stieber A., Gonatas N.K., Brain Res., 188 : 465-72 (1980)
11- Fiume L., Busi C., Mattioli A., Balboni P.G., Barbanti-Brodano G., FEBS Lett., 129 : 261-4 (1981)
12- Uchida T., Miyake Y., Yamaizumi M., Mekada E., Okada Y., Biophys. Biochem. Res. Comm. 87 : 371-9 (1979)
13- Bizzini B., Grob P., Glicksman M.A., Akert K., Brain Res., 193 : 221-7 (1980)
14- Remy M.H., Poznansky M.J., Lancet 8 Juillet: 68-70 (1978)
15- Marshall J.J., Trends Biochem. Sci. 3 : 79-83 (1978)
16- Neufeld E.F., Ashwell G. in "Biochemistry of Glycoproteins & Proteoglycans" pp. 241-66 (ed. Lennarz) Plenum Press London (1980)
17- Welinder K.G., Eur. J. Biochem. 96 : 483-502 (1979)
18- Avrameas S., Ternynck T., Immunochem. 8 : 1175-9 (1971)
19- Wilson M.B., Nakane P.K. in "Immunofluorescence and related staining techniques" pp. 215-24 (eds.Knapp, Holubar and Wick) Elsevier, Amsterdam (1978)
20- Schwartz B.A., Gray G.R., Arch. Biochem. Biophys. 181 : 542-9 (1977)
21- Ullberg S., Acta Radiol. Suppl. 118 : 1-110 (1954)
22- Graham R.C. Jr., Karnovsky M.J., J. Histochem. Cytochem. 14 : 291-302 (1966)

SYNTHESIS AND CHARACTERIZATION OF COVALENTLY BOUND POLYMER-

HORMONE CONJUGATES FOR THE CONTROLLED RELEASE OF HORMONES

W.A.R. van Heeswijk, G.J. Brinks and J. Feijen

Dept. of Chem. Technology, Twente University of
Technology, P.O. Box 217, 7500 AE Enschede
The Netherlands

INTRODUCTION

A promising approach in the release of hormones from polymeric
systems is the use of covalently bound polymer-hormone compounds
which are biodegradable and do not need removal after implantation[1,2]
Anderson et al[3,4,5], Mitra et al.[6] and Marck et al.[7] have shown
that synthetic poly-(α-amino acids) which are biodegradable can be
developed and the rate of *in vivo* biodegradation can be controlled
by varying the hydrophylicity of the side chain groups. Thus, these
polymers can be used, in principle, as carriers in polymer-drug
conjugates.

Recently, we described the coupling of steroids to a biodegradable poly(hydroxyalkyl)-L-glutamine by interpolation of a carbonate
bond. *In vivo* release studies using microparticles of norethindrone
coupled to poly-N^5-(3-hydroxypropyl)-L-glutamine (**1**), which were
implanted subcutaneously in rats, showed a near zero-order release
rate for more than 200 days[1,8].

The rate of the release of the bioactive agent from such systems is governed by a number of parameters: (i) initial molecular
weight, degree of substitution, hydrophylicity of the conjugate
and (ii) the nature of the chemical bonds used between the carrier
and the drug, (iii) the use of so-called spacer groups and (iv)
geometrical form of the device. However, the applicability of these
promising drug delivery systems has been hampered seriously by both
the limited accessibility of the carrier **1** and the lack of efficient methods to attach bioactive agents to the side chains of the
polymer. We now wish to report an improved method for the preparation of **1** and the facile preparation of polymeric mixed succinic

147

acid diesters derived from **1** and either testosterone (**2**) or norethindrone (**3**).

RESULTS AND DISCUSSION

Poly-N[5]-(3-hydroxypropyl)-L-glutamine (**1**) has been prepared[9]
by solvolysis of poly-γ-benzyl-L-glutamate with 3-amino-propan-1-
-ol. However the aminolysis is frequently accompanied by (i) extensive gel formation, presumably *via* interchain amide bonds, and
(ii) chain scission due to aminolysis of the poly-α-L-amino acid
amide linkages giving polymers with molecular weights ranging from
$2.10^4 < \overline{M}_w < 1.4.10^5$ daltons. Both chain scission and concomitant
gel formation can be reduced to some extent by a modification of
the former procedure as devised by Okita et al.[10] However, un-
assisted aminolysis of **4** to prepare **1** (scheme 1) is still limited
by extensive scission of the poly-α-amino acid chain as well as
the formation of gel formed by transesterification.

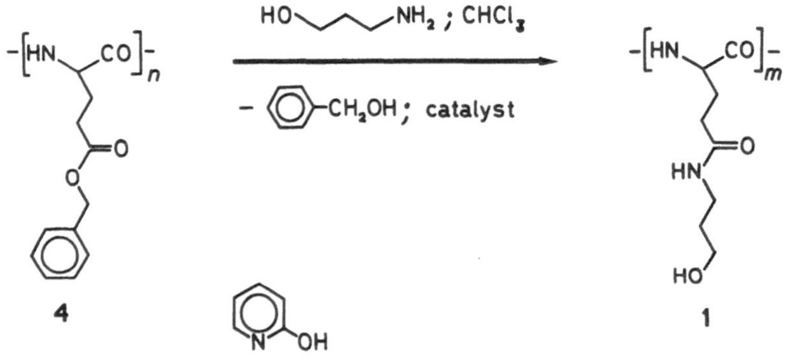

Scheme 1.   Aminolysis of poly-γ-benzyl-L-glutamate **4**.

2-Hydroxypyridine (**5**) is known to enhance aminolysis of esters
by bifunctional catalysis[11,12]. Application of **5** in the aminolysis
of **4** proved to be very succesful both with respect to yield and
molecular weight. Thus, **4** with molecular weights ranging from
$1-4.10^6$ failed in the aminolysis according to Lupu-Lotan[9] and
Okita[10], whereas the addition of **5** to the reaction mixture gave **1**
in typical yields of 80-85% within 24 hrs. Residual benzyl groups
mounted up to ~3%. Materials having intrinsic viscosities of

$[\eta] = 1.42$ dl.g^{-1} (corresponding to $\overline{M}_v$ $3.7 \cdot 10^5$ using the equation $[\eta] = 1.4 \cdot 10^{-5} \overline{M}_v^{0.9}$)[9] have been prepared. The large differences found in the enhancement of aminolysis of low and high molecular weight **4** using **5** cannot be explained solely in terms of mechanistic assistance in the actual displacement of the benzyloxy groups[11]. More likely, it may be concluded that **5** also plays a pivotal role in solubilization of the polymer in the heterogeneous reaction mixture and/or induces conformational changes in **4**, thereby rendering the pendant ester groups more accessible towards attack by the amine.

To attach the drugs testosterone (**2**) and norethindrone (**3**) to this carrier we selected the use of mixed succinic diesters on the basis of the following considerations: 1. It is expected that ester bonds between the drug and spacer are cleaved prior to degradation of the backbone amide bonds. 2. Hemisuccinates, in turn, are more rapidly hydrolysed than the corresponding diesters from which they are generated. Rapid formation of the parent drug from the hemiester is known to occur by intramolecular general acid catalysis. 3. The degradation products are not toxic. 4. The hemisuccinates of steroids have been used as a, so called, disposable solubilizing moiety[13]. 5. Hemisuccinates of steroids are generally readily accessible and stable compounds.

Testosterone hemisuccinate (**6**) was readily prepared by direct acylation of testosterone using succinic anhydride and N-methylimidazole as the acid acceptor in 79% yield (scheme 2). However the preparation of norethindrone hemisuccinate (**7**) by this method gave low yields, presumably due to steric hindrance and poor nucleophylicity of the hydroxyl function. In contrast, symmetrical 2,2,2-trichloroethyl hemisuccinate anhydride (**8**) was found to acylate the tertiary hydroxyl function in excellent yields using 4-dimethylaminopyridine as the acylation catalyst. The latter anhydride was generated *in situ* with the use of dicyclohexyl carbodiimide (scheme 3). Finally, the trichloroethyl protecting group was removed with the use of zinc in acetic acid (scheme 4). Along this route which comprises three steps norethindrone hemisuccinate **7** was prepared in 58%.

In order to attach the hemisuccinates of both steroids **6** and **7** onto the polymeric carrier **1** we have developed a facile coupling technique. This technique comprises the use of an arylsulphonyl halide[14] to generate symmetrical anhydrides of the acids **6** and **7** which readily acylates the hydroxyl groups of **1** in the presence of 4-dimethylaminopyridine. 2,4,6-Triisopropylbenzenesulphonyl chloride (**9**) was selected to prevent O-sulphonylation of the polymer (c.f. ref. 15) (scheme 5). A partial structure of the testosterone polymer conjugate is visualized in scheme 6.

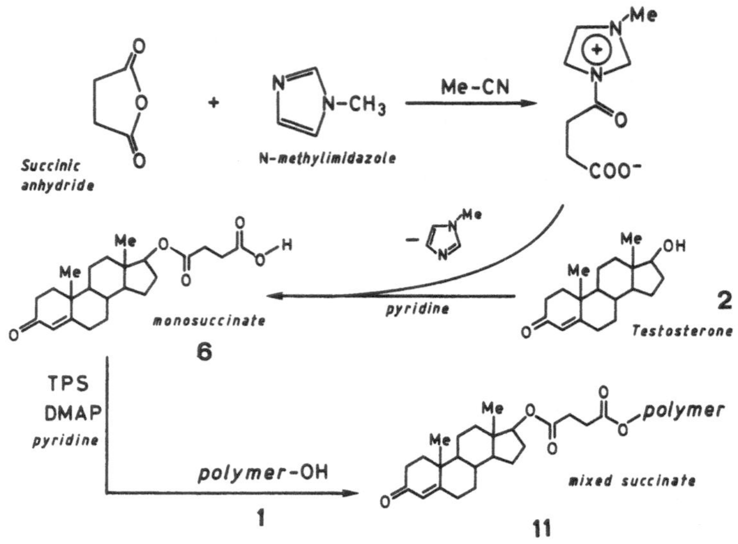

Scheme 2.   Preparation of covalently bound testosterone-
            polymer conjugates.

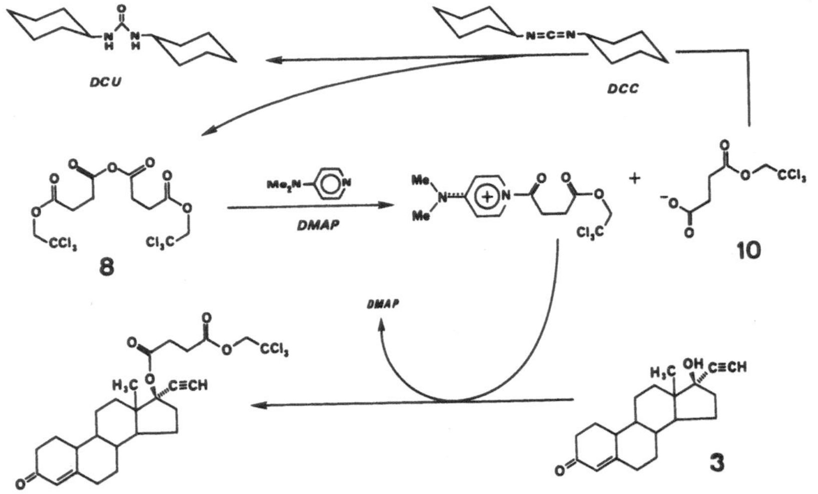

Scheme 3.   Synthesis of 2,2,2-trichloroethyl norethindrone
            succinate diester.

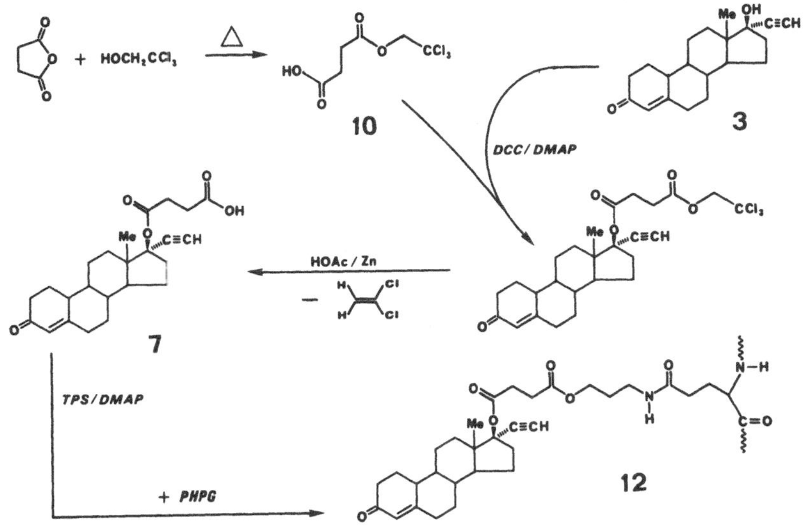

Scheme 4.   Preparation of covalently bound norethindrone –
polymer conjugates.

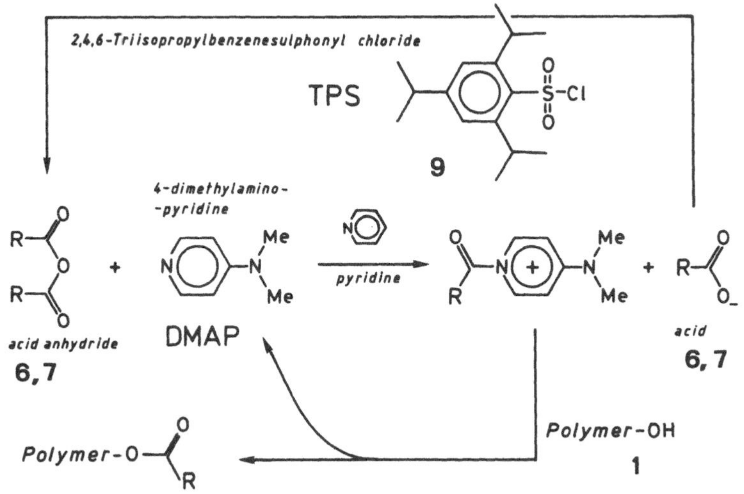

Scheme 5.   Synthetic method for the coupling of drugs
containing carboxyl groups with polymers
containing hydroxyl functions.

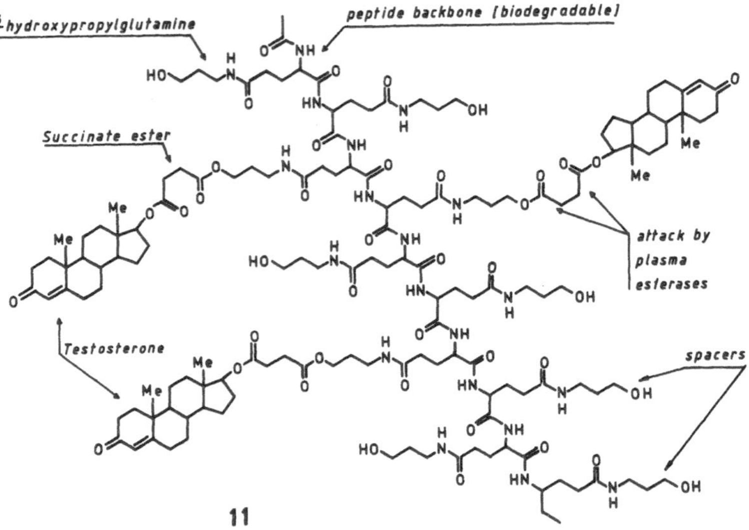

Scheme 6. Partial structure of polymeric mixed succinate
ester derived from 1 and testosterone.

The structures of both chloroform soluble steroid polymer
conjugates were confirmed by [1]H-n.m.r. spectroscopy. Degrees of
substitution of steroid onto available hydroxyl groups mounted
up to 72% (molar) as determined by gravimetrical methods and
[1]H-n.m.r. spectrocopy using H-4, ethynyl-H, and other skeleton
proton integrals. Release studies using these polymers are currently under investigation.

EXPERIMENTAL

General methods

Pyridine (1L) was distilled from chlorosulphonic acid (10 ml)
and kept under nitrogen. All acylations were performed under a
blanketing nitrogen atmosphere. TLC analysis was performed on
Merck Kieselgel 60 F 254 plates using chloroform-methanol 85:15
(v/v) as the solvent and UV and/or iodine vapor visualization.
Mobilities of compounds were expressed relative to reference
compounds. [1]H-n.m.r. spectra (80 MHz) and I.R. spectra were recorded
on a Bruker WP-80 and Beckman IR-33 apparatus, respectively.

Poly-N^5-(3-hydroxypropyl)-L-glutamine (**1**).

Freshly prepared γ-benzyl-L-glutamate *N*-carboxyanhydride[16]

(11 g, 50 mmole) was dissolved in freshly distilled dioxane
(550 ml) and triethylamine (150 µl) in dioxane (5 ml) was added
with stirring. After 100 hr the viscous solution was concentrated
in vacuo until 60-80% of the dioxane had been removed. The resulting
glassy material was freeze dried to afford a porous white mass of
poly γ-benzyl-L-glutamate (**4**).

A solution of 2-hydroxypyridine (**5**, 2.5 g, 26 mmole) in 3-
amino-propan-1-ol (25 ml) was added with the aid of a syringe
through a rubber septum into a flask containing spongy **4** (2 g,
freeze dried from 20 ml of dioxane) *in vacuo*. Then, the amine was
forced by atmospheric pressure (N$_2$) into the matrix of **4**. A number
of repetitive evacuations will remove virtually all of the air
pockets present in **4**. The mixture was kept for 3 hr at 60°C. Then,
chloroform (8 ml) was added to the reaction mixture and the aminolysis was allowed to continue with occasional stirring at 60°C
until a sample was soluble in water. The polymer was precipitated
by slow addition of the reaction mixture to chloroform (300 ml),
filtered, washed with ether and dissolved in water. The aqueous
solution was dialysed exhaustively against distilled water, filtered through a millipore filter (0.45 µ, pore size) in a stirred
cell, concentrated to a small volume using an Immersible Ultrafiltration Unit CX-10 (Millipore, Badford, Mass. USA) and freeze
dried (yield 80-85%). The ^{1}H-n.m.r. data (80 MHz, D$_2$O) were in
agreement with those given by Joubert et al.[17]. No resonances
attributable to residual benzyl groups could be observed (< 3%).
Polymers with molecular weights $\bar{M}_v$, up to 3.7·10^5 daltons were
calculated from intrinsic viscosity data using the equation $[\eta]$ =
1.4·10$^{-5}\bar{M}_v$0.9 , H$_2$O, 28.5°C)[9].

*2,2,2-Trichloroethyl hemisuccinate (**10**)*

A solution of succinic anhydride (6 g, 60 mmole) in 2,2,2-
trichloroethanol (7.2 ml, 75 mmole) was refluxed under a nitrogen
atmosphere for 8 hr. The mixture was cooled, *n*-hexane (100 ml)
was added and boiling was continued for 25 min. The solution was
kept at 4°C for 16 hr and the crystals of **10** were collected,
washed with cold hexane and dried. Yield 12 g (80%)., m.p. 87-88°C
(86-88°)[18].

*17-β-Hydroxy-4-androstene-3-one-17-hemisuccinate (testosterone
hemisuccinate **6**)*

A mixture of succinic anhydride (3.12 g, 31.2 mmole), *N*-
methylimidazole (2.31 ml, 31.2 mmole), and testosterone (**2**, 5.77 g,
20 mmole) in acetonitrile (20 ml) was stirred for 30 hr at 65°C
and concentrated *in vacuo*. The sirup was taken in chloroform (100
ml) and the solution was washed successively with aqueous hydro-
cloric acid (100 ml, 1.2 N, 0°C), water, an aqueous solution of sodium

dihydrogen phosphate (100 ml, 0.25 M) and water. The organic layer
was dried (Na$_2$SO$_4$), concentrated to a small volume and subjected
to flash chromatography[19] using a 2-4% methanol gradient in chloroform. Appropriate fractions (TLC, R$_2$ 0.7; A) were pooled, concentrated to a small volume and induced to crystallization by the
addition of hexane yielding **6** (6.14 g, 79%); m.p. 175-176°C
(173 - 174)[20]. I.R. data for **6**; ν 1740 (ester), 1650 (ketone)
^{1}H-n.m.r. data (CDCl$_3$): δ 5.74 (m, 1H, H-4); 4.62 (m, 1H, H-17);
2.65 (s, 4H, succinate); 1.19 (s, 3H, Me-19), 0.84 (s, 3H, Me-18),
0.8-2.6 (other protons) (c.f. ref. 20).

Norethindrone (17-α-ethynyl-17β-hydroxyestr-4-en-3-one)hemi-
succinate (7)

    a. By direct acylation. A mixture of succinic anhydride (400
mg, 4 mmole), norethindrone (**3**, 298 mg, 1 mmole) and 4-dimethyl-
aminopyridine (550 mg, 4.5 mmole) in pyridine (2 ml) was stirred
at room temperature in the dark up to 11 days at which time the
conversion of **3** into **7** was still far from complete (TLC, R$_3$ 0.7,
A). The dark brown mixture was taken up in chloroform and further
processed as described for **6**. Crude **7** was subjected to flash
chromatography[19] using a 0-6% methanol gradient in chloroform
giving crystals of **7** in very moderate yields (15%).
    b. A solution of 2,2,2-trichloroethyl hemisuccinate (**10**,
1.4 g, 6 mmol), dicyclohexylcarbodiimide (1.38 g, 6 mmole), norethindrone (**3**, 1.47 g, 5 mmole) and 4-dimethylaminopyridine
(122 mg, 1 mmole) in methylene chloride (15 ml) was stirred for 16
hr in the dark under rigid exclusion of water. The mixture was concentrated *in vacuo* and the residue was taken up in methylene
chloride, filtered and subjected to flash chromatography[19] as
described above. The fractions containing **7** were pooled (TLC) and
concentrated to an oil. This oil was taken up in acetic acid (90%,
10 ml) and zinc powder (1.2 g) was added and the suspension was
vigorously shaken for about 6 hr. Removal of the protecting group
was monitored by TLC analysis. The mixture was filtered and the
zinc was washed with acetic acid and the filtrate was concentrated
in vacuo. The residue was taken up in chloroform (5 ml) and the
solution was washed successively with ice cold hydrochloric acid
(0.1 N, 10 ml), three times with water, dried (Na$_2$SO$_4$) and concentrated to an oil which crystallized *in vacuo*, yield 1.15 g,
58%, m.p. 179-181°C. IR data (KBr): 3200-3600 (broad, COOH), 3310
(stretch, C≡CH), 1745 (COOR), 1715 (COOH), 1640 (ketone), 635-
660 (waggle, C≡CH) cm^{-1}. ^{1}H-n.m.r. (CDCl$_3$) data: δ 5.85 (1H, H-4),
2.65 (4H, succinate), 2.60 (s, 1H, C≡CH), 0.93 (s, 3H, CH$_3$), 1.0-
3.0 (other protons).

Poly N^5-(3-hydroxypropyl)-L-glutamine testosterone mixed succinate
(11).

4-Dimethylaminopyridine (275 mg, 2.25 mmole) and **6** (275 mg, 2.25

mmole) were dissolved in pyridine under a blanketing atmosphere and **9** (2.73 g, 9 mmole) was added with stirring. After 20 min polymer **1** (930 mg, 5 mmole) was added. The polymer dissolves in the course of a few hours and the reaction can be monitored by TLC analysis. Addition of more **9** regenerates more acid anhydride to account for any water present in the reaction mixture (scheme 5). The mixture was stirred for an additional 70 hr at room tempe-rature and then added dropwise to an aqueous solution of potassium carbonate (1%) to remove arylsulphonic acid and any **6**. The fibrin-ous mass precipitated was washed with aqueous hydrochloric acid (0.12 N) to remove pyridine (the less toxic ethyl nicotinate ester suits equally well as a substitute for pyridine). In con-trast to the carrier polymer, the steroid conjugates **11** and **12** dissolve in chloroform due to the added lipophylic area in the conjugate. The structure of the conjugate (scheme 6) was confirmed by ^{1}H-n.m.r. spectroscopy and synthetical evidence. ^{1}H-n.m.r. ($CDCl_3$) data: δ 5.73 (m, 1H, H-4); 3.8-4.7 (H-17 and resonances of **1**); 2.58 (4H, succinate); 1.19 (3H, Me-19), 0.82 (3H, Me-18) ppm. I.R. data (KBr): ν 3300-3450 (NH, stretch), 1720 (esters C=O), 1640-1660 (ketone, steroid; amid) cm^{-1}.

Poly N^5-(3-hydroxypropyl)-L-glutamine norethindrone mixed succinate (12)

was prepared in the same manner as described for **11**. I.R. data (KBr): ν 3450 (OH, NH); 1730 (esters) 1640-1660 (ketone, steroid; amid); 670 (C≡CH, waggle) cm^{-1}. ^{1}H-n.m.r. ($CDCl_3$) data: δ 5.82 (m, 1H, H-4); 3.22 (s, 1H, C≡CH); 0.93 (s, 3H, Me); 1-3 (other protons) ppm. Degrees of substitution in the chloroform soluble conjugates **11** and **12** were determined by the proton integrals of the ^{1}H-n.m.r. resonances of appropriate diagnostic signals (H-4, ethynyl-H, H-17). Both conjugates should be stored in the dark and in the cold.

REFERENCES

1. D.E. Gregonis, J. Feijen, J.M. Anderson and R.V. Petersen, Polymer Preprints 20 (2) 612 (1979).
2. J. Feijen, D.E. Gregonis, C.E. Anderson, R.V. Petersen and J.M. Anderson, J. Pharm. Sci. 69, 871 (1980).
3. J.M. Anderson, A. Hiltner, K. Schodt and R. Woods, J. Biomed. Mater. Res. Symp. 3, 25 (1972).
4. J.M. Anderson, D.F. Gibbons, R.L. Martin, A. Hiltner and R. Woods, J. Biomed. Mater. Res. Symp. 5 197 (1974).
5. H.R. Dickinson and A. Hiltner, J. Biomed. Mater. Res. 15 591 (1981).
6. S. Mitra, N. van Dress, J.M. Anderson and R.V. Petersen, D.E. Gregonis and J. Feijen, Polymer Preprints 20, 32 (1979).
7. K.W. Marck, Ch.R.H. Wildevuur, W.L. Sederel, A. Bantjes and J. Feijen, J. Biomed. Mater. Res. 11, 405 (1977).

8. R.V. Petersen, C.G. Anderson, S.M. Fang, D.E. Gregonis,
    S.W. Kim, J. Feijen, J.M. Anderson and S. Mitra,
    Controlled Release of Bioactive Materials, R. Baker (ed.),
    Ac. Press. N.Y.(1980), p.45.
9. M. Lupu-Lotan, A. Yaron, A. Berger and M. Sela, Biopolymers
    $\underline{3}$, 625 (1965).
10. K. Okita, A. Teramoto and H. Fujita, Biopolymers $\underline{9}$, 717,
    (1970).
11. H.C. Beyerman, W.M. van der Brink, Proc. Chem. Soc. 266,
    (1963).
12. L.M. Litvinenko, N.M. Oleinik, Russ. Chem. Revs. $\underline{47}$ (5),
    401 (1978).
13. E.J. Ariens, Modulation of Pharmacokinetics by Molecular
    Manipulation, Drug Design Vol. II, E.J. Ariens, ed.
    Acad. Press. N.Y. (1971), p. 80.
14. J.H. Brewster, C.J. Ciotty, J. Am. Chem. Soc. $\underline{77}$, 6214
    (1955).
15. K. Itakura, N. Katagiri, C.P. Bahl, R.H. Wightman and
    S.A. Narang, J. Am. Chem. Soc. $\underline{97}$, 7327 (1975)
16. W.D. Fuller, M.S. Verlander, M. Goodman, Biopolymers $\underline{15}$,
    1869 (1976).
17. F.J. Joubert, M. Lupu-Lotan and H.A. Scheraga, Biochemistry
    $\underline{9}$, 2197 (1970).
18. C.W. Cambridge, C. Fitzmaurice, Brit. Patent 1.060.073;
    Chem. Abstr. $\underline{66}$, 88638k (1967).
19. W.C. Still, M. Kahn and A. Mitra, J. Org. Chem. $\underline{43}$, 2923
    (1978).
20. K. Holmberg, B. Hansen, Acta Chem. Scand. $\underline{B33}$, 410 (1979).

IMPROVED DRUG DELIVERY TO TARGET SPECIFIC ORGANS USING LIPOSOMES AS

COATED WITH POLYSACCHARIDES

Junzo Sunamoto*, Kiyoshi Iwamoto*, Masahiro Takada†,
Teruaki Yuzuriha†, and Kouichi Katayama†

* Department of Industrial Chemistry, Faculty of
Engineering, Nagasaki University, Nagasaki 852, Japan
† Eisai Co. Ltd., Toyosato-machi, Tsukuba-gun, Ibaragi
300-26, Japan

An assembly of a cell wall-like structure on the outmost surface of liposomes has been constructed, which makes liposomes
tough against chemical and physicochemical lyses of liposomal
membranes caused by external stimuli. In accordance with this
idea, we have prepared partly modified polysaccharides, O-palmitoyl-
pullulan (OPP) and O-palmitoylamylopectin (OPA), and coated the
outmost surface of egg phosphatidylcholine liposomes by them. The
efficiency in coating liposomes with the artificial cell wall has
been ascertained by four different methods: (1) isolation of
polysaccharide-coated liposomes by gel-filtration, (2) reduced
permeability for a water soluble material, carboxyfluorescein,
encapsulated in the interior of liposomes, (3) increased resistance
against the enzymatic lysis with phospholipase D for the coated
liposomes, and (4) decreased probability in the enzymatic digestion
with pullulanase of the polysaccharide strongly bound to the surface of liposomes. These results suggest a wider usage of the
polysaccharide-coated liposomes as an improved drug carrier.
    When conventional liposomes are administrated, they are highly
distributed in liver and kidney in general because of their hydrophobic (lipophilic) character. In this work, however, $[^{14}C]$-CoQ$_{10}$
encapsulated in the OPA-coated liposomes was found to be more
highly distributed in spleen and lung after intravenous injection
through the femoral vein of male ginea pigs.

INTRODUCTION

    Bacterial and plant cell membranes are well known to be
covered with cell walls mainly composed of polysaccharide derivatives.[1,2] The gram-positive bacterial cell walls are chiefly

157

composed of peptideglycan, while the cell wall of plant cells consists of cellulose. The cell walls are required to maintain the shape and stiffness of cells and to protect the cell membranes against the chemical and physical forces such as osmotic stimulation from the exterior. On the other hand, the saccharide determinants such as glycolipids and glycoproteins at the cell surface are considered to play an important role in various biological recognition processes involving an antigen-antibody interaction, toxin recognition, and cell-cell adhesion.[3,4]

Liposomes have gained wide acceptance in chemotherapy as potential carriers in introducing drugs and macromolecules into cells.[5,6] However, several problems to be overcome still remain before the liposomal drug-delivery system achieves its usefulness in biological and medicinal usages.[5,6] First of all is to depress the permeability of water soluble materials entrapped in the liposomes, and the second is to access an ability for liposomes to specifically approach the target cells or tissues. Hence, in order to increase the physicochemical and chemical stability of liposomes and to achive specific direction of liposomes to target cells or tissues, we have newly prepared chemically modified polysaccharides, O-palmitoylpullulan (OPP) and O-palmitoylamylopectin (OPA) (Fig. 1), and tried to coat the outmost surface of liposomes with them just like a cell wall. In this article we would like to emphasize the usefulness of the polysaccharide-coated liposomes as an improved drug carrier to target specific tissues or organs.

MATERIALS AND METHODS

Materials

Egg phosphatidylcholine (egg PC) was isolated and purified according to the method described by Singleton et al.[7] Pullulan-50 (weight average molecular weight, 50,100), pullulan-230 (M.W., 230,000), and amylopectin-112 (M.W., 112,000) were gifts from Hayashibara Biochemical Laboratories, Inc., Okayama. Fluoresceinylisothiocyanate (FITC, Dojin Laboratories, Kumamoto, Japan), carboxyfluorescein (CF, Eastman Kodak, Rochester, N.Y.), and sodium 1-anilino-8-naphthalenesulfonate (ANS, Tokyo Kasei Co. Ltd., Tokyo) were commercially available, respectively. Phospholipase D (*Streptomyces chromofuscus*) was a gift from Toyo Jyozo Co. Ltd., Tokyo. Pullulanase (*Aerobacter aerogenes subtilis*) was a gift from Hayashibara Biochemical Laboratories, Inc., Okayama. OPP and OPA were prepared according to the procedures for preparing O-palmitoyldextran.[8,9] The substitution degree of the palmitoyl residue per 100 glucose units in these polysaccharides was determined by ^{1}H-nmr and found to be 1.0 for OPP-230, 1.8 and 5.6 for OPP-50, and 4.9 for OPA-112, respectively. They were abbreviated such as OPP-230(1.0), OPP-50(1.8), OPP-50(5.6), and OPA-112(4.9). Fluoresceinylthiocarbamoyl(FITC)-O-palmitoylpolysaccharides were prepared according to the method for preparing FITC-dextran.[10]

The substitution degree of FITC group per 100 glucose units was fluorometrically determined to be 0.54 for OPP-50(1.8) and 0.48 for OPA-112(4.9). We abbreviate them totally such as FITC(0.54)-OPP-50(1.8) and FITC(0.48)-OPA-112(4.9). Other organic and inorganic chemicals were commercially available as analytical grade and were used without further purification.

Preparation of the polysaccharide-coated liposomes

Polysaccharide-coated liposomes were prepared by essentially the same procedure as that adopted for the conventional liposomes.[11,12]

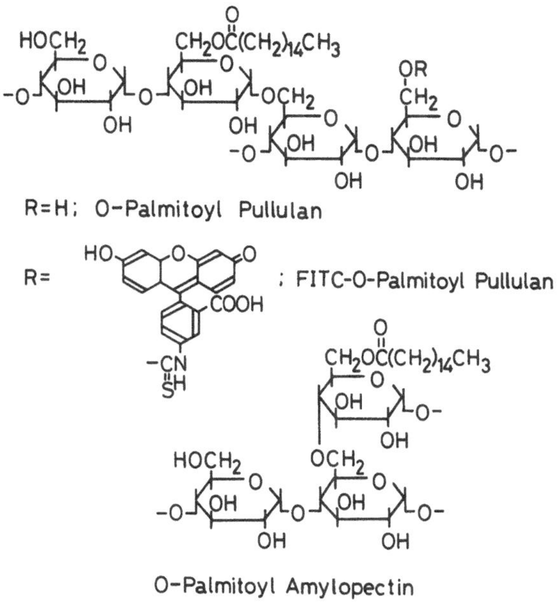

Fig. 1.   Structures of the modified polysaccharides employed in this work.

A 1.0 ml aqueous solution of 20 mM Tris-HCl buffer (pH 8.6)
containing 5 mg of OPP or OPA and 200 mM NaCl was added to the
sonicated liposome suspension (4.0 ml) of egg PC (30.0 mg) containing CF in their interior core of vesicles. After stirring for 30
min at 20 °C, the resulting mixed suspension was passed through a
Sepharose 4B column (ϕ1.6× 55 cm) pre-equilibrated in 20 mM Tris-
HCl aqueous buffered solution (pH 8.6) containing 200 mM NaCl, by
which procedures free polysaccharides, polysaccharide-coated
multilamellar, single-walled liposomes, and free CF were isolated,
respectively. CF-Unloaded liposomes were also prepared and coated
with polysaccharides by essentially the same procedure. The
concentration of liposomes was determined as an inorganic phosphate
according to Allen's procedure.[13]

Fluorescence measurement for CF release

Fluorescence measurements were carried out on a Hitachi 650-
10S fluorospectrophotometer equipped with a thermoregulated cell
compartment. The trianionic species of CF (above pH 7.5) emits at
520 nm by excitation at 470 nm. However, when concentrated up to
about 200 mM, its emission intensity diminished about 90% by the
concentration quenching mechanism.[14,15] The release of CF from
liposomes was followed by monitoring an increase in the fluorescence
intensity at 520 nm upon the liberation of CF from the concentration
quenching.[14,15] The total amount of CF encapsulated in liposomes
was determined by destroying the liposomal membranes with 30 μl of
10% (v/v) aqueous Triton X-100 solution for 1.0 ml of the liposome
suspension.

Reaction with phospholipase D

ANS emits at 490 nm by exciting at 380 nm in liposomal membranes, but in an aqueous solution because of strong quenching with
water. The ANS-loaded liposomes were prepared and isolated by
essentially the same method as described. When lecithins are
specifically destructed by phospholipase D upon the hydrolytic
cleavage of the bond between the phosphate and choline moieties,
ANS is released to the bulk aqueous phase along with the choline
group and its fluorescence is strongly quenched by water. Hence,
the destruction of lecithins by phospholipase D could be quantitatively followed by monitoring the fluorescence intensity of ANS at
490 nm.[16] For this purpose, to a 3.0 ml of single-walled liposome
suspension of egg PC (2.03×10^{-4} M as lecithin) containing $1.78\times$
10^{-5} M of ANS in bilayers was added 2.0 mg of OPP. After preincubation for 10 min at 37 °C, the reaction was initiated by injecting
200 mM Tris-HCl (pH 8.0) containing 15 mM Ca^{2+} ion and phospholipase
D (4.14 unit).

Hydrolytic digestion of polysaccharide with pullulanase

Hydrolytic digestion of pullulan with pullulanase and subsequent release of FITC moiety from the surface of liposomes were examined by a decrease in the steady-state fluorescence depolarization. Measurements were run on a Union Giken fluorescence polarization spectrophotometer FS-501S, where the FITC fluorescence (520 nm) was detected by exciting the sample at 440 nm using a sharp cut-off filter Y46 (Hoya Glass Works, Tokyo). To a 3.0 ml solution of liposome (1.0×10^{-4} M as egg PC) coated with a given amount of FITC(0.54)-OPP-50(1.8) was added 20 µl of pullulanase solution (2.0 unit) after preincubation for 5 min in 200 mM Tris-HCl (pH 8.0) at 25.0 °C.

Preparation of polysaccharide-coated liposome encapsulating [^{14}C]-CoQ$_{10}$

Small single-walled liposomes encapsulating [^{14}C]-CoQ$_{10}$ were prepared according to the method described previously.[11,12] Gel-filtration was carried out on a Sepharose 4B column ($\phi 1.6 \times 45$ cm) pre-equilibrated in 10 mM phosphate buffer (pH 7.4) containing 150 mM sodium chloride. The obtained liposome suspension of egg PC was concentrated to 2 ml by ultrafiltration. The concentrated liposome suspension containing 5.2 mg of egg PC was incubated with 5.2 mg of OPP or OPA under stirring for 30 min at room temperature. The resulting polysaccharide-coated liposomes were used for the following animal experiments.

Tissue distribution of polysaccharide-coated liposome containing [^{14}C]-CoQ$_{10}$

Polysaccharide-coated and -uncoated liposomes encapsulating [^{14}C]-CoQ$_{10}$ were intravenously administrated to the femoral vein of male guinea pigs (300-350 g) at a dose of 0.6 mg/kg of body weight. Blood samples were obtained by ear vein puncture at appropriate time interval. Guinea pigs were killed by decapitation 24 hr after intravenous injection. Just before decapitation, the heart and brain were perfused with cold saline, and then heart, brain, lung, liver, spleen, adrenal gland and kidney were removed for scintillation counting. The radioactivity was measured by the following manner. One hundred milligrams of each tissue was incubated with 0.5 ml of Soluene 350 for 2 hr at 50 °C. After adding 5 ml of scintillation fluid consist of Instagel/0.5 N HCl (9/1, by vol.) to the solubilized tissue sample, the radioactivity of samples was measured by using a Aloka LSC 653 liquid scintillation counter.

RESULTS AND DISCUSSION

Figure 2 shows a typical example of gel-filtration (Sepharose 4B) for isolating multilamellar (fr. 9-13) and single-walled

liposomes coated with FITC(0.54)-OPP-50(1.8) (fr. 14-21) and the
free polysaccharide (fr. 22-32). The fractionings were monitored
by both the optical density (turbidity) at 220 nm and fluorescence
intensity of FITC at 520 nm. Similar elution diagrams were obtained
also for other FITC-*O*-palmitoylpolysaccharide-coated liposomes.
When liposomes were incubated with simple FITC-OPP carrying no

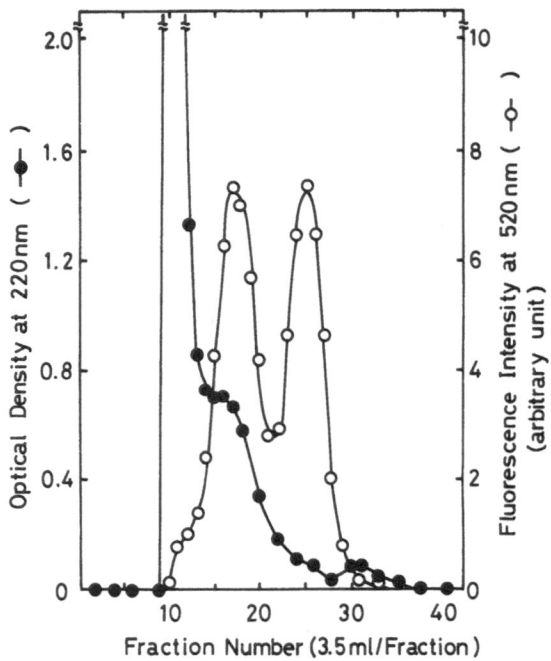

Fig. 2.   A typical example of gel-filtration of liposomes coated
          with FITC(0.54)-OPP-50(1.8) from the free FITC-OPP
          developed with 200 mM Tris-HCl (pH 8.0) as an eluant.
          Liposomes were detected by turbidity at 220 nm (−●−) and
          fluorescence intensity of FITC moiety at 520 nm (−○−).
          Coating with the polysaccharide was performed by incubat-
          ing 3.0 ml of a sonicated liposome suspension prepared
          from 30 mg of egg lecithin with 1 ml of 200 mM Tris-HCl
          containing 5 mg of the FITC-OPP at 25 °C (see text).

hydrophobic legs under the controlled condition, no polysaccharide
bound liposomes were isolated in the liposome fraction during gel-
filtration. First, thus, this finding shows apparently that polysaccharides bearing convenient acyl residues can bind tightly to
the outer surface of liposomal membranes.

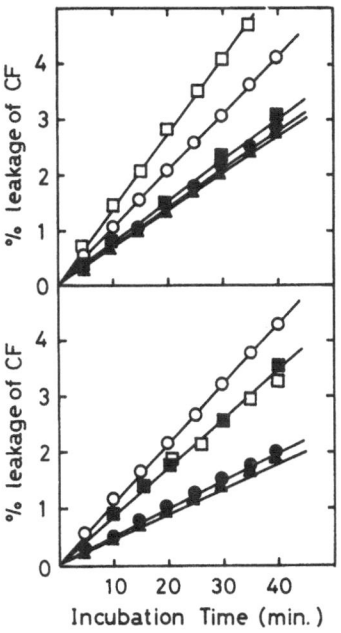

Fig. 3. Spontaneous release of carboxyfluorescein (200 mM) encap-
         sulated in the interior core of liposomes with and without
         artificial cell wall as a function of time at 50 °C and
         $\mu = 0.20$ M (NaCl) in 20 mM Tris-HCl (pH 8.6); top, small
         single-walled liposomes (30 mg, 2.4×10^{-4} M as lecithin
         in the cuvette) without polysaccharides ($-\bigcirc-$) and as
         coated with 5 mg of OPP-50(5.6) (■), OPP-50(1.8)
         ($-●-$), OPP-230(1.0) ($-▲-$), and OPA-112(4.9) ($-\square-$),
         respectively; bottom, large multilamellar liposomes
         themselves ($-\bigcirc-$) and as coated with OPP-50(5.6) ($-■-$),
         OPP-50(1.8)($-●-$), OPP-230(1.0) ($-▲-$), and OPA-112(4.9)
         ($-\square-$), respectively, in the same manner as that applied
         for small liposomes.

Secondly, we investigated the effect of coating with these polysaccharides on the barrier function of liposomal membrane. The spontaneous release of carboxyfluorescein (CF) encapsulated in the interior of liposomes as a function of time was investigated at 50.0 °C (Fig. 3). Coating the outer surface of liposomes with these polysaccharides brought about several fold decrease in the permeability for CF. The substitution degree of acyl residues to behave as an anchor rather than the molecular weight of polysaccharide seems more important to coat conveniently the surface of liposomes.

Thirdly, we inspected the resistance of liposomes to the enzymatic lysis. When phospholipase D was added to the preincubated suspension of liposomes which contain 1-anilino-8-naphthalene-sulfonate (ANS) close to the membrane surface, the fluorescence from ANS was effectively quenched upon the release of ANS to the bulk aqueous phase (Fig. 4).[16] This quenching was effectively

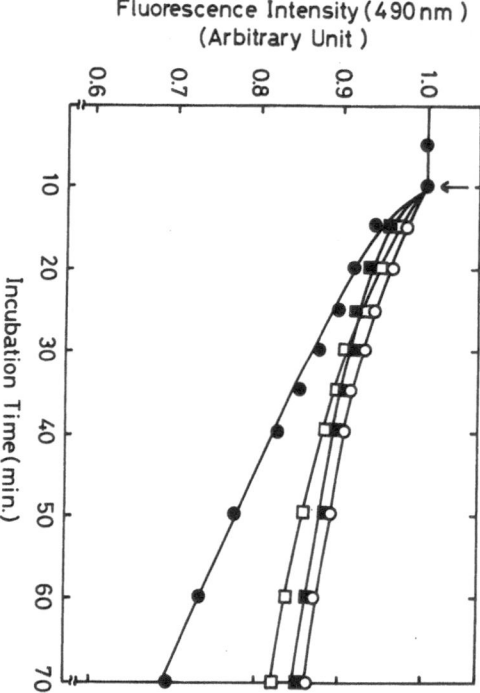

Fig. 4. Fluorescence quenching of ANS upon the enzymatic lysis of single-walled liposomes without coating (−●−) and as coated with OPP-230(1.0) (−O−), OPP-50(1.8) (−■−), and OPA-112(4.9) (−□−) as a function of incubation time. For detail, see text. Vertical arrows indicate the point when phospholipase D was injected after preincubation.

depressed for OPP- or OPA-coated liposomes compared with free
liposomes, which means that coating liposomes with polysaccharides
can protect liposomes effectively from the enzymatic lysis.

The fourth investigation was the enzymatic digestion of FITC-
OPP bound to liposomes with pullulanase.[17] The steady-state fluorescence depolarization (p) about the FITC(0.54)-OPP-50(1.8)-coated
liposomes showed the restriction of mobility of polysaccharides,
where the higher p-value were obtained for FITC-OPP with liposomes
than for free FITC-OPP without liposomes (Fig. 5). When pullulan
was digested by pullulanase, which is the enzyme to cleave specifically the $\alpha 1 \rightarrow 6$ linkage of pullulan (see Fig. 1),[17] a drastic decrease in the p-value was observed. This is caused by gains of the

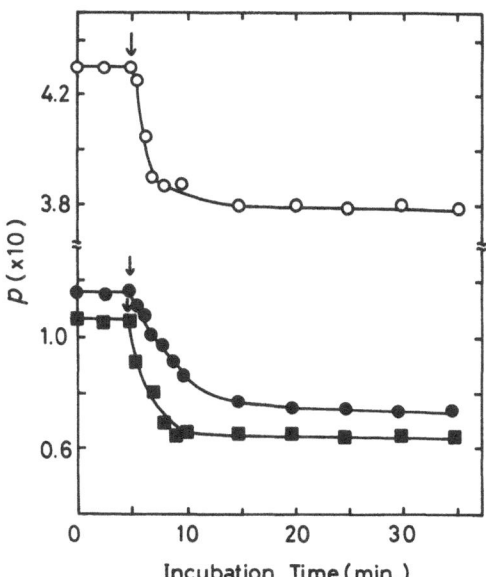

Fig. 5.   Changes in fluorescence depolarization (p) upon the enzymatic digestion of FITC(0.54)-OPP-50(1.8) as a function
of time: —□— , the free polysaccharide; —●— , the polysaccharide bound to small single-walled liposomes; and
—O— , the polysaccharide bound to large multilamellar
liposomes, respectively. For detail, see text. Vertical
arrows indicate the point when pullulanase was injected
after preincubation.

Table 1.   Tissue Distribution of $[^{14}C]$-CoQ$_{10}$ (%) 24 hr after
           Intravenous Injection of Small Single-walled
           Liposomes into the Femoral Vein of Male Guinea Pigs[a]

Liposome[b] Tissue	I	II	III	IV	V
Liver	30	40	49	52	50
Adrenal gland	0.5	1.3	1.5	2.0	2.2
Heart	0.19	0.5	0.05	0.03	0.2
Spleen	2.0	0.3	3.4	1.6	2.1
Kidney	0.24	0.25	0.23	0.3	0.3
Lung	0.27	0.19	13.1	0.3	0.5
Brain	0.03	0.03	0.02	0.03	0.06

I: Egg PC/Chol (MLV)   II: Egg PC/Chol
III: Egg PC/Chol/OPA-112(4.9)   IV: Egg PC/Chol/OPP-50(1.8)
V: Egg PC/Chol/OPP-230(1.0)

a) Values in the Table show the percentages of radioactivity
   of an injected dose.
b) Liposomes were composed of egg PC and cholesterol (3:1,
   by mol.).

Abbreviation; Chol, cholesterol; MLV, multilamellar vesicle

freedom of mobility upon the release of the fluorescent moiety from liposomal surface to bulk aqueous phase (Fig. 5). Interestingly, in addition, a fast decrease (15-35%) in the p-value ceased within several minutes just after adding the enzyme to the liposome suspension, and a continuous decrease up to the zero level, which is corresponding to the value of free polysaccharide in the absence of liposomes, was not observed. Even if more enzyme was added to the system, further decrease in the p-value was never observed. This may mean that the enzyme can attack only the part as branched and/or directed to an aqueous phase in the polysaccharide molecule and hardly digest the main chains which are strongly bound onto the surface of liposomes. Though data were not shown, simple pullulan which did not bring any hydrophobic anchor such as the palmitoyl residue did not show any significant effect as an artificial cell-wall for liposomes.

In any event, we could first success to assemble a cell wall-like coat on the outer surface of liposomes using partly modified polysaccharides and consequently to make liposomes tough against the permeability of water soluble drugs and the enzymatic lysis of lipids as well. These results encouraged us to employ this cell-wall-assembled liposome as an improved drug carrier. Hence, we investigated the fate of the cell-wall-assembled liposomes encapsulating [^{14}C]-CoQ$_{10}$ after intravenous injection into the male guinea pigs. Table 1 shows the tissue distribution (%) of [^{14}C]-CoQ$_{10}$ encapsulated in several polysaccharide coated liposomes and conventional liposomes 24 hr after intravenous injection. Clearly from Table 1, [^{14}C]-CoQ$_{10}$ is highly distributed in adrenal gland, spleen, liver, and lung. In the case of OPA-coated liposomes, most interestingly, the distribution of [^{14}C]-CoQ$_{10}$ in lung drastically was increased compared with those of other liposome systems. Similar tissue distribution pattern after intravenous injection has been previously observed also in the case of drug delivery system using glycolipid contained liposomes.[18] In any event, these present findings suggest that modifying the surface of liposomes enables to alter the tissue distribution of liposomal drug carrier. Further and more detailed investigations are in progress in our laboratories using different kinds of polysaccharide.

ACKNOWLEDGMENT

We are grateful to Dr. Takeshi Fujita, Eisai Co. Ltd., Tsukuba Institute, for his valuable suggestion and discussion.

REFERENCES

1. H. J. Rogers, H. R. Perkins, and J. B. Ward, "Microbial Cell Walls and Membranes," Chapman and Hall, London, New York (1980).
2. A. L. Lehninger, "Biochemistry," Worth Publisher Inc., New York (1975).

3. R. C. Hughes, The Complex Carbohydrates of Mammalian Cell Surfaces and Their Biological Roles, Essays Biochem. 11:1 (1975).
4. N. Sharon and H. Lis, Glycoproteins: Research Booming on Long-
   Ignored, Unbiquitous Compounds, Chem. Eng. News March 30:21
   (1981).
5. D. A. Tyrell, T. D. Heath, C. M. Colley, and B. E. Ryman, New
   Aspects of Liposomes, Biochim. Biophys. Acta 457:259 (1976).
6. G. Gregoriadis, Liposomes in Therapeutic and Preventive Medicine: The Development of Drug-Carrier Concept, Ann. N. Y. Acad.
   Sci. 308:343 (1978).
7. W. S. Singleton, M. S. Gray, M. L. Brown, and J. L. White,
   Chromatographically Homogeneous Lecithin from Egg Phospholipids,
   J. Am. Oil. Chemists' Soc. 42:53 (1965).
8. U. Hämmerling and O. Westphal, Synthesis and Use of O-Stearoyl
   Polysaccharides in Passive Hemagglutination and Hemolysis,
   Eur. J. Biochem. 1:46 (1967).
9. M. Suzuki, T. Mikami, T. Matsumoto, and S. Suzuki, Preparation
   and Antitumor Activity of O-Palmitoyldextran Phosphates, O-
   Palmitoyldextrans, and Dextran Phosphate, Carbohydr. Res. 53:
   223 (1977).
10. A. N. de Belder and K. Grahath, Preparation and Properties of
    Fluorescein-Labelled Dextrans, Carbohydr. Res. 30:375 (1973).
11. J. Sunamoto, H. Kondo, and A. Yoshimatsu, Liposomal Membranes.
    I. Chemical Damage of Liposomal Membranes with Functional
    Detergent, Biochim. Biophys. Acta 510:52 (1978).
12. J. Sunamoto, T. Hamada, and H. Murase, Liposomal Membranes. IV.
    Fusion of Liposomal Membranes Induced by Several Lipophilic
    Agents, Bull. Chem. Soc. Jpn., 53:2773 (1980).
13. R. J. L. Allen, The Estimation of Phosphorus, Biochem. J. 34:
    858 (1940).
14. J. N. Weinstein, S. Yoshikami, P. Henkart, R. Blunmenthal, and
    W. A. Hagins, Liposome-Cell Interaction: Transfer and Intra-
    cellular Release of a Trapped Fluorescent Marker, Science
    195:489 (1977).
15. F. C. Szoka, Jr., K. Jacobson, and D. Papahadjopoulos, The Use
    of Aqueous Space Markers to Determine the Mechanism of Interac-
    tion between phospholipid Vesicles and Cells, Biochim. Biophys.
    Acta 551:295 (1979).
16. M. Nakagaki and I. Yamamoto, The Interfacial Reaction of
    Lecithin and Phospholipase D at the Micellar Surface, Yakugaku
    Zasshi 101:1099 (1981).
17. J. Sunamoto, K. Iwamoto, and H. Kondo, Liposomal Membranes.
    VII. Fusion and Aggregation of Egg Lecithin Liposomes as
    Promoted by Polysaccharides, Biochem. Biophys. Res. Commun.
    94:1367 (1980).
18. M. R. Mauk, R. C. Gamble, and J. D. Baldeschwieler, Targeting
    of Lipid Vesicles: Specificity of Carbohydrate Receptor
    Analogues for Leukocytes in Mice, Proc. Natl. Acad. Sci. USA
    77:4430 (1980).

SYNTHESIS AND RELEASE OF CONTRACEPTIVE STEROIDS

FROM BIOERODIBLE POLY(ORTHO ESTER)S

J. Heller, D.W.H. Penhale,
B. K. Fritzinger, and J. E. Rose

Polymer Sciences Department
SRI International
Menlo Park, CA 94025

INTRODUCTION

Because of possible adverse effects or a desire to terminate therapy, implanted bioerodible drug delivery systems should be easily removable at any time. For this reason, solid devices that maintain their mechanical integrity throughout the major portion of their delivery regime are particularly attractive. A further desirable feature is drug release that is close to zero order.

In developing such devices, two fundamentally different approaches are possible. In one, mechanism of drug release is by diffusion from a reservoir through a rate-limiting bioerodible polymer membrane,[1] and in the other, drug release is controlled by matrix erosion.[2] However, to achieve zero order drug delivery from monolithic erosional devices the erosion process must be confined to the surface of the solid device.[3]

The principal focus of our work is the development of a sub-dermally implantable, bioerodible monolithic device that would release contraceptive steroids by close to zero order kinetics for at least six months. Also, polymer erosion and drug release should be coupled so that no polymer remains in the tissue after all the drug has been released.

The major difficulty in developing such a system is the requirement of surface erodibility: to achieve this, the hydrolytic erosion process at the surface of the device must occur at a very much faster rate than the hydrolytic erosion process in the matrix interior. To develop such a system we have synthesized polymers that contains linkages in the polymer backbone that are very labile in acid and very stable in base and then use pH-modifying agents incorporated into the matrix that will produce a pH at the outer surface of the device that is lower than that in the interior of the matrix.

The polymers are poly(ortho esters)[4] and in our initial approach we have used finely micronized sodium carbonate as the pH modifier with the expectation that polymer erosion would only occur at the outer surface of the device, where the sodium carbonate is sufficiently neutralized by the external environment.[5]

In this chapter we present _in vitro_ and _in vivo_ norethindrone release rate data (from experiments in which we used the basic salt sodium carbonate and the neutral salt sodium sulfate), discuss the effect of thermal treatment on polymer properties, and finally discuss problems that arise when an extremely water-insoluble drug such as levonorgestrel is used.

EXPERIMENTAL PROCEDURES

Polymers were prepared as described previously[4] by condensation of equimolar amounts of a diketene acetal and a diol

where R = H for 3,9-bis(methylene 2,4,8,10-tetraoxaspiro[5,5] undecane) and R = CH_3 for 3,9-bis(ethylidene 2,4,8,10-tetraoxaspiro [5,5]undecane).

Polymer Fabrication

Drug-containing devices were either circular discs 6.3 mm in diameter and 0.6- or 1.2-mm thick or rods 2.3 mm in diameter and 10-mm long. The discs were punched from sheets fabricated by first blending the polymer with micronized drug and salt in a Teflon-coated pan, in a dry box, at temperatures between 135° and 200°C and then pressing the blend in a Carver press at the same temperature and 10,000 psi. Before the discs were punched, the sheets were warmed to 50° to 60°C. Rods were prepared by using a screw-activated extruder and cutting the rods to size with a heated knife. Current fabrication techniques involve blending the drug and polymer at temperatures not exceeding 130°C in a modified Brabender mixer.

Drug Release

Drug-containing devices were enclosed in a stainless steel mesh bag and moved vertically in a pH 7.4 phosphate buffer contained in test tubes thermostated at 37°C. Drug released into the buffer was analyzed by high pressure liquid chromatography. Care was taken to ensure that total drug concentration in the buffer remained below 10% by changing solutions at frequent intervals.

Polymer Characterization

Polymer molecular weight and molecular weight distribution were determined with a Waters Gel Permeation Chromatograph (GPC) using 10^4, 10^3, and 10 Å microstyrogel columns in tetrahydrofuran at room temperature. The GPC effluent was passed through a Chromatix KMX-6 laser, low-angle light-scattering apparatus.[6] Melt viscosities were measured with a parallel plate rheometer attachment on a thermomechanical analyzer and a Du Pont 990 Controller.[7]

RESULTS AND DISCUSSION

Drug Release Mediated by Water-Soluble Salts

Figure 1 shows a comparison of rate of norethindrone release from circular discs containing 10 wt% drug and either 10 wt% Na_2CO_3 or 10 wt% Na_2SO_4. The numbers below the arrows indicate total weight loss.

As described previously[5,8-10] these data can be explained by an osmotic imbibing of water mediated by the incorporated water-soluble salt as described by Fedors.[11] According to this mechanism, when a hydrophobic polymer containing water-soluble inclusions is placed in an aqueous environment, water will begin to diffuse into the polymer until it comes into contact with an included salt particle that will then begin to dissolve. Because the polymer acts as a semipermeable membrane, an osmotic pressure will begin to act radially outwards as a result of the increasing volume of water drawn into the inclusion.

As a consequence of this process, a swelling front develops and drug release occurs by the swelling-induced relaxation of the polymer chains. Because poly(ortho esters) are stable in base, virtually no polymer erosion occurs when Na_2CO_3 is used, but when the neutral salt Na_2SO_4 is used, polymer erosion does take place. However, because erosion significantly lags drug release, erosion is not the dominant process and drug release again occurs by the swelling-induced relaxation of polymer chains.

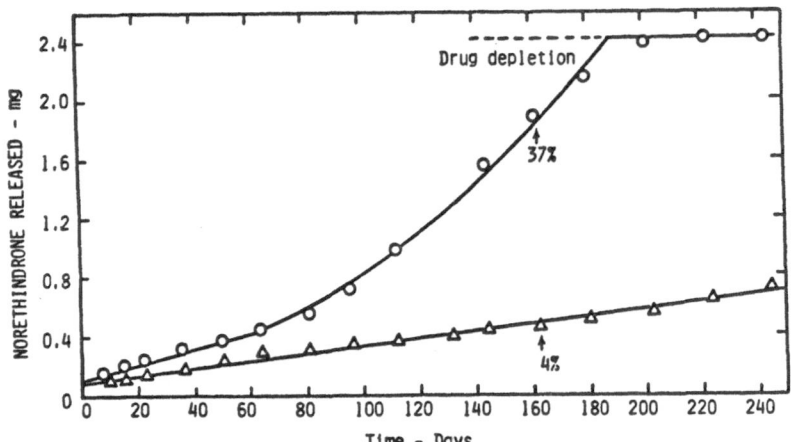

Fig. 1.   Norethindrone (NE) release from 3,9-bis(methylene 2,4,8,
           10-tetraoxaspiro[5,5]undecane)/1,6-hexane diol poly(ortho
           ester), 6.3-mm diameter discs at pH 7.4 and 37°C.
           o 10 wt% NE, 10 wt% Na₂SO₄
              disc thickness 0.6 mm, total drug content 2.4 mg
           ▲ 10 wt% NE, 10 wt% Na₂CO₃
              disc thickness 1.2 mm, total drug content 4.0 mg

        Because no polymer erosion takes place when the basic salt
Na₂CO₃ is used, surface area of the swelling front remains constant
and drug release is also constant. However, when the neutral salt
Na₂SO₄ is used polymer erosion takes place within the water inclusions, and as a consequence of the erosion process, surface area
increases with a consequent increase in drug release rate.

Effect of Polymer Fabrication Temperature on Drug Release

        An effect not realized in our previous work was a dependence
of polymer erosion and drug release rate on fabrication temperature.
Figure 2 shows the effect of fabrication temperature on levonorgestrel release rate.

        Although at the present time the effect of fabrication tempera-
ture on erosion rate and drug release is not completely understood
and is currently under investigation, the following polymer
degradation mechanism appear reasonable

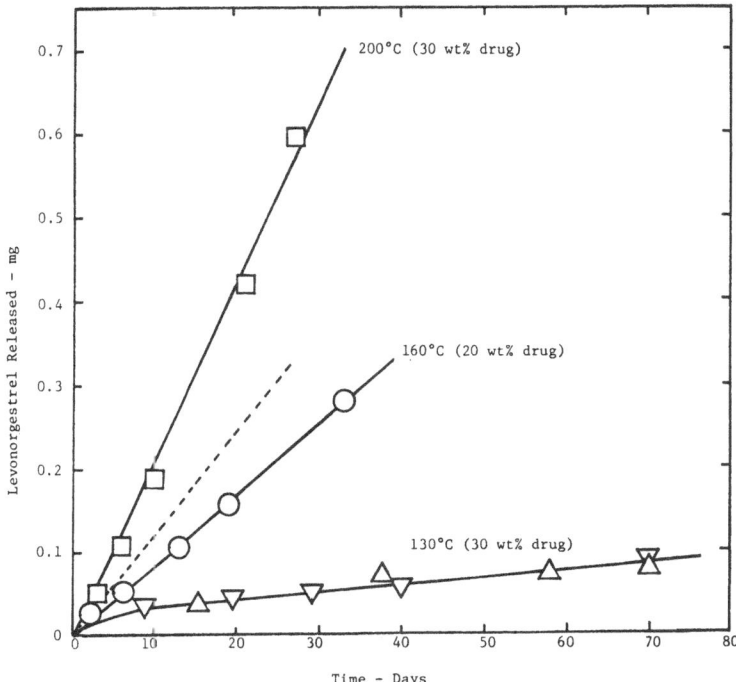

Fig. 2.   Cumulative release of levonorgestrel from 6.3 mm
poly(ortho ester) discs fabricated at different
temperatures. Polymers prepared from 3,9-bis(ethylidene
2,4,8,10-tetraoxaspiro[5,5]undecane) and various diols.
▲ 1,6-hexanediol at pH 7.4
▼ 1,6-hexanediol at pH 7.0
O 70/30 t-cyclohexanedimethanol/1,6-hexanediol at pH 7.4
□ 70/30 t-cyclohexanedimethanol/1,6-hexanediol at pH 7.4
--- polymer fabricated at 160°C corrected to 30 wt% drug

where R is $-(CH_2)_4-$ for 1,6-hexanediol and $-\bigcirc-$ for t-cyclohexane-
dimethanol. Preliminary data[12] indicate that this degradation only
occurs at significant rates at temperature exceeding 130°C so that
currently materials are fabricated at or below that temperature.

A similar degradation mechanism has been proposed for the
thermal degradation of polyurethanes which also dissociate into
their isocyanate and diol constituents.[13]

Because the necessity to fabricate at high temperatures is
related to the molecular weight dependent polymer melt viscosity,
we are currently controlling stoichiometry so that polymers having
molecular weights no higher than about 20,000 are produced. The
effect of stoichiometry on polymer molecular weight is shown in
Figure 3, and polymer melt viscosity measured at 150°C as a function
of molecular weight is shown in Figure 4.

Drug Solubility Considerations

Because a number of contraceptive steroids, and particularly
levonorgestrel, have water solubilities of only a few parts per
million, the effect of drug water solubility on rate of release
from bioerodible polymers into an aqueous medium becomes an important
consideration and can, in fact, become rate limiting.

This effect is illustrated in Figure 5, which shows rate of
release of levonorgestrel from a 3,9-bis(ethylidone 2,4,8,10-
tetraoxaspiro[5,5]undecane)/(70/30) t-cyclohexane dimethanol-1,6-
hexanediol polymer containing 20 wt% drug. Total weight percent
loss of the device is shown above the arrows.

Clearly, matrix erosion has little effect on rate of drug
release since after 50 days rapid acceleration of erosion rate without a concomitant increase in rate of drug release is noted. The
release rate data also seem to indicate two regimes, one up to
about day 48 and one from day 48 on. Thus, up to day 48, rate of
levonorgestrel release is about 8 mcg/day and subsequent to that
day it doubles to about 16 mcg/day. Furthermore, weight loss data
show that up to day 48 erosion and drug release is coupled (4.3%
weight loss, 7% drug released) but after day 48 erosion significantly
leads drug release and at day 143 weight loss is 75%, while the
amount of drug released is only 41%.

According to these data, during the first stage levonorgestrel
release is controlled by matrix erosion, but during the second stage
the rate of levonorgestrel release is controlled by the rate of
dissolution of the drug particles exposed by the eroding polymer.
Because polymer erosion occurs more rapidly than rate of drug
solubilization, polymer erosion leads drug release.

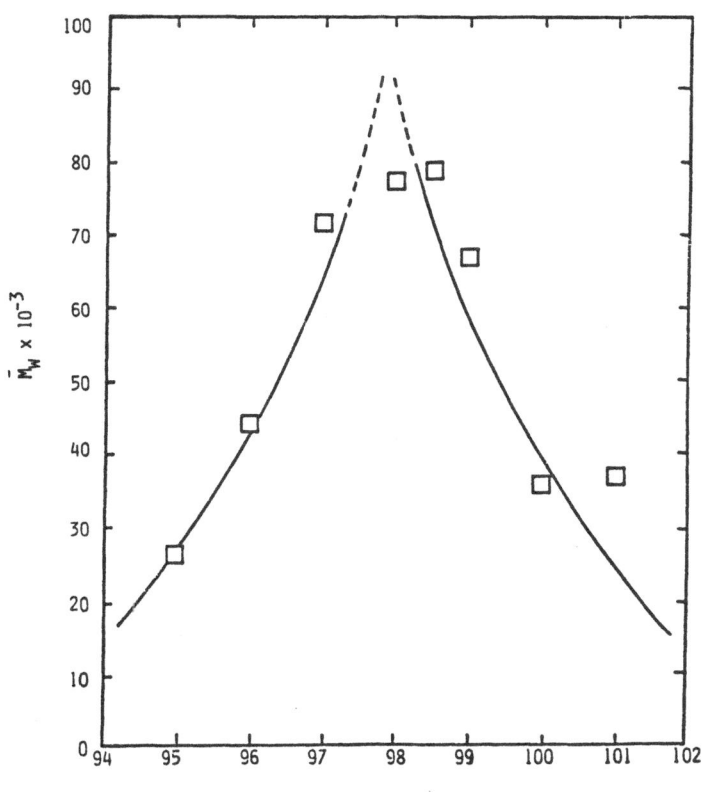

Fig. 3.   Effect of stoichiometry on weight average molecular weight of poly(ortho ester) prepared from 3,9-bis(ethylidene 2,4,8,10-tetraoxaspiro[5,5]undecane) and trans-cyclohexanedimethanol.

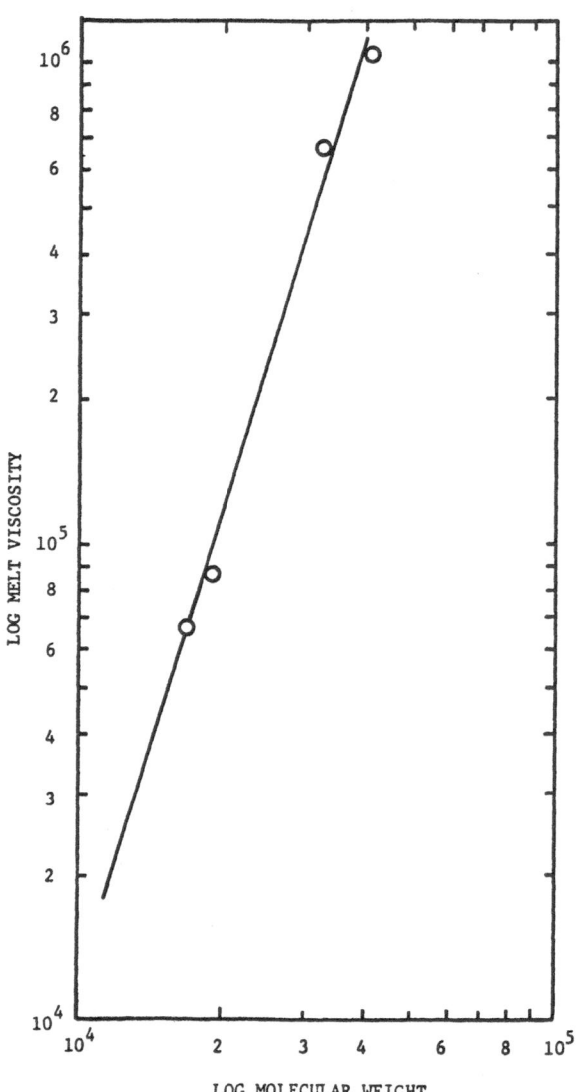

Fig. 4.   Melt viscosity versus molecular weight for a 3,9-
           bis(ethylidene 2,4,8,10-tetraoxaspiro[5,5]undecane/
           (70/30) t-cyclohexanedimethanol, 1,6-hexanediol polymer
           measured at 150°C.

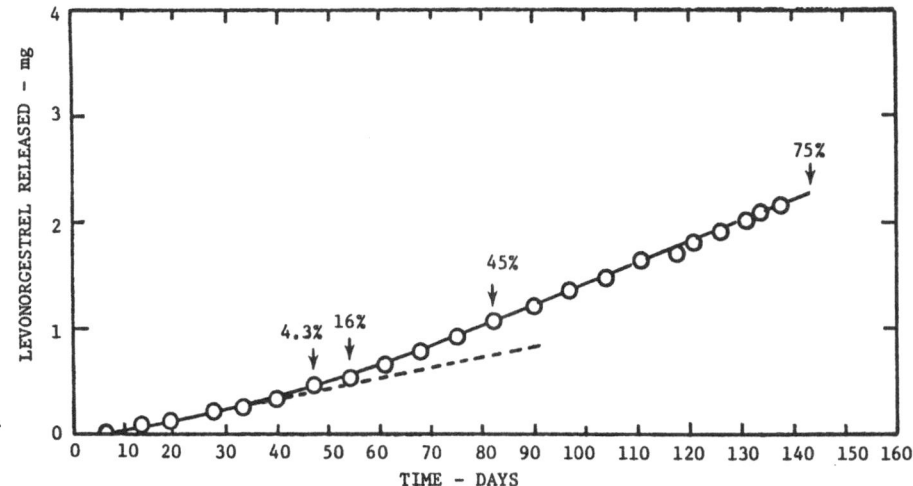

Fig. 5.   Cumulative release of levonorgestrel from a 3,9-
bis(ethylidene 2,4,8,10-tetraoxaspiro[5,5]undecane/
(70/30) t-cyclohexanedimethanol/1,6-hexanediol polymer
containing 20 wt% levonorgestrel. Total drug 5.6 mg.
Discs 6 mm x 0.5 mm at pH 7.4 and 37°C. Numbers
indicate weight loss.

ACKNOWLEDGMENT

We wish to acknowledge the contributions of Dr. D. B. Cotts who
supervised polymer characterization work and provided valuable advice.

This work was supported by the Contraceptive Development Branch,
Center for Population Research, National Institute of Child Health
and Human Development, NIH under Contract No. 1-HD-7-2826,

REFERENCES

1.   C. G. Pitt, T. A. Marks, and A. Schindler, Biodegradable Drug
      Delivery Systems Based on Aliphatic Polyesters: Application to
      Contraceptives and Narcotic Antagonists, in: "Controlled
      Release of Bioactive Materials," R. W. Baker, ed., Academic
      Press, New York (1980).
2.   J. Heller, Drug Release from Bioerodible Systems, in: "Medical
      Applications of Controlled Release Technology," R. S. Langer
      and D. C. Wise, eds., CRC Press, Boca Raton, Florida, in
      press.

3. J. Heller and R. W. Baker, Theory and Practice of Controlled
    Drug Delivery from Bioerodible Polymers, in: "Controlled
    Release of Bioactive Materials," R. W. Baker, ed., Academic
    Press, New York.

4. J. Heller, D.W.H. Penhale, and R. F. Helwing, Preparation of
    Poly(Ortho Esters) by the Reaction of Ketene Acetals and
    Polyols, J. Polymer Sci., Polym. Lett. Ed., 18:619 (1980).

5. J. Heller, D.W.H. Penhale, R. F. Helwing, B. K. Fritzinger,
    and R. W. Baker, Release of Norethindrone from Polyacetals
    and Poly(Ortho Esters), AIChE Symposium Series No. 206,
    77:28 (1981).

6. D. B. Cotts, Dependence of Chain Conformation and Solution
    Thermodynamics on Composition for Poly(Ortho Esters) Prepared
    for Ketene Acetals and Mixtures of Diols, J. Polymer Sci.,
    Polym. Lett. Ed., in press.

7. G. J. Dienes and H. F. Klemn, Theory and Application of the
    Parallel Plate Plastometer,   J. Appl. Phys., 17:458 (1946).

8. J. Heller, D.W.H. Penhale, R. F. Helwing, and B. K. Fritzinger,
    Release of Norethindrone from Poly(Ortho Esters), Polymer Eng.
    Sci., 21:727 (1981).

9. J. Heller, D.W.H. Penhale, R. F. Helwing, and B. K. Fritzinger,
    Controlled Release of Steroids from Poly(Ortho Esters), in:
    "Controlled Release Delivery Systems," T. J. Rosemand and
    S. F. Mansdorf eds., Marcel Dekker, New York, in press.

10. J. Heller, D.W.H. Penhale, B. K. Fritzinger, J. E. Rose, and
    R. F. Helwing, Controlled Release of Contraceptive Steroids
    from Biodegradable Poly(Ortho Esters), Contracept. Deliv.
    Systems, 4:000 (1983).

11. R. F. Fedors, Osmotic Effects 1. Water Absorption by Polymers,
    Polymer, 21:207 (1980).

12. J. Heller and J. E. Rose, Thermal Degradation of Poly(Ortho
    Esters) prepared from Ketene Acetals and Diols, to be published.

13. H. J. Fabris, Thermal and Oxidative Stability of Urethanes in :
    "Advances in Urethane Science and Technology," Volume 4,
    Technomic Publishing Co., Westport, Connecticut (1976).

PRESS-COATED SYSTEMS FOR DRUG RELEASE CONTROL

Ubaldo Conte, Paolo Colombo, Carla Caramella
and Aldo La Manna

Department of Pharmaceutical Chemistry
University of Pavia - Italy

INTRODUCTION

The final aim of controlled delivery systems is to obtain drug release kinetics meeting the therapeutic need. Since a constant plasma level is adequate in many forms of therapy, a zero-order release of drug from dosage form is often desirable.

Among drug delivery systems, those based on a diffusion-controlled release through a polymeric membrane appear to be the most reliable in order to achieve a zero-order release [1-4].

These systems are based on the concept of a constant surface for mass transport. After an initial lag time this process is described by the following equation:

$$\frac{dQ}{dt} = \frac{DSC_S}{h}$$

where dQ/dt is the release rate, D the diffusion constant through the membrane surface, C_S the concentration at the internal membrane surface and h is the membrane thickness (Figure 1).

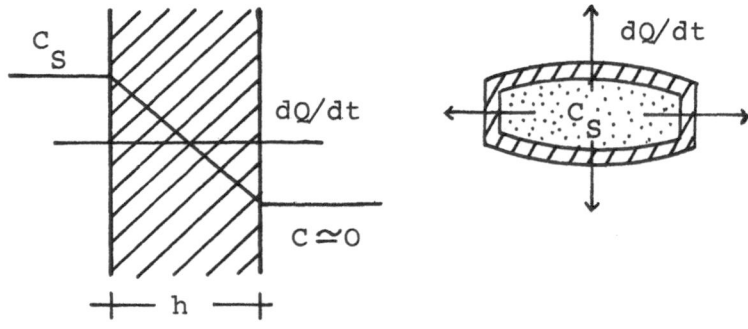

Figure 1

Most of these systems raise problems due to feasibility of industrial preparative procedures.

The aim of the present work was to employ the dry
coating technique to prepare a delivery system composed
by an active core coated and a polymer coat acting as a
diffusion controlling membrane.

EXPERIMENTAL

Materials

Polymer: Polyvinylalcohol ELVANOL 71-30 (Dupont)

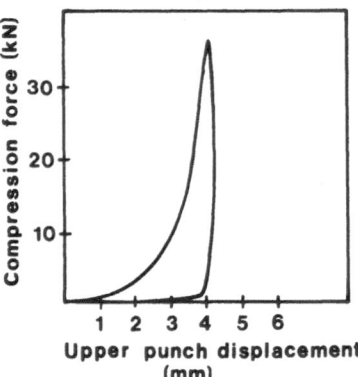

Figure 2. Elvanol compression cycle.

TABLE I

Typical properties

Hot water soluble	
Resistant to cold water	
Form	granular
Color	white
Molecular weight	medium
Hydrolysis,mole %[a]	99.0-99.8
Saponification number[b]	3-12
Residual polyvinyl acetate, weight %	0.5-1.8
Viscosity, cP(mPa.s)[c]	28-32
Solution pH	5.0-7.0
Volatiles, % max	5
ASH (as Na_2O), % max[a]	5
Bulk density (kg/m^3)	400-432
Specific gravity	1.30
Resin density (kg/m^3)	1294
Specific volume (m^3 kg)	$7.7.10^{-4}$
Refractive index	1.54
Hardness, shore unplasticized	greater than 100
Specific heat (j/kg.K)	1674
Swelling (in water at 25°)	100%
Packing fraction	0.388
Repose angle	23°
Particles dimensions (Sieving)	226 ± /µm

----

[a] Dry basis

[b] Milligrams potassium hydroxyde per gram polymer

[c] Viscosity of a 4% aqueous solution at 20°C, determined by Hoeppler falling ball method.

Active principle: Diprophylline

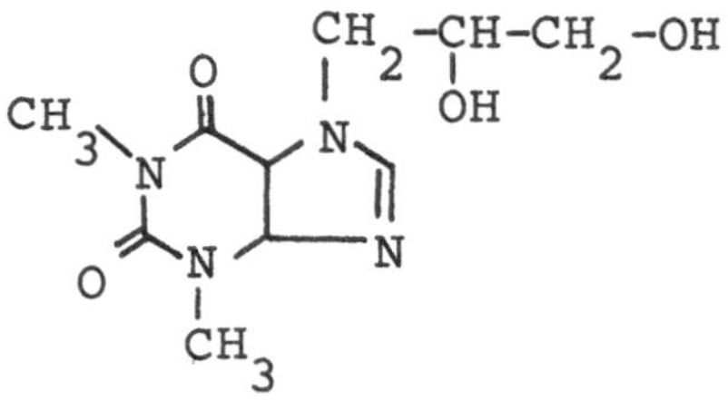

Dose: 200 mg thrice daily [5]

Half-life: 1.7 ± 0.4 h

Water solubility: 1 gr/3ml

The delivery systems were prepared from Diprophylline cores of 7.00 and 9.00 mm diameter.
The Elvanol 71-30 polymer coat was applied using die
punch sets of 9.0 and 12.0 mm respectively. This polymer
was chosen on the basis of the previously assessed
favourable flow and compression properties.

The quantity of polymer applied ranged from 75 to
300 mg and the compression force employed for coating
was between 12 and 35 kN.

Some series of these systems were prepared at
different compression force levels and drug release
rate was checked by U.S.P. XX paddle dissolution method.

RESULTS

Release kinetics of press-coated systems fits well
the mechanism previously described (Figure 3).
The curve is characterized by a delay due to hydration
of the Elvanol coat (lag time), a linear portion corresponding to a constant concentration gradient across
the Elvanol barrier and an exponential final portion
after the complete core dissolution.

Release constants and lag times were calculated by
linear regression analyses and best fitting was accomplished by progressively discarding experimental data
within the initial ninety minutes.

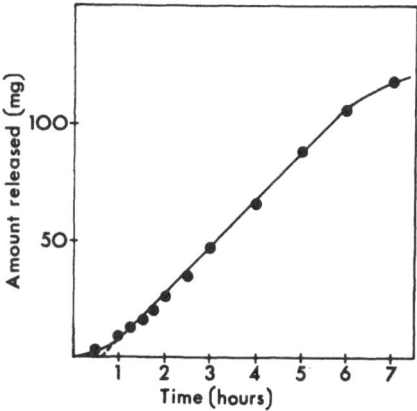

Figure 3. Release kinetics of press-coated systems.
           Core: 123 mg, Ø 7 mm. Coat: 81 mg, Ø 9 mm.
           Compression force: 22 kN.
           K = 19.2 mg/h.   Lag time = 36 min

The release constant obtained for several series of press-coated systems is independent of compression force level within the range examined (12-35 kN) (Figure 4).

     This suggests that the hydration and swelling properties of the Elvanol coat produce an hydrated barrier having always the same permeability characteristics which do not depend on the initial hydrated barrier porosity.

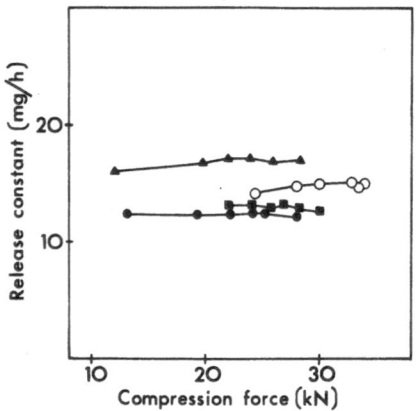

Figure 4. Release constant versus compression force.
          Punch Ø: core 7 mm; coating 9 mm
          Coat (mg):   ▲ 78;   ○ 99;   ■ 128;   ● 145.

On the contrary the lag time seems related to the compression force which would indicate that the porosity of the polymer barriers determines the hydration time.

On increasing the applied polymer quantity, a decrease of release constant and a simultaneous increase of lag time are observed (Figure 5).
The larger surface area of the 9-12 mm systems determines release constant values generally greater than those of the 7-9 systems.

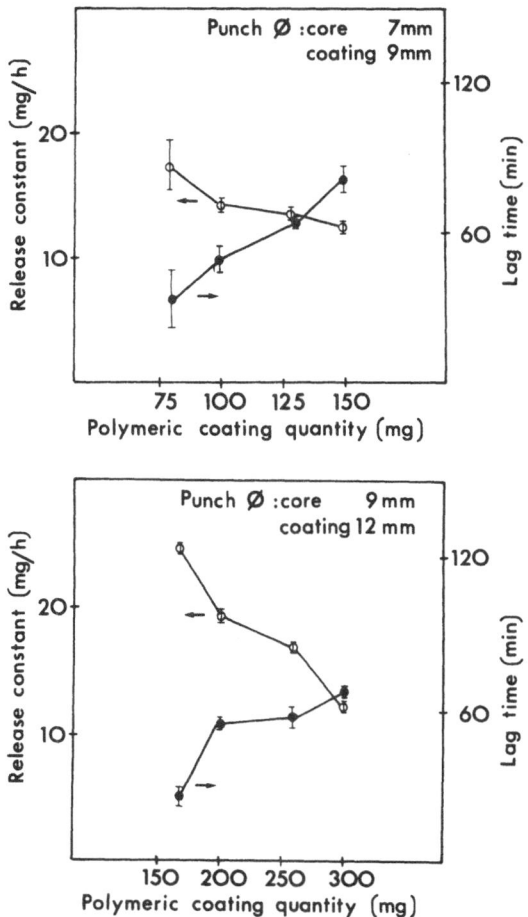

Figure 5. Release constant and lag time versus polymeric coating quantity.

CONCLUSIONS

The results obtained confirm that the system described releases the active principle with zero-order
kinetics, that the release rate can be controlled by
the area and thickness of the coat and, at least with
Elvanol, it is not strictly dependent on compression
force level.

Such an approach, although depending on an appropr-
riate choice of drug-polymer combination, would allow
to obtain advantages as for ease of preparation and
standardization of industrial production.

REFERENCES

1. H. Lapidus, N.G. Lordi, Drug relesae from compressed
        hydropilic matrices, J.Pharm.Sci. 57:1293 (1968)
2. S. Borodkin, F.E. Tucker, Linear drug release from
        laminated hydroxypropylcellulose - polyvinyl
        acetate films, J.Pharm.Sci. 64:1289 (1975)
3. C.F. Lerk, W.J. Bolink, K. Zuurman, Solid dosage
        form with constant release, Pharm. Ind. 38:561
        (1976)
4. J.L. Salomon, E. Doelker, P. Buri, Sustained release
        of a water-soluble drug from hydrophilic compres-
        sed dosage form, Pharm.Ind. 41:799 (1979)
5. J. Zuidema, F.W.H.M. Merkus, Pharmacokinetics and
        pharmacodynamics of diprophylline, Pharm.Weekblad
        Sc. Ed. 3:216 (1981)

The Authors wish to thank Dr. Elisa Pisoni for
experimental work and Mrs. M.C.Sacchi for poster preparation.

POLYMERIC OPHTHALMIC DRUG DELIVERY SYSTEMS: PREPARATION AND
EVALUATION OF PILOCARPINE-CONTAINING INSERTS

Marco F. Saettone, Boris Giannaccini, Patrizia Chetoni,
Giancarlo Galli* and Emo Chiellini**

Istituto di Chimica Farmaceutica, *Istituto di Chimica
Organica Industriale, and **Istituto di Chimica Generale
(Facoltà di Ingegneria)
Università di Pisa, 56100 Pisa, Italy

INTRODUCTION

Ophthalmic drugs are traditionally administered as solutions
(collyria) or as anhydrous (petrolatum) ointments. However, it is
well known that solutions do not result in administration of a precise
and consistent dosage, since the induced lacrimation and reflex blink-
ing that follow instillation quickly remove the drug from the precorneal area. Furthermore, eyedrop instillation typically results in a
pulse entry mechanism, with undesirable transient peaks of drug concentration in the aqueous humour, and periods of nonmedication in the
time intervals between pulses. Ointments, on the other hand, are somewhat better retained, but induce discomfort and blurred vision. Moreover, on account of their greasy nature they do not release effectively
all types of drugs, and show poor patient compliance. These factors
have stimulated the search for more sophisticated ophthalmic dosage
forms, capable of delivering a precise drug dosage, and of maintaining
an optimal concentration in the precorneal area for an extended period
of time, thus rendering unnecessary a frequent administration of eye-
drops.

In the recent years, several types of solid ophthalmic vehicles
(inserts) have been developed. These are intended for a direct application into the conjunctival sac, and can be basically categorized

187

according to the following scheme:

```
                                     ⎧ Reservoir (insoluble)
          DIFFUSIONAL INSERTS        ⎨                  ⎧(insoluble)
                                     ⎩ Monolithic ⎨
                                                        ⎩(soluble)
```

BIOERODIBLE INSERTS

Diffusional reservoir inserts consist of a core of drug surrounded by an inert rate-controlling polymeric barrier. A typical example of this system is the Ocusert® [1], which delivers pilocarpine at a nearly zero-order rate for one week, and, being insoluble, necessitates removal from the conjunctival sac after exhaustion.

Diffusional monolithic insoluble systems usually consist of hydrophilic soft contact lenses (e.g., hydroxyethyl methacrylate type polymer) presoaked in drug solutions[2]. The latter systems, on account of their many disadvantages, have found only limited applications beyond the experimental field.

Erodible inserts contain drug dispersed throughout a polymeric matrix (polylactic, polyglycolic, alginic acid, polypeptides, etc.): the drug is gradually leached from the matrix as it slowly erodes and disintegrates in the cul-de-sac. Devices of this kind have been mainly described in the literature[3], but have not found commercial applications to the present date.

Diffusional monolithic soluble inserts, such as those described in the present report, may offer some advantages over the systems described before. In spite of a drug release with first-order kinetics, rather than with the more desirable zero-order kinetics, they have been reported to favour penetration and to prolong the effect of the medicinal agent. Furthermore, they neither necessitate removal from the eye at the end of the therapy, nor suffer an easy expulsion from the conjunctival sac during the sleep of the patient, as is the case of insoluble inserts. Forms of this type have been described in relatively old publications (glycerinated gelatin "lamellae" of the 1948 British Pharmacopoeia). Soluble inserts based on synthetic polymers, e.g., poly(acrylamide), poly(ethyl acrylate), poly(vinyl alcohol), poly(vinyl pyrrolidone) have been developed and experimented on humans mainly in the U.S.S.R. by Maichuk and coworkers[4] with drugs such as antibiotics, sulfonamides, pilocarpine, atropine, dexamethasone, etc.

The present preliminary study had the purpose of evaluating, on both a physico-chemical and a biological basis, a series of commercially available polymers as possible substrates for soluble inserts containing pilocarpine. It was considered an essential prerequisite to the synthesis of new polymers, specifically tailored as ophthalmic drug carriers, now underway.

EXPERIMENTAL

The following commercially available polymer samples were used as received: poly(vinyl alcohol), PVA (Polyviol, Wacker Chemie GmbH); hydroxypropylcellulose, HPC (Klucel, Hercules Inc.); poly(acrylic acid), PAA (Carbopol 940, Goodrich Chemical Co.). Some properties of the polymers are reported in Table 1.

Table 1.   Physico-chemical properties of the polymeric products employed

Sample	Polymer	$\overline{M}_w$ $(\cdot 10^{-3})$	$\overline{HD},\%^a$	T_m, °C [b]
A		25	86–89	185
B	PVA	72	99–100	225
C		90	97–99	215
D		72	97–99	215
E	HPC	500	---	210
F		100	---	200
G	PAA	1000	---	nd [c]

[a] Degree of hydrolysis as referred to parent poly(vinyl acetate). [b] By DSC. [c] No DSC endothermic peaks up to 270 °C.

Pilocarpine nitrate (m.p. 176–178 °C) was used as received from the manufacturer (E. Merck, Darmstadt). Pilocarpine base was obtained from the nitrate by alkalinization to pH 8–9 of the aqueous solution with ammonium hydroxide, and extraction with chloroform.

A 2.0% w/v solution of pilocarpine nitrate in pH 5.5 isotonic phosphate buffer (hereafter referred to as S), and a viscous gel containing 1.54% w/v pilocarpine base and 0.77% w/v poly(acrylic acid)

G (hereafter referred to as GS) were used as reference standards for the biological tests.

Transparent, flexible films containing pilocarpine nitrate were obtained by slow evaporation (50 °C) of 5.0% w/v solutions of polymers containing the appropriate amount of pilocarpine nitrate. The films (0.4 - 0.5 mm thickness) were cut in the form of small disks (4.0 mm diameter), each containing 1.0 ± 0.05 mg pilocarpine nitrate. PVA films containing pilocarpine base salified by PAA-G were prepared by adding to the 5.0% w/v PVA solutions the appropriate amount of pilo- carpine base and a small excess over the equivalent amount of PAA-G. The resulting viscous gels were liquified by ultrasonic treatment, then were poured into petri dishes and evaporated as specified above. The final inserts contained in this case an amount of pilocarpine base (0.768 mg) corresponding to 1.0 mg of pilocarpine nitrate. All inserts were routinely analyzed for pilocarpine by HPLC [5], after thorough extraction with methanol. HPC inserts containing the pilocarpine PAA salt could not be prepared, since a precipitate was formed on admix- ture of the two polymeric materials.

Miosis-time data on rabbits (male albino, 2-2.5 Kg) were col- lected by placing 50 μl of solution S, or gel GS, or one insert into the lower conjunctival sac of one eye of the animals, the other eye serving as control. The measurements were made at intervals, under standard light conditions, by estimating to the nearest 0.1 mm, with a micrometer, the horizontal diameter of the pupil. Each type of preparation was tested at least on a set of 10 different animals. In no case eye irritation, or expulsion of the insert was observed.

In vitro release tests were carried out on 200 mg samples of polymer films, by determining pilocarpine release to a stirred aque- ous medium (pH 6.98 phosphate buffer, 10.0 ml) at 30 °C. Solution samples (2.0 ml) were withdrawn at appropriate intervals, and were replaced with an equal amount of fresh buffer. Pilocarpine was ana- lyzed spectrophotometrically by the ferric hydroxamate method de- scribed by Gibbs and Tuckerman [6].

Differential scanning calorimetry (DSC) investigations were performed on a Perkin-Elmer DSC-2 apparatus equipped with scanning autozero. Polymer samples and inserts (8-10 mg) and equivalent amounts of pilocarpine salts (ca. 1.0 mg) were analyzed at heating-cooling rates of 20 °C /min under dry nitrogen flow. Indium standards were employed for temperature calibration and enthalpy change evaluation.

RESULTS AND DISCUSSION

"In vivo" Release Studies

Typical miosis-time data for some preparations under study are illustrated in Fig. 1. Administration of 50 µl of the 2.0% pilocar-pine nitrate solution S (corresponding to 1.0 mg of drug) produced miosis of relatively short duration (2 hours), declining rapidly after reaching the peak effect. Conversely, a PVA-C insert containing 1.0 mg pilocarpine nitrate produced a more prolonged effect (ca. 4 hours), that was practically identical with the effect produced by adminis-tration of 50 µl of the gel GS, containing a corresponding amount of pilocarpine base. As shown in the Figure, when the pilocarpine-PAA salt was administered as dispersion in a PVA-C insert, a stronger miotic response with an activity plateau, and an overall 5-hour du-ration were observed.

The areas under the curves illustrating the miotic response versus time (AUC) for all preparations tested were evaluated from

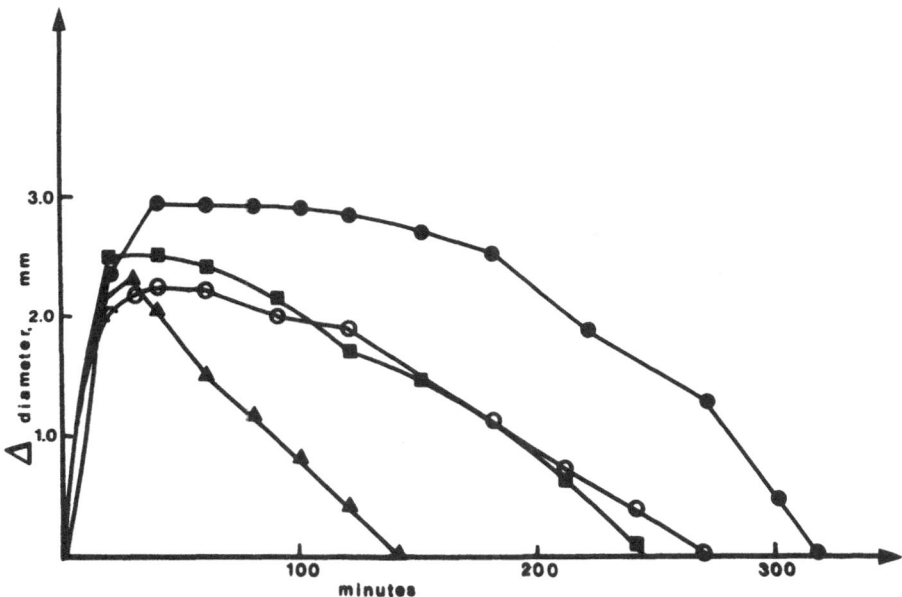

Fig. 1. Mean change in pupillary diameter versus time for some preparations under investigation. Key: ▲, solution S; ■, gel GS; O, PVA-C insert with pilocarpine nitrate; ●, PVA-C insert with pilocarpine PAA-G salt.

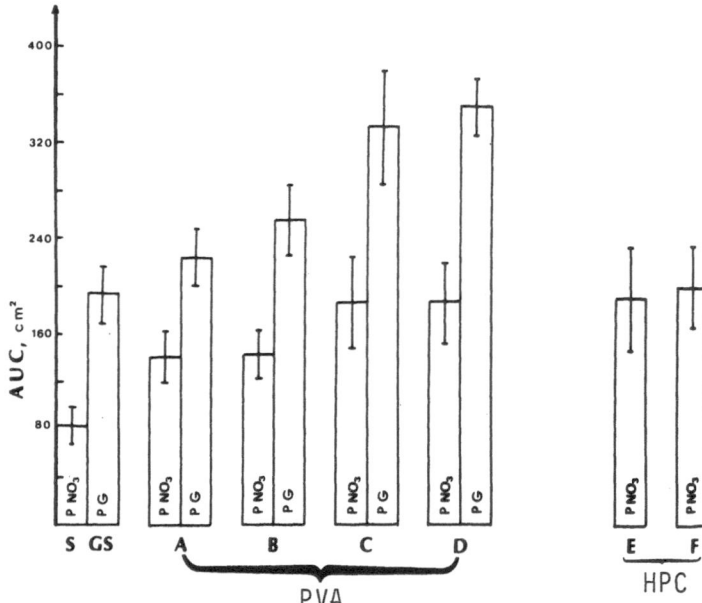

Fig. 2. Areas under the miotic activity versus time curves for all
preparations described in the present study. Key: PNO_3,
pilocarpine nitrate; PG, pilocarpine PAA-G salt. Vertical
lines over bars indicate 95% confidence limits. See text
for further details.

graphs such as those illustrated in Fig. 1, and are reported in Fig.
2 with the corresponding 95% confidence limits. The AUC values should
reflect the aqueous humour concentration of the drug, thus being
indicative of the bioavailability of pilocarpine from the vehicles
under study. It is immediately apparent from the data in Fig. 2 that
administration of 1.0 mg pilocarpine nitrate dispersed in the solid
inserts (PVA A, B, C and D, and HPC E and F) produced a statistically
significant ($P < 0.05$) bioavailability increase over the aqueous solution (1.5 to 2.5 times). The AUC values determined for polymers A, B,
C, D and E containing pilocarpine nitrate were not statistically
different from one another. The HPC-F insert showed a slightly but
significantly higher activity with respect to PVA A and B. The fact
that some inserts, shortly after being placed in the conjunctival
sac of the animals, lost their solid-state integrity, tending to
become viscous liquids (A and F) or gel-like structures (E), while
others (B, C and D) maintained their disk shape throughout the experiment, apparently had no relevance towards their specific activity.

Interestingly, the GS gel showed a significantly higher bioavailability with respect to PVA inserts A and B containing pilocarpine nitrate, while it had practically the same activity as the other inserts (C, D, E and F) containing nitrate. However, the highest AUC values were obtained, as shown in Fig. 2, with the PVA inserts containing the pilocarpine-PAA G salt. Within this series, the best results were given by PVA inserts C and D (producing a 4.5 times bioavailability increase over the aqueous solution S) followed by PVA B and A, respectively, which showed significantly lower AUC values with respect to the first two inserts. It should be mentioned that, out of the four inserts containing the pilocarpine-PAA salt, only one (PVA-A) assumed a gel-like structure when placed in the conjunctival sac, while all the others preserved their integrity.

"In vitro" Release Studies

Typical "in vitro" release data from the present inserts are reported in Fig. 3, where the fractions of drug released (F) from the PVA-C inserts containing pilocarpine nitrate and the drug PAA-G salt are plotted versus the square root of time. A linear relationship between fraction of drug released and square root of time up to 70-80% release was observed in all cases, indicating diffusion as the most important mechanism contributing to drug release from these systems under the "in vitro" experimental conditions [7]. The release "rates" from the inserts containing the PAA-salt, corresponding to the slopes of the initial portions of the F vs. $\sqrt{t}$ plots, were somewhat lower in all cases with respect to the release "rates" from the inserts containing pilocarpine nitrate. Furthermore, the former inserts still retained 10-15% bound drug after a 24-hour desorption time, whereas the nitrate-containing inserts had completely released their drug content after the same time interval. "In vitro" release, however, was comparatively fast in all cases, 60-80% of the total amount of drug being released to the aqueous phase within 30 minutes.

DSC Studies

Raw polymer materials, drug-free and pilocarpine-containing inserts, pilocarpine nitrate and PAA-salt were also investigated by differential scanning calorimetry with the aim of establishing possible interactions between polymer matrix and drug.

The heating curves of all polymer samples were characterized by

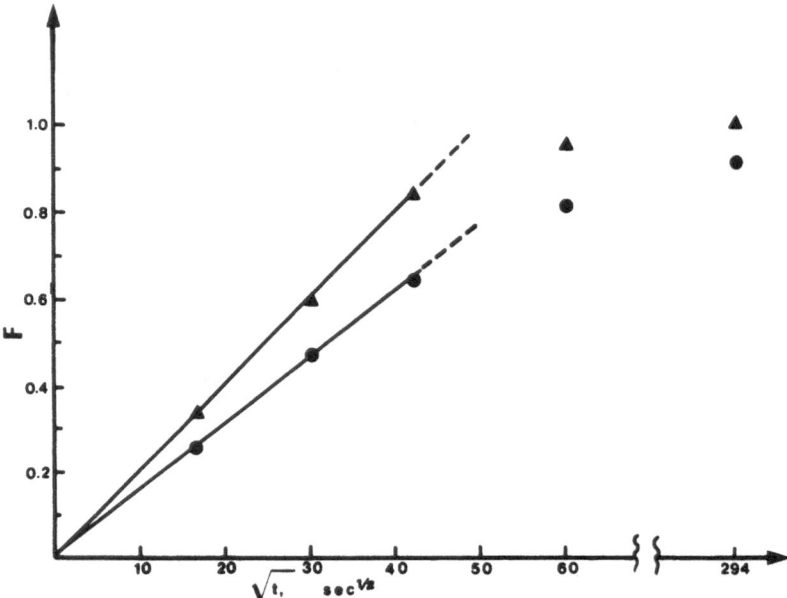

Fig. 3.   Graphs illustrating "in vitro" release of pilocarpine
from PVA-C inserts containing pilocarpine nitrate (▲)
and pilocarpine PAA-salt (●).

a rather sharp, intense melting peak, and the temperature values
corresponding to the maximum of the endotherm are reported in Table
I. Glass transition temperatures in the range 60–80 °C were also
determined for poly(vinyl alcohol) polymers.

The inserts prepared from polymer solutions had DSC traces
very similar to those of the corresponding parent homopolymers,
as far as both temperature and enthalpy changes are concerned. This
implies that no DSC-appreciable physico-chemical modifications
occur during the preparation of the films.

The thermograms of pilocarpine nitrate show a melting endotherm
at 176 °C with an overwhelming exotherm centered at 190 °C. A further
exothermic transition is located at 220 °C. On heating, pilocarpine
(I) is known to undergo isomerization to the more stable, pharmacologically less active isopilocarpine (II)[8]. Therefore, it is plau-
sible to ascribe the strong exotherm immediately after melting at
176 °C to the formation of isopilocarpine. The rather strong and

$$I \qquad\qquad\qquad\qquad II$$

sharp exotherm present at 220 °C is tentatively assigned to the conversion of isopilocarpine to some other structural isomer. The heating
curve of the pilocarpine PAA--salt, on the contrary, show only a broad
melting transition at 260 °C. On heating the PAA-G no transitions
were observed in the full range investigated, thus indicating that
the ionic binding of pilocarpine to the anionogenic polymer matrix
helps inhibiting the isomerization process. A very complex thermal
behaviour was exhibited by the inserts containing pilocarpine nitrate.
In fact, exothermic transitions due to pilocarpine isomerization are
present along with the expected endothermic transitions attributable
to the melting of the polymer (Figs. 4 and 5). Interestingly, the
occurrence of such exotherms was found to be somehow dependent on
structural parameters of the poly(vinyl alcohol) matrix, such as molecular weight and apparent degree of hydrolysis.

In the thermograms of the doped films prepared from PVA characterized by high molecular weight and an almost quantitative degree
of hydrolysis (PVA-C), a sharp exotherm occurs at 230 °C, analogous
to that observed with pilocarpine nitrate (Fig. 4). On the contrary,
in the case of films prepared from PVA having a lower molecular weight
and a degree of hydrolysis below 90% (PVA-A), a broad exotherm appears
at 190 °C, intermediate between those observed with pilocarpine nitrate (Fig. 5). On cooling from the melt, in all cases broad exothermic peaks with two distinct minima were observed in the range
160–110 °C.

The results seem to point out that host polymer and drug behave
rather independently of each other, even though non–bonded or weak
hydrogen bonding interactions must play a role in determining the
conformational assembling of pilocarpine nitrate in the matrix.

A completely different behaviour was observed in the case of the
inserts containing pilocarpine PAA-salt. All inserts, except B,

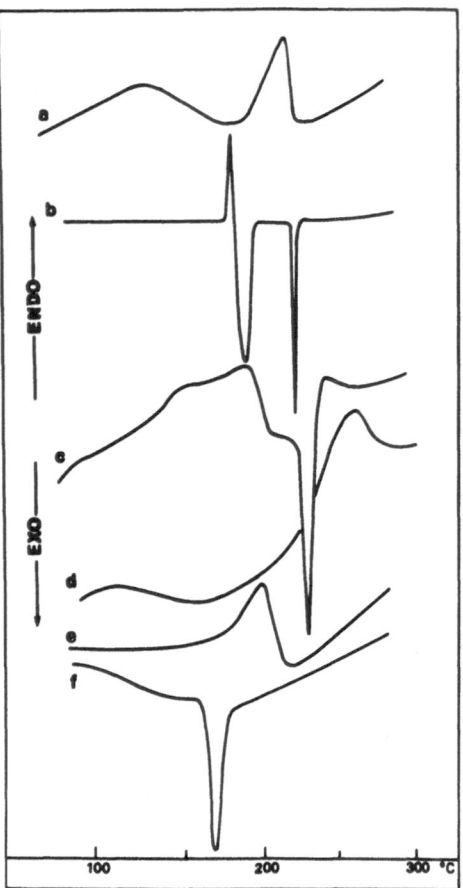

Fig. 4.   DSC curves of inserts prepared with poly(vinyl alcohol)
           PVA-C. Key: a) polymer film; b) pilocarpine nitrate; c)
           polymer insert containing pilocarpine nitrate; d) pilo-
           carpine PAA-salt; e,f) polymer insert containing pilo-
           carpine PAA-salt: heating and cooling, respectively.

had crystalline melting temperatures lower than those of the corresponding films, whereas no peaks attributable to the melting of
the drug could be detected. On cooling, polymer crystallization occurred in a narrow interval of temperatures with a relatively small
degree of supercooling (~30 °C; Cf. Figs. 4 and 5).

     From these thermal features it may be conceived that chemical
interactions among matrix, polymeric anion and drug concur in establishing a tight binding of pilocarpine, that in turn may result in

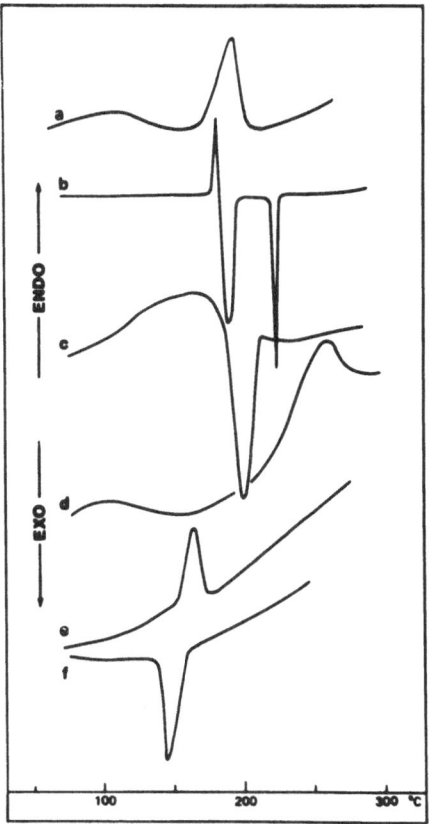

Fig. 5.   DSC curves of inserts prepared with poly(vinyl alcohol)
           PVA-A. Key: a) polymer film; b) pilocarpine nitrate; c)
           polymer insert containing pilocarpine nitrate; d) pilo-
           carpine PAA-salt; e,f) polymer insert containing pilocar-
           pine PAA-salt: heating and cooling, respectively.

more restrained mobility and release.

    While these suggestions are still tentative, it is worth stress-
ing the sensitivity and potential of thermal analysis in the assessment of basic matrix-drug parameters for controlled delivery systems.

CONCLUSIONS

The following essential points may be indicated as the main out-
come of the present preliminary study. (a) Out of over 20 commercial,
pharmaceutical or food-grade polymers tested in the investigation,
only a handful were found suitable as insert-formers, both on a technological and on a biological basis. While it cannot be excluded that
other polymers, yet to be tested, may show superior properties, two
PVA grades (C and D) appeared to be the most promising, also in view
of their compatibility with the PAA-G polymer. (b) Salification of the
drug base with the polyanionic polymer G exerted a profound influence
on overall drug activity, that was enhanced when the salt was administered as dispersion in a solid insert. This effect had been partially anticipated in the literature. Some authors[9-11] had indeed
indicated the activity enhancement resulting from ophthalmic administration of pilocarpine—polyanionic polymer salts. The further
availability enhancement resulting from administration of the salt
dispersed in an appropriate PVA matrix appears interesting, and
worthy of further investigation.

The reported data seem to point to the relevance to ophthalmic
availability both (i) of the type of polymer forming the insert
matrix, and (ii) of the diffusive properties of the drug base, that
may be regulated by the structure of the polyanion. Concerning the
first point, a previous study[12] in which several iso-viscous preparations were compared, had shown that the chemical nature of the
polymer, and not its solution viscosity per se may play an essential
role in determining the ophthalmic availability of a drug. In the
same investigation, the superiority of PVA over other thickening
agents was also established. In the case of the present inserts, the
PVA polymeric matrix is probably endowed with some degree of affinity
for the corneal and conjunctival structures, thus forming a stable
drug depot in the cul-de-sac, and a film over the precorneal (absorption) area, from which pilocarpine diffuses at an optimal rate. As
for what point (ii) is concerned, evidence of interactions between
the drug and the PAA-G polymer, presumably resulting in a decreased
"in vivo" diffusion rate, has been gathered both from "in vitro"
release studies and from DSC data. A slower delivery rate of pilocarpine from the PVA/PAA-salt matrix is in all evidence due to the
restraint imposed by the polyanionic macromolecular structure on
diffusing material. Work is now in progress with the aim of clari-
fying the above mentioned points.

REFERENCES

1.   J. W. Shell and R. W. Baker, Diffusional Systems for Controlled
      Drug Release of Drugs to the Eye, Ann. Ophthalmol. 6:1037 (1974).
2.   S. Podos, B. Becker, C. Asseff and J. Harstein, Pilocarpine
      Therapy with Soft Contact Lenses, Amer. J. Ophthalmol. 73:336
      (1972).
3.   A. Zaffaroni, Selective Administration of Drug with Ocular
      Therapeutic System, U.S. Pat. 4,186,184 (Jan 29, 1980) to Alza
      Corp.;   through Chem. Abstr. 92:185937b (1980).
4.   Y. F. Maichuk, Ophthalmic Drug Inserts, Invest. Ophthalmol.
      14:87 (1975).
5.   L. D. Dunn, B. S. Scott and E. D. Dorsey, Analysis of Pilo-
      carpine and Isopilocarpine in Ophthalmic Solutions by Normal-
      Phase High-Performance Liquid Chromatography, J. Pharm. Sci.
      70:446 (1981).
6.   I. S. Gibbs and M. M. Tuckerman, Optimal Ferric Hydroxamate
      Method for Determination of Intact Pilocarpine, J. Pharm. Sci.
      59:395 (1970).
7.   G. L. Flynn, S. H. Yalkowsky and T. J. Roseman, Mass Transport
      Phenomena and Models, Theoretical Concepts, J. Pharm. Sci. 63:
      479 (1974).
8.   R. K. Hill and S. Barcza, Stereochemistry of the Jaborandi
      Alkaloids, Tetrahedron 22:2889 (1966).
9.   S. P. Loucas and H. M. Haddad, Solid-State Ophthalmic Dosage
      Systems in Effecting Prolonged Release of Pilocarpine in the
      Cul-De-Sac, J. Pharm. Sci. 61:985 (1972).
10.   S. P. Loucas and H. M. Haddad, Solid State Ophthalmic Dosage
      Systems II. Use of Polyuronic Acid in Effecting Prolonged
      Delivery of Pilocarpine in the Eye, Metabol. Ophthalmol. 1:27
      (1976).
11.   R. D. Schoenwald and R. E. Roehrs, Sustained Release Ophthalmic
      Drug Dosage, U.S. Pat. 4,271,143 (May 9, 1979) to Alcon Labo-
      ratories, Inc.
12.   M. F. Saettone, B. Giannaccini, A. Teneggi, P. Savigni and
      N. Tellini, Vehicle Effects on Ophthalmic Bioavailability:
      the Influence of Different Polymers on the Activity of Pilo-
      carpine in Rabbit and Man, J. Pharm. Pharmacol. 34:464 (1982).

BASIC PHYSICAL PARAMETERS OF POLYMERIC MATRICES

INFLUENCING DRUG RELEASE

Fabio Carli

Physical Pharmacy Laboratory, Pharmaceutical
Development
Farmitalia Carlo Erba
via Imbonati 24, Milan, Italy

SUMMARY

The basic physical parameters of polymeric matrices prepared by simple physical mixing of drug and polymer powders are reviewed. The wettability of the matrix, its pore structure and the permeability of the polymer particles resulted to be the controlling factors in the overall penetration process and in the subsequent drug release mechanism. A simple experimental approach to the identification of the penetration mechanism is proposed, which allows to distinguish between a capillary driven penetration and a diffusion controlled permeation process.

INTRODUCTION

The embedding of a drug into an inert porous polymeric matrix, first proposed by Higuchi[1-4], is still one of the most used techniques to obtain a controlled release. This technique is usually applied to sustained release tablets aimed at releasing the drug along the gastro-intestinal tract[5-8], but it has been proposed also for other special applications, e.g., as microreservoirs for drug delivery systems based on the osmotic pump principle[9].

201

The drug release pattern from such porous matrices is strongly influenced by the penetration process of the external liquids inside the microstructure of the matrix. If the liquid can wet the matrix (contact angle below 90°) the penetration takes place basically via a capillary motion through the pores. In this case the basic physical factors influencing the penetration are the contact angle, the pore size and volume, the surface tension of liquid and its viscosity[10-11]. If the liquid does not wet the matrix (contact angle above 90°), penetration can take place only by dissolution od the drug particles or by permeation of the polymeric particles themselves. In this case the basic factors controlling the diffusion of the liquid molecules through the polymer are the crosslinking density, the crystallinity of the polymer etc.[12] These different penetration mechanisms are schematized in figure 1.

It is the object of this paper to show how the physical factors outlined above can exert a strong influence on the drug release and to show how it is possible to experimentally identify the mechanism of penetration taking place in the matrices under study.

Polymer
particles

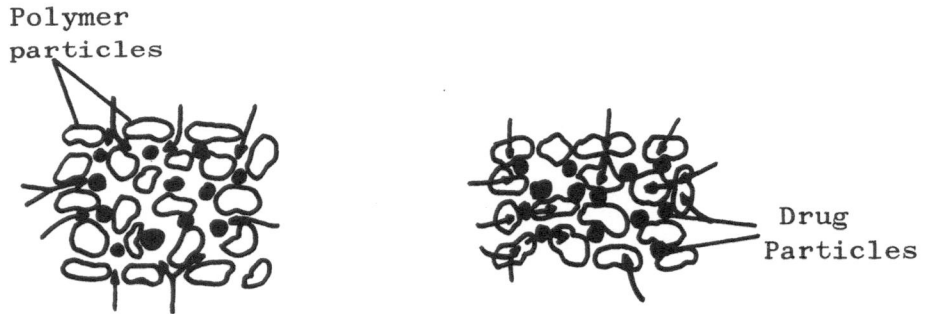

Drug
Particles

CAPILLARITY   PERMEABILITY

Fig. 1.   The different penetration mechanisms into porous
           polymeric matrices.

MATERIALS AND METHODS

Materials

The polymers used were: polyvinylchloride (Sicron 230, Montedison, Italy); vinylchloride/vinylacetate copolymer (Sicron 822, Montedison, Italy); polytetrafluoroethylene (Fluon, I.C.I.,U.K.); polyethylene (Montedison,Italy); acrylic polymer (Paraloid K 120 N, Rohm & Haas, Italy); acrylic/methacrylic acid esters copolymer with 10 % of quaternary ammonium groups (Eudragit RL, Rohm Pharma, Germany) and with 5 % of these groups (Eudragit RS, Rohm Pharma, Germany).

Acetylsalycilic acid was chosen as the drug model (F.U., Farmitalia Carlo Erba, Italy).

Methods

Matrices were made by directly compressing an uniform aspirin-polymer powders mixture (10% w/w of the drug) using an instrumented compaction machine (Nassovia, Germany), with a single 13 mm flat face punch and die.

Penetration measurements were carried out using a simple apparatus, elsewhere described[13], and demineralized water saturated with aspirin as penetration liquid, at room temperature.

Total pore volume and pore size distribution were determined using mercury porosimeter (Mod. 225, Carlo Erba Strumentazione, Italy), with intrusion pressures ranging from 0 to 2000 atm.

Liquid/solid contact angles were measured by a Wettability Tester (Lorentzen-Wettre, Sweden) and the method of Mack[14].

Aspirin release rates in 500 ml of distilled water thermostated at 37° were determined using the beaker method with mild agitation (40 r.p.m.), as suggested by Sjogren[15] and exposing a single surface to the dissolution medium. Aspirin was assayed spectrophotometrically (Pye Unicam 1800) following the method of Javaid and Cadwallader[16].

RESULTS AND DISCUSSION

Wettability

The influence of wettability characteristics of the
matrices on the drug release rate is illustrated by the
data presented in table 1. As the wettability decreases,
i.e., the contact angle increases, becoming higher than
90°, the drug release rate decreases, although the pore
volume is not significantly different. This means that
in the case of hydrophobic matrices (e.g. polyethylene)
the pore network cannot be penetrated by capillarity,
so that the pore volume effectively available for drug
diffusion is only that left by the already dissolved drug
particles.

Table 1.   Influence of Wettability on Aspirin Release
           Rate from Matrices.

Matrix	Pore Volume ml/g	Water/Matrix Contact Angle	Aspirin Release Rate $mg\ cm^{-2}\ sec^{-1/2}$
CPVC	0.14	70°	0.16
Acrylic	0.18	51° 01'	0.15
CPVC-PFTE	0.12	100° 01'	0.10
Polyethylene	0.13	95° 58'	0.02

Pore Volume

    The relevance of the pore structure on the drug re-
lease rate from porous matrices is shown in table 2 and
figure 2. For a wettable polymer such as the acrylic, the
reduction in the total pore volume caused by any process,
e.g. sintering[17], brings about a correspondent reduction
of the volume penetrated by capillarity; thus the pore volume effectively available for drug diffusion is reduced
and drug release rate becomes slower. Infact, as shown in
figure 2, there is a linear relationship between the release rate and the penetration volume: this allows to
choose the correct pore structure characteristics in order
to obtain the desired release pattern.

Table 2.   Influence of Pore Volume on Aspirin Release
           Rate from Acrylic Matrices Sintered for Dif-
           ferent Times.

Sintering Time hrs	Pore Volume ml/g	Water penetration Volume ml/g	Aspirin Release Rate mg cm^{-2} sec$^{-1/2}$
0	0.176	0.167	0.073
0.5	0.169	0.104	0.033
1-5	0.156	0.094	0.028
15-24	0.139	0.077	0.018

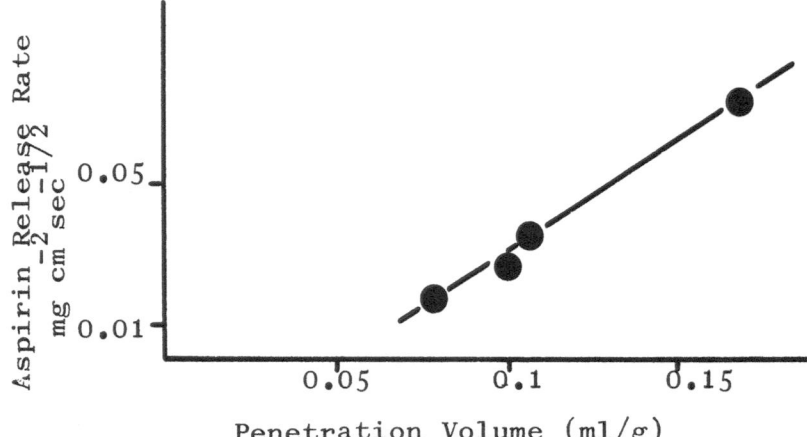

Fig. 2. Influence of water penetration volume on the
        aspirin release rate from acrylic matrices,
        sintered for different times.

Permeability

If the matrix exhibits a water contact angle higher
than 90°, no capillary penetration can take place; but
solvent molecules can diffuse through the polymer itself,[12]
bringing about the penetration of the matrix anyway. This
possibility is illustrated by polymers such as the copolymers of acrylic and methacrylic acid esters commercialized
under the trade names Eudragit RS and RL: the former con-
taines 5% of hydrophilic quaternary ammonium groups, the
latter 10%. Although they are non-wettable, they present
a strong water penetration rate proportional to the percentage of hydrophilic groups present. As shown in table 3,
the higher permeability rate of Eudragit RL matrices originates an aspirin release rate much larger than that
from the Eudragit RS matrices. In the case of Eudragit RL
matrices, the permeation process is accompanied by the
swelling of the polymer particles and a consequent brea-
kage of the pore structure of the matrix, which further
contributes to the higher release rate.

Table 3. Influence of Quaternary Ammonium Groups Percentage on the Characteristics of Eudragit Matrices.

Polymer	$\geq N-$ %	Water/Matrix Contact Angle	Permeation Rate[a] $ml^{1/m}/g^{1/m}/sec$	Aspirin Release Rate $mg\ cm^{-2}\ sec^{-1/2}$
EudragitRS[b]	5	98° 08'	1.95×10^{-6}	0.036
EudragitRL[b]	10	97° 56'	1.40×10^{-4}	0.130

[a] Derived by the intercept of a double logarithmic plot of penetration data, with m equal to the slope of this plot[10].

[b] Prepared at ∼500 MNm^{-2} of compaction pressure.

Identification of Penetration Mechanism

From the data presented above, it is evident that in the design of a porous matrix tablet with specific drug release rate, is strictly necessary to prelyminarly identify the process of penetration taking place in the polymeric system under examination.

A simple approach is offered by the double logarithmic plotting of penetration data. The author has shown that in the case of capillary penetration, the slope m of this double logarithmic plot depends on the pore structure of the matrix being penetrated[18]. Thus, if a polymeric matrix prepared at two different compaction pressures shows two different m values, one can reasonably conclude that penetration takes place via a capillary process. This case is illustrated by the CPVC matrices, which gave different penetration patterns (figure 3 and table 4) related to the different porous structures shown in figure 4.

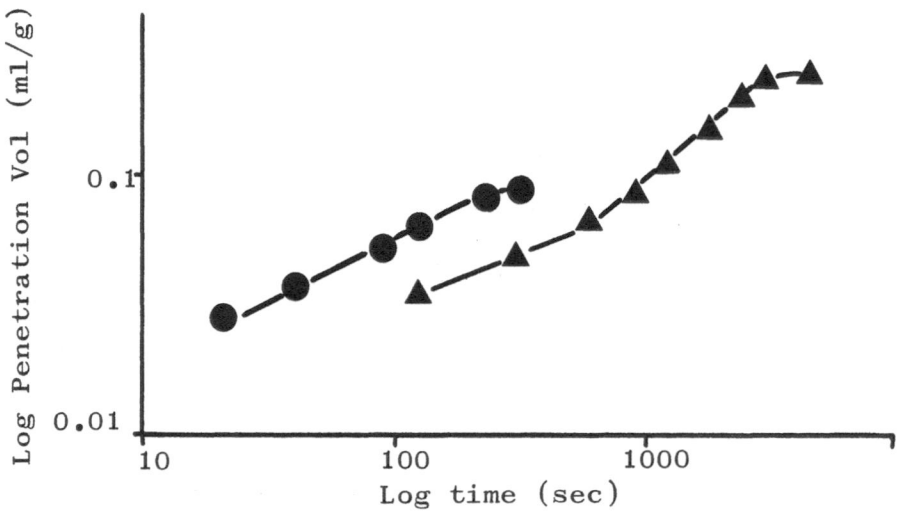

Fig. 3.   Effect of compaction pressure on water penetration into CPVC matrices. Compaction pressure:
$-\blacktriangle-$ 53.6 MNm^{-2}; $-\bullet-$ 441.2 MNm^{-2}.

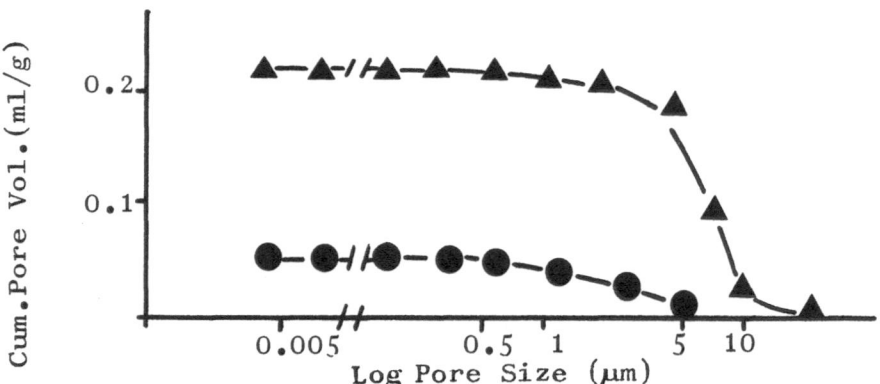

Fig. 4.   Effect of compaction pressure on cumulative pore size distribution of CPVC matrices. Compac-
tion pressure: $-\blacktriangle-$ 53.6 MNm^{-2}; $-\bullet-$ 441.2 MNm^{-2}

On the contrary, if the penetration is only due to the permeation of the polymeric particles, also dramatic changes in the interparticle porous network do not cause any change in the parameters of double logarithmic plots of the penetration data. This point is illustrated by ERL matrices prepared at different compaction pressures: the penetration pattern (figure 5 and table 4) is practically the same, although the porous structure (figure 6)is largely different.

Table 4.   Influence of Compaction Pressure on the Penetration Characteristics of Polymeric Matrices.

Polymer	Compaction Pressure MNm^{-2}	Penetration Rate[a] $ml^{1/m}/g^{1/m}/sec$	m[b]	Mechanism of Penetration
CPVC	53.6	5.0×10^{-5}[c]	0.799[c]	
				Capillarity
CPVC	441.2	3.3×10^{-5}	0.451	
EudragitRL	54.3	1.3×10^{-4}	0.940	
				Permeability
EudragitRL	516.1	1.5×10^{-4}	0.982	

[a] Derived by the intercept of double logarithmic plot of penetration data[10].

[b] Slope of double logarithmic plot of penetration data[10].

[c] Relative to predominant penetration region[18].

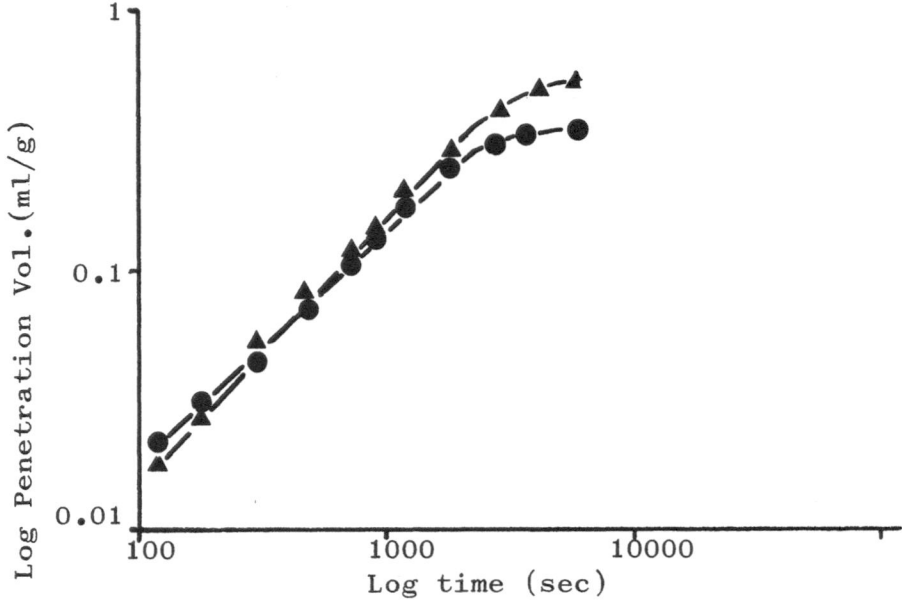

Fig.5. Effect of compaction pressure on water penetration into Eudragit RL matrices. Compaction pressure: $-\blacktriangle-$ 54.3 MNm^{-2}; $-\bullet-$ 516.1 MNm^{-2}.

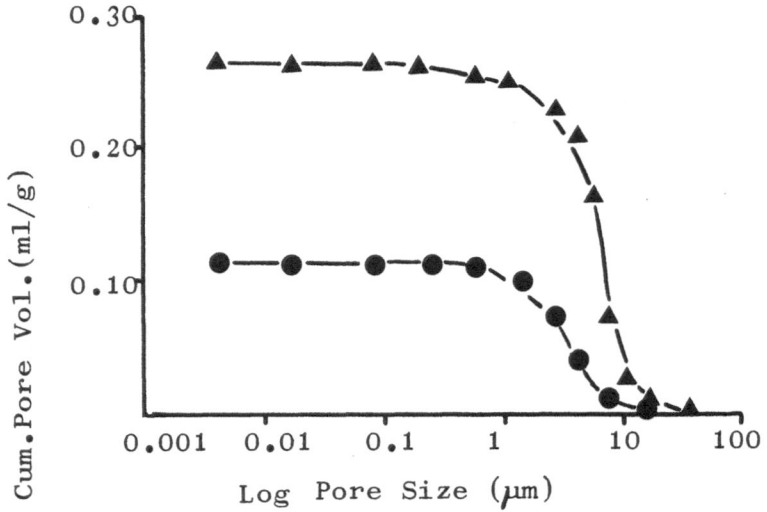

Fig. 6. Effect of compaction pressure on cumulative pore size distribution of Eudragit RL matrices. Compaction pressure:$-\blacktriangle-$54.3MNm^{-2};$-\bullet-$516.1MNm^{-2}.

REFERENCES

1. T. Higuchi, J.Pharm.Sci. 52:1145(1963)
2. S.J.Desai et al.,J.Pharm.Sci. 54:1459(1965)
3. S.J.Desai et al., J.Pharm.Sci.55:1224(1966)
4. P.Singh et al., J.Pharm.Sci. 57:217(1968)
5. N.H.Choulis and H.Papadopoulos, Pharmazie 30:233(1975)
6. R.Huttenrauch and U.Scmeiss, Pharmazie 30:229(1975)
7. R.Rowe, Manufact.Chemist.Aeros.News 46:23(1975)
8. P.F.D'Arcy et al., J.Pharm.Sci. 60:1028(1971)
9. F. Theeuwes, U.S.Patent, 3-977-404,(1976)
10. F.Carli and L.Simioni, J.Pharm.Pharmac. 31:128(1979)
11. F.Carli and L.Simioni, Pharm.Acta Helv. 53:320(1978)
12. E.Drioli, in Proceedings of the Symposium"Employ des
    Polymers dans l'Elaboration de Nouvelles Formes Medi-
    camenteuses",2-3 October 1980,Geneve
13. P.Couvreur et al., Ann.Pharm.Franc. 34:123(1976)
14. G.L.Mack, J.Phys.Chem. 40:159(1936)
15. J.Sjogren, Acta Pharm. Suecica 8:153(1971)
16. K.A.Javaid and D.E.Cádwallader, J.Pharm.Sci. 61:1370,
    (1972)
17. F.Carli and L.Simioni, Int.J.Pharm.Tech.&Prod.Mfr.
    2:23(1981)
18. F.Carli et al., J.Pharm.Pharmac. 33:129(1981)

SECTION II

POLYMERS AS BIOMATERIALS

POLYMERS IN MEDICINE - AN OVERVIEW

Donald J. Lyman

Department of Materials Science and Engineering
University of Utah
Salt Lake City, Utah 84112, USA

The history of biomaterials (i.e. the use of natural or artificial materials in restoring the function of tissue or parts of the body) dates further back than even the most ancient ruins on this beautiful island of Sardinia. The Edgar Smith papyrus papers of over 4000 B.C. describe sutures and other wound closure devices. Later work around 2000 B.C. describe metals in bone repair. Somewhat later, the use of goose quills in vascular repair was mentioned. However, it was not until the 1800's that we saw the real beginning of metals used in bone reconstruction surgery and in the 1930's (with the beginning of plastic industries) to see the beginning use of polymers in a variety of reconstruction applications.

Yet for the most part, it was the aftermath of the heavy casualties of World War II that sparked the search for needed implants and extracorporeal devices. In almost every case, it was a material development for industrial use that made the first workable device possible. For example, the artificial kidney of Dr. Kolff was made possible because of cellophane sausage casings. Dr. Scribner's application of the artificial kidney in 1960 was possible because polysiloxane and polytetrafluroethylene materials were available to make the A/V shunts. The first polymer vascular grafts were made by Hufnagel in 1945 of polymethyl methacrylate and were further modified by Voorhees in 1952 to be made of synthetic polymer fabric. Voorhees simply sewed two pieces of Vinlon sail cloth together to make the grafts. From that came the specially woven and knitted tubular fabric of Dacron that we can use today in large artery repair.

I started in the field of biomaterials in 1962 working with
Dr. Scribner on new synthetic polymer membranes. It has been an
exciting 20 years with many changes occurring. It has certainly
not been 20 years of frustrations as some people have suggested.
From the standpoint of meetings, in the early 1960's there were
possibly only four non-M.D.s at the ASAIO meetings (Ed Leonard from
Columbia, Bob Leininger from Battelle, Les Babb from Seattle and
myself from Stanford Research Institute). Now look at the expansion as indicated by the number of people meeting here today on
just biomaterials. Also, 20 years ago we met in places like
Atlantic City, Philadelphia, and Chicago. Who could not say that
this beautiful seeting in Porto Cervo is not an advancement.

We have also seen many medical advancements: a workable hip
joint with its metal shaft set in a PMMA grout and the acetabular
cup of ultrahigh molecular weight polyethylene; polymeric finger
joints that enable the reconstruction of an arthritic hand to
reduce pain and allow it to function; and, a functioning artificial heart. In 1970 we had still not broken the 100-hour survival
and now the group at Utah is ready for the first human implantation.

In other advances, we see polymers and natural tissue in competition. For example, in small diameter blood vessel repair, the
saphenous vein is still the best. However, we are now approach-
ing the first human implantation of small diameter synthetic polymer
vascular prostheses to help those patients who have no useable
saphenous vein. The ball and disc type of heart valve prostheses
are still not as good as the processed natural tissue porcine
valve. Yet even here we use a polymeric or metal stint to support
the porcine tissue. The artificial kidney has been miniaturized,
but in terms of patient well-being it has improved very little
since the 1940's. Also, the A/V shunts have given way to A/V
fistulas since we do not have good polymers for long-term access.
Peritoneal dialysis is becoming a better dialysis procedure for
the patient, primarily because of the improvement of a polymeric
access to the peritoneal cavity.

In new areas of work which have great potential are the
artificial skin for helping burn patients, new membrane oxygena-
tion, and, of course, the drug delivery devices which is the major
focus of this Conference. As you can see, it has been an exciting
20 years.

That doesn't mean we don't have problems. Many of these old
and new devices are still being developed empirically and not
developed on the basis of knowledge. We talk about materials
specifications for devices but what do we mean by purity? And how
much purity is needed? For example, PVC containing certain industrial additives clot blood. In contrast, the low molecular weight
impurity in Biomer may be what makes it nonthrombogenic.

What are the right chemical, physical and mechanical properties the polymer must have to make a successful device? To determine this, we need to understand how the material interacts at the
molecular, cellular and macroscopic functional levels.

We must understand that the polymer properties can be altered
by how it is fabricated; for example, by contamination from molds,
by surface oxidation, by conformational changes in surface and
bulks morphology, etc. Steam sterilization can be regarded as an
annealing step and chemical and radiation sterilization a polymer-
modifying step, with all possibly altering the properties of the
polymer. Also, the body is a very corrosive environment to
materials and can alter the performance of an implant.

How do we get knowledge of these material requirements? We
need to conduct good tests in vitro which are clean, reproduceable
and allow accurate measurements. However, they may give no real
knowledge. Thus we need good in vivo tests even though they can
be messy, show variations from animal to animal, and species to
species, and present difficulties getting accurate data. Correlation of the in vitro and in vivo can lead to the advancement of
knowledge. We also need to interact better with people in the
biological sciences and medicine so that tests are conducted in
the best possible manner. There are also many things that we as
polymer chemists can do better. We have to more accurately characterize our materials in the form in which it will be used. For
example, Figure 1 shows a film cast on a glass plate. The surface
morphology is different and its response in cell culture is different. Thus, the same material would be considered good and bad.

We must also consider the final application and whether or not
the design is proper for the proposed use. The best material will
fail in a poor design. We must also consider if the device is
being handled properly on implantation. To achieve this we need
good interaction between the medical scientist or surgeon and the
physical scientist.

Finally, keep in mind that we are doing this research to help
the patient. To do this we need good teamwork between a variety
of disciplines so that the problems can be adequately defined, the
in vitro and in vivo research done at the highest level, and
definitive animal studies conducted to prove that the studies can
progress to our ultimate goal - the human animal.

The true coupling of disciplines requires trust and close
interaction of individuals. This has not been helped by the current
practices of competitive federal funding nor by the less than pro-
fessional ethical behavior of some individuals and some industries.
However, this week we have a chance to learn from each other and to
expand our minds and, further, to possibly set up future

collaborative interactions. I know that we are all eager to begin
this excellent conference arranged by our hosts from the University
of Pisa.

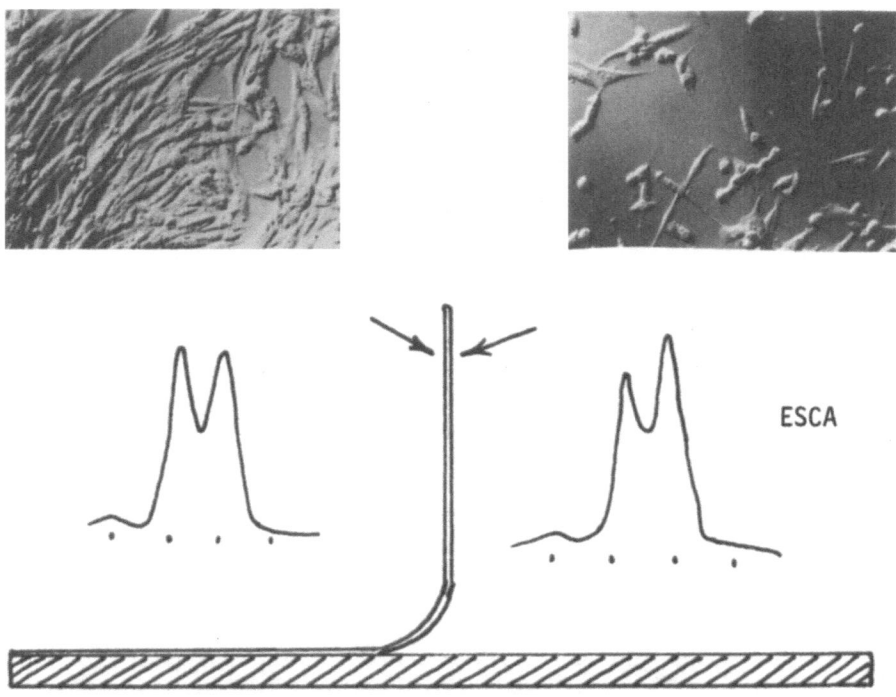

Figure 1.   The effect of mold surface on the surface morphology of
            a polymeric film and cell growth on the two surfaces.

FUNCTIONAL POLYESTERS AND POLYAMIDES FOR MEDICAL APPLICATIONS

OF BIODEGRADABLE POLYMERS

Robert W. Lenz and Philippe Guerin

Chemical Engineering Department
University of Massachusetts
Amherst, MA   01003

INTRODUCTION

For many medical, agricultural or veterinary uses it would be desirable to have a functionalizable, biodegradable polymer which could be fabricated in the molten state by either injection molding, melt spinning or melt extrusion. These polymers could serve as carriers for biomedically active agents, which would be chemically bound to the polymer and, after release of the agent, the polymer would degrade to non-toxic, low molecular weight residues capable of being eliminated from the body by normal metabolic pathways.

This goal was the objective of the present investigation, which is concerned with the preparation of polymers from naturally-occurring, multifunctional hydroxyacids and aminoacids. On degradation such polymers could revert in the body to their original natural products. The natural products chosen for this purpose were malic acid (I), which is an intermediate in the citric acid cycle, and aspartic acid (II), which is a common component of many proteins. High yield synthetic procedures were developed for the conversion of both compounds into cyclic monomers, which could be polymerized to polymers of controlled molecular weights. Malic acid was converted to the β-lactone and aspartic acid to the β-lactam, and each was polymerized to its respective polyester and polyamide, as outlined in the equations below:

---

*Present address: Ecole Normale Superieure, Laboratoire de Chimie, Paris, France.

$$
\begin{array}{cccc}
\begin{matrix} COOH \\ | \\ Y-CH \\ | \\ CH_2 \\ | \\ COOH \end{matrix}
& \xrightarrow{}\xrightarrow{}
\begin{matrix} COOR \\ | \\ Y-CH \\ | \\ CH_2 \\ | \\ COOH \end{matrix}
& \xrightarrow{}
\begin{matrix} COOR \\ | \\ CH-X \\ | \quad | \\ CH_2-C\!\!\diagdown_O \end{matrix}
& \xrightarrow{}
\left[XCH-CH_2-\overset{\displaystyle O}{\overset{\|}{C}} \atop | \atop COOR \right]_n
\end{array}
$$

(I) X = 0,   (II) X = NH

These two polymers are of particular interest for several reasons: (1) polyesters and polyamides, in general, show good biocompatibility; polyglycolic and polylactic acids and their copolymers for the former, and poly-L-lysine and poly-L-glutamate for the latter, are all reportedly biodegradable in the body; (2) the pendant carboxylic groups on the malic and aspartic acid polymers can be used either to impart water solubility, and hence plasma solubility, or to attach drugs containing alcohol on amine groups (or spacers connected to drugs) for sustained release of bioactive materials; (3) polyanions from carboxyl-containing polymers can be pharmacologically active in themselves;[2] (4) stable ester derivatives of the polymers can be prepared for evaluation as biomaterials in applications such as sutures, prostheses and surgical implants;[3] and lastly (5) the presence of an asymmetric carbon atom in the repeating units of these polymers is of potential use because its configurational order or stereoregularity can be modified by changing the distribution of the two enantiomer units which, in turn, would cause changes in polymer crystallinity and, thereby, changes in the mechanical and biological properties. It is also important to study the effect of chirality on biological activity in itself.

Both sets of monomers and polymers were prepared in their optically active and racemic forms, and the comparative physical properties of the two polymers were determined. The ring-opening polymerization reactions of the lactone and lactam were readily carried out by a variety of catalysts or initiators as described in the following sections.

MALIC ACID POLYMERS

Monomer Synthesis

Racemic Malolactonate.[1] The first cyclic monomer prepared was the benzyl ester of the β-lactone of malic acid. Initial attempts to make the lactone directly from malic acid itself were unsuccessful, so bromosuccinic acid was chosen as the starting material. This compound was converted to its anhydride by refluxing in acetyl chloride, and the anhydride was reacted with benzyl alcohol to yield a mixture of the two bromosuccinic acid monobenzyl ester isomers. Only one of these esters, IIIB, is capable of being converted to the lactone by the ring closure, nucleophilic substitution reaction. These reactions are shown in the following sequence:

$$\begin{array}{c} COOH \\ | \\ CHBr \\ | \\ CH_2 \\ | \\ COOH \end{array} \longrightarrow \begin{array}{c} Br-CH-C{<}^{O}_{O} \\ | \\ CH_2-C{\lessgtr}_O \end{array} \xrightarrow{C_6H_5CH_2OH} \begin{array}{c} COOH \\ | \\ CHBr \\ | \\ CH_2 \\ | \\ COOCH_2C_6H_5 \end{array} + \begin{array}{c} COOCH_2C_6H_5 \\ | \\ CHBr \\ | \\ CH_2 \\ | \\ COOH \end{array}$$

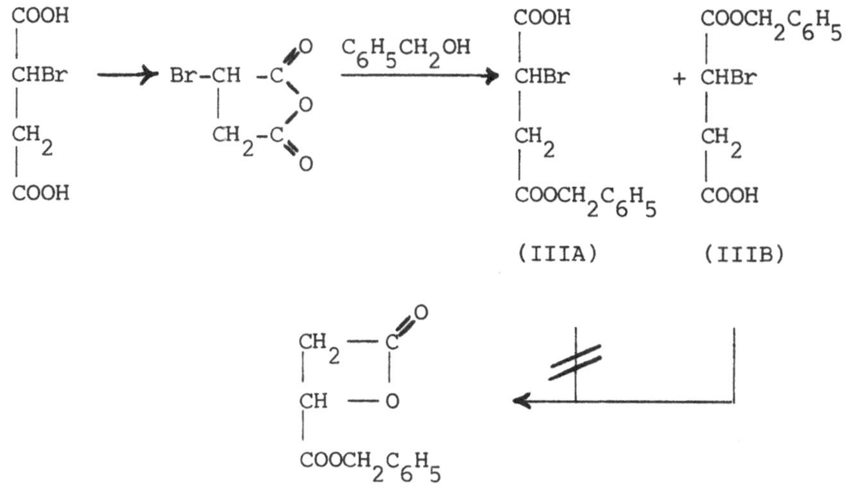

$$\qquad\qquad\qquad (IIIA) \qquad\qquad (IIIB)$$

$$\begin{array}{c} CH_2 - C{\lessgtr}^O \\ | \qquad\quad | \\ CH\ -\ O \\ | \\ COOCH_2C_6H_5 \end{array} \quad \nleftarrow$$

The lactonization reaction was carried out on the crude mixture of the two half esters in aqueous solution, and the lactone formed was separated by extraction with an organic solvent. Further purification was carried out by preparative HPLC followed by high vacuum distillation (B.P. $105°C/10^{-3}$ mm Hg). The overall yield of the monomer was about 20%. The lactone was characterized by infrared spectroscopy and ^{1}H and C^{13} NMR spectroscopy.

Optically Active Benzyl Malolactonate.[4] The route used to prepare the racemic monomer was not applied to the preparation of the optically active monomer because the optically active form of bromosuccinic acid is difficult to prepare and racemization could occur in the S_N2 ring-closure reaction. The starting material instead was L-malic acid itself, which is readily available, and the following scheme shows the method of preparation of the optically active malolactone benzyl ester monomer. It should be noted in this scheme that at no time, during the different steps of the synthesis sequence, were bonds made or broken at the chiral carbon atom:

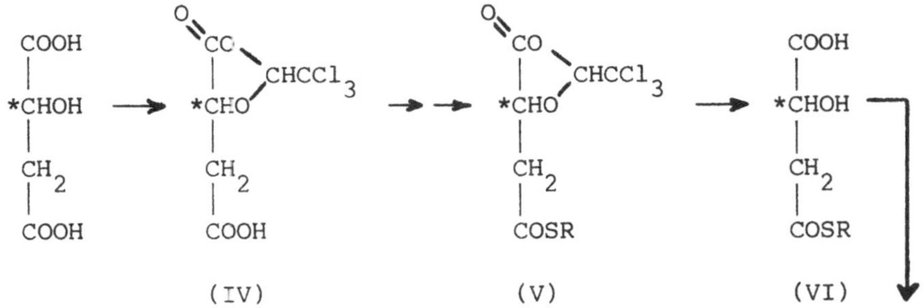

$$\begin{array}{c} COOH \\ | \\ {*}CHOH \\ | \\ CH_2 \\ | \\ COOH \end{array} \longrightarrow \begin{array}{c} O{\ll}^{CO}_{\diagdown CHCCl_3} \\ | \\ {*}CHO \\ | \\ CH_2 \\ | \\ COOH \end{array} \longrightarrow\longrightarrow \begin{array}{c} O{\ll}^{CO}_{\diagdown CHCCl_3} \\ | \\ {*}CHO \\ | \\ CH_2 \\ | \\ COSR \end{array} \longrightarrow \begin{array}{c} COOH \\ | \\ {*}CHOH \\ | \\ CH_2 \\ | \\ COSR \end{array}$$

$$\qquad (IV) \qquad\qquad\qquad (V) \qquad\qquad\qquad (VI)$$

(next page)

$$
\begin{array}{ccc}
 & \overset{\displaystyle COOCH_2C_6H_5}{\underset{\displaystyle \ \ |}{\ }} & \overset{\displaystyle CH_2-C\!\!\nearrow^{\!\!O}}{\underset{\displaystyle |\ \ |}{\ }} \\
\text{(previous} & \longrightarrow \ \ *CHOH & \longrightarrow \ \ *CH\ -O \\
\text{page)} & | & | \\
 & CH_2 & COOCH_2C_6H_5 \\
 & | & \\
 & COSR & \\
 & & \\
 & (VII) & (VIII)
\end{array}
$$

The choice of this synthesis route[4] was guided by the final reaction in the sequence, the formation of the lactone from the hydroxy-
thioester, which was done by the same procedure as that reported for
the preparation of macrolides.[5] This intramolecular esterification
reaction was catalyzed by a mercuric salt catalyst. The first goal
of the sequence, therefore, was to block one of the two carboxylic
acid groups of malic acid with a protective group, in this case the
chloralide group. A crystalline white product, IV, was isolated
which had a melting point of 142-145°C and a specific rotation,
$(\alpha)_D$ = +27° in dioxane. This compound was converted to malic acid
chloralide acid chloride (M.P. 83°C, $(\alpha)_D$ = +12° in dioxane) and
the latter to the thioester, V, by reaction of IV with a thallous
thiolate. S-octadecyl malic acid chloralide (M.P. 81°C, $(\alpha)_D$ = +35°
in dioxane) was prepared in this step and then converted to S-octadecyl malic acid, VI (M.P. 93°C, $(\alpha)_D$ = -45° in dioxane). This compound was esterificated with benzyl alcohol to form the S-octadecyl
malic acid benzyl ester product, VII (M.P. 52-54°C, $(\alpha)_D$ = -75° in
dioxane). The lactonization step was carried out with mercuric
methanesulfonate as the catalyst, and the optical purity of the L-
malolactone benzyl ester, VIII (M.P. 35-37°C, $(\alpha)_D$ = -37° in dioxane), after purification by HPLC and distillation, was determined
to be 91% by hydrolyzing this compound to malic acid and measuring
the rotation of the latter.

Polymer Preparation and Properties[1,4,6]

The benzyl ester of malolactone was used as the monomer instead
of malolactonic acid to eliminate potential problems in the ionic
polymerization reactions. The carboxylic acid group, if present,
could react with the initiators or cause either chain transfer or
termination reactions during the polymerization. The polyester of
this pendant ester was readily converted to poly-β-malic acid by
hydrogenolysis without change in its molecular weight, according to
the following reaction:

$$
\left[OCHCH_2\overset{\displaystyle O}{\overset{\|}{C}} \right]_n \xrightarrow[\text{cat}]{H_2} \left[OCHCH_2\overset{\displaystyle O}{\overset{\|}{C}} \right]_n + nC_6H_5CH_2OH
$$

$$
\begin{array}{cc}
| & | \\
COOCH_2C_6H_5 & COOH
\end{array}
$$

β-Lactones are known to polymerize by both anionic and cationic mechanisms depending upon the substituents on the lactone ring.[7] In general, however, β-substituted-β-lactones are polymerized only with cationic initiators, presumably because an anionic polymerization occurs by an S_N2 reaction at the β-carbon atom and steric hindrance from the substituents at the β-position would interfere with the propagation step.

In the present study, it was important that the L-malolactone benzyl ester monomer be polymerized to the polyester in sufficiently high molecular weights for the biomedical applications in mind.[8] To achieve this goal, a number of different types of initiators and catalysts were evaluated for the polymerization of both the racemic and optically active monomers, including macrozwitter ions, nucleophilic initiators, cationic catalysts and initiators, organometallic catalysts, and nucleophilic initiator-crown ether complexes. The results obtained in these polymerization reactions are collected in Tables 1 and 2 for the optically active[4] and racemic monomers,[1,6] respectively.

As seen in both tables, anionic initiators were effective in this polymerization, which may be the first example of a β-substituted-β-lactone being successfully polymerized through an anionic

Table 1. Polymerization of Racemic Malolactone Benzyl Ester[4]

'Initiator	Reaction Conditions[a]		Polymer Product		
	Temp. °C	Time days	Yield %	$\overline{M}_{GPC}$[b]	$\overline{M}_w/\overline{M}_n$[b]
Anionic					
$(C_2H_5)_3N$	70	5	65	3100	1.5
$C_6H_5COO^{\ominus}(C_2H_5)_4N^{\oplus}$	70	5	80	2900	1.4
$C_6H_5COO^{\ominus}Na^{\oplus}$ DBCE[c]	50	1	80	2800	1.3
Cationic					
$AlCl_3$	70	20	10	1000	2.1
$(C_6H_5)_3C^{\oplus}SbF_6^{\ominus}$	30	1	10	400	2.0

[a]Reactions carried out on pure monomer in absence of solvent.

[b]Determined by gel permeation chromatography, GPC, relative to polystyrene standards.

[c]DBCE: dibenzo-18-crown-6-ether.

Table 2. Polymerization of Optically Active Malolactone Benzyl
          Ester[4]

Initiator	Reaction Conditions				Polymer Product		
	Medium	Temp. °C	Time days	Yield %	$\overline{M}_{GPC}$ [a]	$\overline{M}_w/\overline{M}_n$ [a]	$[\alpha]_{25}^D$ [b]
Anionic							
$(C_2H_5)_3N$	solution	30	28	20	2600	2.1	+10.6
$(C_2H_5)_3N$	bulk	70	5	80	2900	1.4	+12.0
$C_6H_5COO^{\ominus}(C_2H_5)_4N^{\oplus}$	solution	30	28	25	3500	2.1	+11.9
$C_6H_5COO^{\ominus}(C_2H_5)_4N^{\oplus}$	bulk	70	3	75	7300	1.4	+13.8
Cationic							
$FeCl_3$	solution	30	28	5	1800	2.4	- 7.8
$FeCl_3$	bulk	70	7	40	3200	1.9	- 7.5

[a] Determined by GPC relative to polystyrene standards.

[b] Optical rotation in degrees.

mechanism. It is likely that the electronic activation of the β-
position by the β-carbobenzoxy group offset the steric hindrance of
this substituent and permitted the S_N2 propagation reaction to occur.
But with this monomer, chain transfer can also occur by proton abstraction of the acidic hydrogen atom at the α-carbon atom. Proton
abstraction would most likely be followed by an elimination reaction
to form the salt of maleic or fumaric acid, which could reinitiate
the polymerization. Hence, chain transfer could occur to limit the
molecular weight of the polymer formed.

For the optically active monomer, preliminary results showed
that the optically active polymers which were formed by either anionic or cationic initiators had the same magnitude but the opposite
sign of rotation.[4] This result is consistent with the suggestion
above that the anionic polymerization occurs by an S_N2 reaction.
For an optically active monomer, therefore, anionic propagation at
the β-carbon atom causes inversion of the chiral center, while cationic propagation, which is believed to occur at the ester bond,
occurs with retention of the configuration. The optical properties
of these polymers are being studied further to confirm these results
and interpretations.

The physical properties of the polyesters obtained from the racemic monomer were quite different for polymerization reactions based on either anionic initiators or coordination initiators.[1] All of the homogeneous anionic initiators gave amorphous polymers, but the heterogeneous triethylaluminum-water initiator gave a crystalline polymer as determined by differential scanning calorimetry, DSC. The latter had a broad melting point extending over a temperature range from 165 to 185°C. This result is consistent with previous observations of the ability of this initiator to cause a stereospecific polymerization of racemic β-alkyl-β-propiolactones.

DSC analysis of the optically active poly-L-malolactone benzyl ester also showed multiple endotherms in the temperature range of 140-160°C, regardless of the catalyst system used to prepare the polymer.[4] The difference in the melting points of the optically active and racemic polymers could be due to differences in either the degree of stereoregularity or the molecular weights of the polymers.

The extremely important effect of monomer purity on polymer molecular weight can be seen in Table 3.[4] The impurities in the monomer are, most likely, the cause of the very low molecular weights obtained in Tables 1 and 2. Reactive impurities are apparently still present even after careful purification of the monomer and are assumed to be either malolactone carboxylic acid or benzyl alcohol, formed by hydrolysis of the malolactone benzyl ester during the monomer distillation. The results in Tables 1 and 2 are for experiments made with monomer which has been purified through only one purification by HPLC and distillation.

Table 3. Effect of Monomer Purification and Polymerization Temperature in Anionic Polymerization on Molecular Weights[a6]

Purification Method[b]	Reaction Conditions		Polymer Product	
	Temp. °C	Time days	Yield %	$\overline{M}_{GPC}$[c]
I	60	10	80	8,000
II	60	19	80	22,000
II	40	30	99	36,000
II	25	40	84	45,000
III	60	12	95	33,000
III	40	30	93	50,000

[a]Initiated with 0.1 mole% triethylamine in pure monomer.

[b]Methods: I, one pass through HPLC and one distillation; II, two passes through HPLC and one distillation; III, two passes through HPLC and two distillations.

[c]Determined by GPC relative to polystyrene standards.

The molecular weight distributions observed by gel permeation chromatography, GPC, were quite variable with $\overline{M}_w/\overline{M}_n$ ratios varying between 1.2 and 2.7 relative to polystyrene standards. The influence of the temperature of polymerization on polymer molecular weight was also seen to be important. With triethylamine as initiator, the molecular weight of the polymer obtained decreased when the temperature was increased.

ASPARTIC ACID POLYMERS

Monomer Synthesis

The benzyl ester of the β-lactam of aspartic acid, XI, was synthesized by a four-step sequence of reactions involving: (1) esterification of both carboxyl groups to form the diester, IX; (2) replacing one proton of the amine function with the trimethylsilyl protective group to form X; (3) ring closure by formation of an amide anion with a very strong base (a Grignand reagent) to form a lactam, and (4) removal of the protective group to form the monomer, XI, as follows:

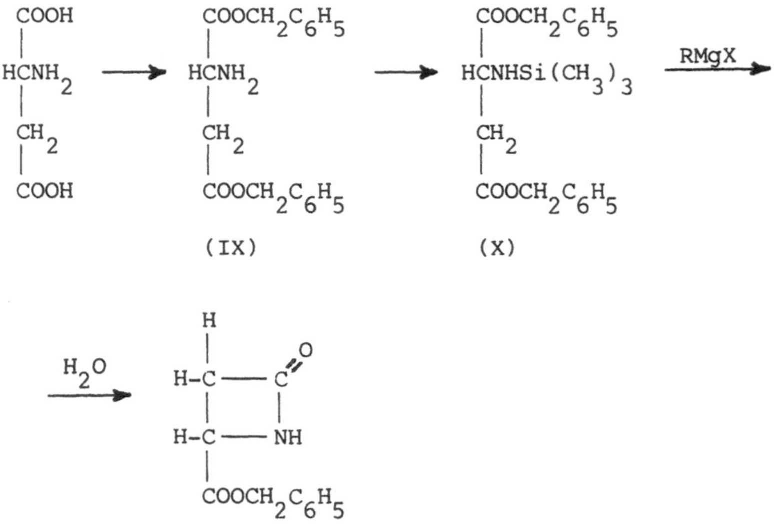

COOH
|
HCNH$_2$ $\longrightarrow$
|
CH$_2$
|
COOH

(IX)

COOCH$_2$C$_6$H$_5$
|
HCNH$_2$ $\longrightarrow$
|
CH$_2$
|
COOCH$_2$C$_6$H$_5$

COOCH$_2$C$_6$H$_5$
|
HCNHSi(CH$_3$)$_3$ $\xrightarrow{\text{RMgX}}$
|
CH$_2$
|
COOCH$_2$C$_6$H$_5$

(X)

$\xrightarrow{\text{H}_2\text{O}}$

H
|
H-C —— C=O
| |
H-C —— NH
|
COOCH$_2$C$_6$H$_5$

(XI)

The recrystallized lactam monomer, XI, was obtained in approximately 10% overall yield and had a melting point of 129–131°C. Elemental analysis and NMR spectroscopy verified the high purity and structure of the monomer.

Polymer Preparation and Properties

   Polymers of aspartic acid were prepared by two methods: (1) the anionic polymerization of the β-benzyl ester of the β-lactam monomer, and (2) the thermal polymerization of aspartic acid, XII, as shown in the equations below:[9]

| (1) | (XI) | (XIII) | (XIV) |

$$x=0.1-1.0$$

| (2) | (XII) | (XV) |

   The anionic polymerization reaction using a potassium pyrrolid-one-acetylpyrrolidone initiator system formed a polymer containing either a single type, XIII, or two types, XIII and XIV, of repeating units, depending on the solvent used for the polymerization reac-tion.[9] A solvent of very low polarity and coordinating ability such as methylene chloride, gave essentially a homopolymer containing only the β-aspartamide units, XIII, while a polar solvent such as DMSO gave a copolymer containing mostly aspartimide units, XIV. In contrast, the thermal polymerization of the bulk monomer gave only the imide homopolymer, XV, as previously shown.[10]   The re-sults for the β-lactam polymerization are collected in Table 4. As shown in Table 4, the cationic polymerization of the β-lactam also gave an amide-imide copolymer. For both types of initiators, however, only relatively low molecular weight polymers were formed.

   In the anionic polymerization reaction, imide formation is be-lieved to occur within the internal β-amide units formed during the polymerization by reaction of their amide groups with either the strong base initiator or activated monomer, XVI, to form the amide anion which can react with the pendant ester group, as follows:

Table 4. Polymerization of the Aspartic Acid Lactam Benzyl Ester[9]

Initiator	Reaction Conditions			Polymer Product		
	Solvent	Temp. °C	Time h	Yield %	Amide Unit Content[a] %	$\bar{M}_n$[b]
Anionic[c]						
KP/AP	DMSO	20	48	50	10	--
KP/AP	CH_2Cl_2	20	30	40	90-100	6,600
Cationic						
CH_3COOH	$C_6H_5NO_2$	160	26	46	40-50	8,000
C_6H_5COOH	$C_6H_5NO_2$	160	26	29	40-50	6,200

[a] Polymer structure determined by NMR and elemental analysis.

[b] Determined by GPC relative to polystyrene standards.

[c] KP/AP - potassium pyrrolidone and acetyl pyrrolidone.

(XIII)   (XVI)

(XIV)

Presumably, the alkoxide ion formed in this reaction can react with the lactam to recreate the activated monomer. However, this reaction may be quite slow compared to the polymerization reaction. Similarly in the cationic polymerization, imide formation probably occurs by an acid-catalyzed amide-ester interchange reaction to yield benzyl alcohol. In both cases, very little is known about the effects of these byproducts on the polymerization reaction.

The polyacids of these polymers can be prepared by either hydrogenolysis of the poly-β-amide, XIII, or hydrolysis of the imide copolymer, XIV, or homopolymer, XV. The hydrolyzed imide polymers are believed to contain both α- and β-amide units, as follows:

CONCLUSIONS

Two new water-soluble polyacids, especially selected for their potential use in biodegradable synthetic drug carriers, have been prepared, using different types of monomers, catalysts and initiators. Further investigations are in progress directed at obtaining high molecular weight polymers in the range of 100,000-200,000, but the particular molecular weights chosen for evaluation will depend upon the final use of the polymer. The combination of controlled crystallinity and controlled number of carboxylic groups can be used to increase or decrease the water swellability and solubility, and thereby the rate of degradation, of these polymers. The degradation rates, in turn, would directly control either the release rate of the drug, when the polymers are used as carriers, or the stability of the polymers as biomaterials.

While effective use in the human body is the goal of this program, many other requirements have to be met, in particular those which are usually referred to as biocompatibility; that is, non-toxicity and non-antigenicity. To date acute toxicity and immunogenicity in mice and rabbits have been evaluated by our collaborators on racemic poly-β-malic acid samples prepared in the same manner as used in this study. The results indicate that this polymer is non-toxic. Previous investigators have also shown this behavior for the polyimide obtained by the thermal polymeriza-[12,13] tion of aspartic acid, and its derivatives, as described above. The drug-containing polymers were obtained by reacting the polyimide with a pharmacologically active amine to form the α- and β-forms of[13] polyaspartamide containing pendant amide groups.

ACKNOWLEDGMENT

The authors are grateful to the National Science Foundation
and to the C.N.R.S. of France for the fellowship provided to
Philippe Guerin and to the N.S.F.-sponsored Materials Research
Laboratory of the University of Massachusetts for the support of
this work.

LITERATURE REFERENCES

1. M. Vert and R. W. Lenz, A.C.S. Polymer Preprints, 20, No. 1,
    608 (1979).
2. L. G. Donaruma, Ed., "Polymers in Medicine and Biology. Vol. I.
    Anionic Polymeric Drugs," John Wiley, N.Y., 1980.
3. M. Vert and F. Chabot, Makromol. Chem., Suppl., 5, 30 (1981).
4. R. Wojcik, Ph.D. Dissertation, University of Massachusetts,
    1982.
5. S. Masamune, Y. Hayase, W. R. Chan and R. L. Sobczak, J. Am.
    Chem. Soc., 98, 7874 (1976).
6. D. Johns, Ph.D. thesis research, University of Massachusetts,
    private communication.
7. D. Johns, R. W. Lenz and A. Luecke in "Ring-Opening Polymeriza-
    tion. Vol. I," K. Ivin, ed., Academic Press, in press.
8. R. W. Lenz and M. Vert, U.S. Patent 4,265,247 (1981) and U.S.
    Patent 4,320,753 (1982).
9. T. Mang, University of Massachusetts, private communication.
10. A. Vegotsky, K. Harada and S. W. Fox, J. Am. Chem. Soc., 80,
    3361 (1958); J. Vlasak, F. Rypacek, J. Drobnik and V. Saudek,
    J. Polymer Sci., Polymer Symposium, 66, 59 (1979).
11. C. Brand, C. Bunel, M. Vert, P. Bouffard, M. Cabaut and B.
    Dalpech, Proceedings of IUPAC Symposium of Macromolecules,
    Amherst, MA, 1982; p. 384.
12. J. Drobnik, V. Saudek, J. Vlasak and J. Kalal, J. Polymer Sci.,
    Polymer Symposium, 66, 65 (1979); J. Drobnik, J. Kalal, L.
    Dabrowska, R. Praus, M. Vachova, and J. Elis, J. Polymer Sci.,
    Polymer Symposium, 66, 75 (1979).

POLYURETHANES AS BIOMATERIALS - ASSESSMENT OF BLOOD COMPATIBILITY

Vera Sá da Costa[*], E.W. Merrill, E.W. Salzman,
D. Brier-Russel, L. Kirchner, D.F. Waugh,
G. Trudell III, S. Stopper and V. Vitale

Departments of Chemical Engineering and Biology, Massachusetts Institute of Technology, Cambridge, U.S.A., Department of Surgery, Harvard Medical School and Beth Is-
rael Hospital, Boston, U.S.A.

INTRODUCTION

Segmented polyurethanes (SPU) are being widely used as biomaterials in blood contacting devices (e.g.angiographic catheters).
Most of the time the results are very encouraging ones, showing
very low blood platelet activation, but substantial variations have
been reported. For this reason it is important to relate blood
platelet activity with chemical composition, specially surface composition since blood "sees" only the surface of a material and will
react accordingly. In fact the surface and the bulk of SPU will not
usually be the same since in these polymers there is phase separation
of two mutually incompatible molecular sequences. One sequence,
which can be a polyether, constitues the continuous phase and is
rubbery at room temperature ("soft" phase). The other sequence,
consists of strongly hydrogen bonding units derived from diisocyanates and (usually) diamines which cluster as glassy or crystalline
micelles, acting as the discontinuous phase ("hard" phase).

With the purpose of relating blood compatibility and surface
composition a series of segmented polyurethanes was synthesised.
Three different α,ω polyether diols of different molecular weights
(Table 1) have been used, as well as two diisocyanates (Table 2).

---

*Presently at Centro de Quimica Fisica Molecular, Instituto Superior
Tecnico, Lisboa, Portugal

Table 1. POLYETHERS USED

Polytetramethylene oxide, PTMO

$$HO(CH_2CH_2CH_2CH_2O)_n - CH_2CH_2CH_2CH_2OH$$

$\overline{M}_n$= 2000, 1000, 650
n = 27, 13, 8

Polypropylene oxide, PPO

$$HO-(CH-CH_2-O)_n CH-CH_2-OH$$
$$\quad\quad |\qquad\qquad\; |$$
$$\quad\quad CH_3\qquad\quad CH_3$$

$\overline{M}_n$= 2000,1200, 400
n = 42, 27, 7

Polyethylene oxide, PEO

$$HO-(CH_2-CH_2-O)_n-CH_2-CH_2-OH$$

$\overline{M}_n$= 8000, 4500, 1500, 1000, 600
n = 176, 101, 33, 22, 13

Table 2. DIISOCYANATES USED

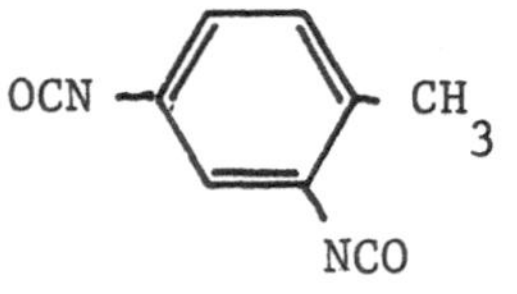

OCN — ⬡ — CH$_3$ / NCO   2,4-tolylene diisocyanate
                                       2,4 TDI

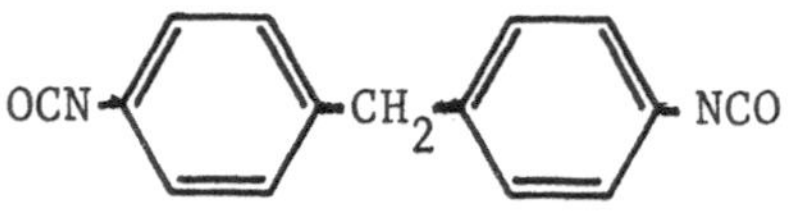

OCN — ⬡ — CH$_2$ — ⬡ — NCO   4,4'-diphenylmethane diiso-
                                             cyanate
                                       MDI

SYNTHESIS

Two different synthetic procedures were used. A two step process shown in Figure 1 and a three step one as in Figure 2. In all cases the chain extension of isocyanate terminated prepolymers was accomplished with ethylene diamine. The synthesis took place in a 2:1 volume mixture of dimethyl sulfoxide (DMSO) and 4-methyl-2-pentanone at 80°C for the first step and at room temperature for the chain extension.

The three step synthesis brings a new variable in the phase separation of segmented polyurethanes. In a polyurethane synthesised by a two step process, we may have a phase separation as shown in Fig. 3. In the three step process, the diisocyanate "connector" either remains in the "soft" phase, in which case the polyether acts as a chain of double molecular weight, with a stuctural "defect" in the midle, or the "connector" may associate with the "hard" phase (Fig. 4), the results of which would be that the polyether chain would be fixed at each end of its primary length to a "hard" segment junction.

The polymers obtained are rather complicated in molecular structure, contrary to what could be thought from the idealized synthesis process. This is due to several reasons as dimerization of diisocyanates, secondary reactions of the -NCO group with the urethane and urea linkages originating allophanate and biuret groups, and the polymerization process itself since the pure "hard segment triad" is a fiction. In reality, according to Flory's theory single diisocyanate residues, pentads and heptads must also form. The same can be said with respect to the single ether chain "soft" segment. Actually we found that "soft" segment chains with two or three ether chains connected by diisocyanate residues will form in Type I polyurethanes.

An added complicating factor arises in what we will call Type III and Type IV polyurethanes. These are similar to Type I and Type II respectively, but instead of using a pure polyether in the first step a 1:1 mixture of PPO 1200 and PEO 1000 is used. In this case the sequences of ether chains connected by a diisocyanate residue or a "hard" segment can be PPO-PPO, PPO-PEO or PEO-PEO.

For all the reasons mentioned the phase separation achieved is widely dependent on synthesis conditions and casting procedures.

SURFACE PREPARATION

Films were cast from 1-5% solutions in DMF, the evaporation of the solvent taking place under a warm (40°C) argon flow.

Films were cast on glass slides, glass microscope slides (for

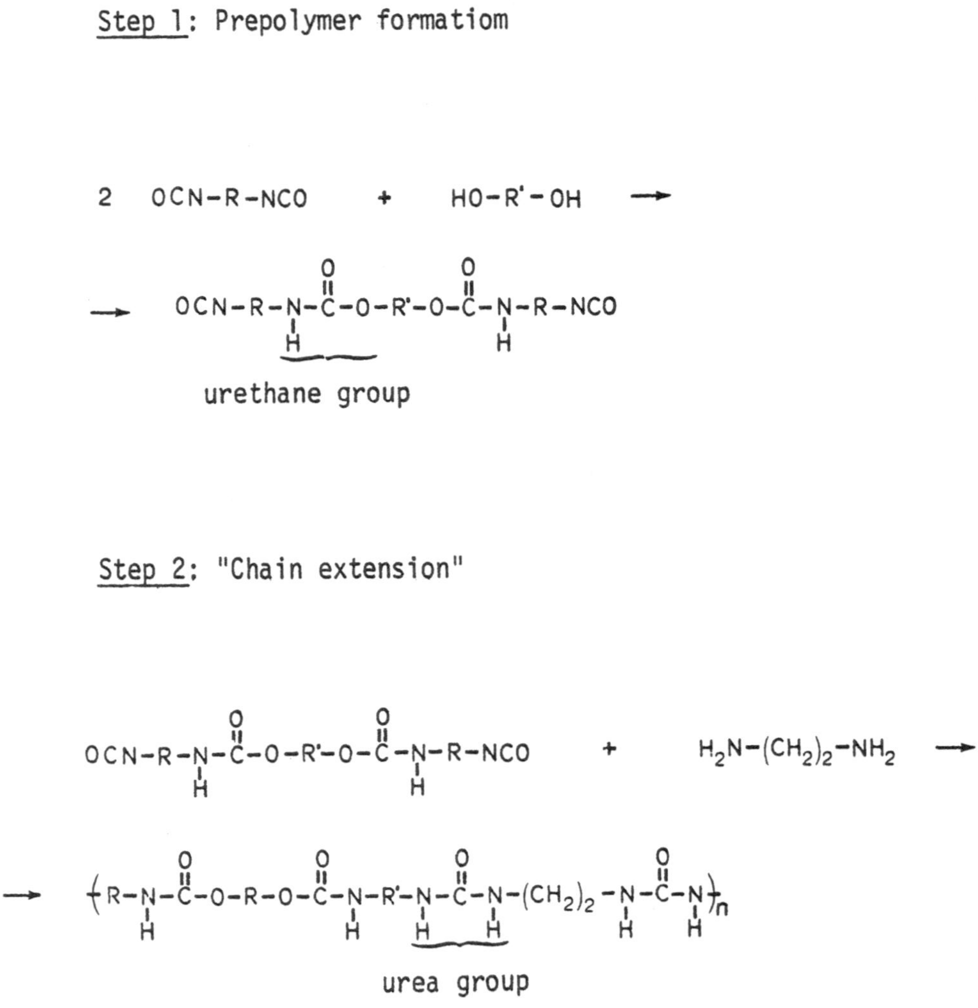

Step 1: Prepolymer formatiom

2   OCN-R-NCO   +   HO-R'-OH   →

$$\longrightarrow \quad \underset{H}{OCN-R-N}-\overset{O}{\overset{\|}{C}}-O-R'-O-\overset{O}{\overset{\|}{C}}-\underset{H}{N}-R-NCO$$

urethane group

Step 2: "Chain extension"

$$\underset{H}{OCN-R-N}-\overset{O}{\overset{\|}{C}}-O-R'-O-\overset{O}{\overset{\|}{C}}-\underset{H}{N}-R-NCO \quad + \quad H_2N-(CH_2)_2-NH_2 \quad \longrightarrow$$

$$\longrightarrow \quad \left(R-\underset{H}{N}-\overset{O}{\overset{\|}{C}}-O-R-O-\overset{O}{\overset{\|}{C}}-\underset{H}{N}-R'-\underset{H}{N}-\overset{O}{\overset{\|}{C}}-\underset{H}{N}-(CH_2)_2-\underset{H}{N}-\overset{O}{\overset{\|}{C}}-\underset{H}{N}\right)_n$$

urea group

Figure 1. Two step synthesis - Type I polymer

Step 1: Couple two diols by diisocyanate: Create a "modified"
polyether with diisocyanate "connector"

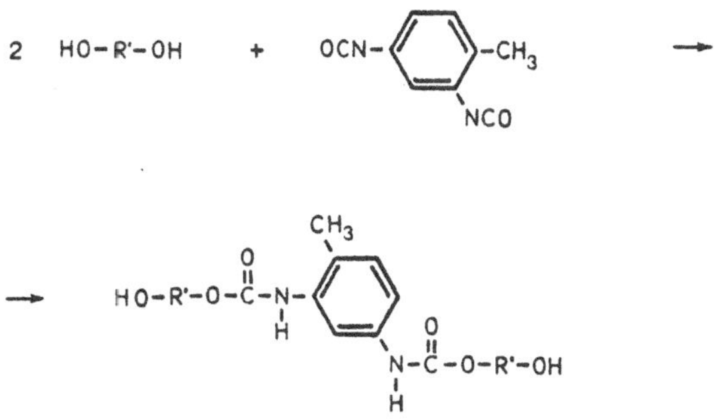

Step 2: Create "modified" prepolymer

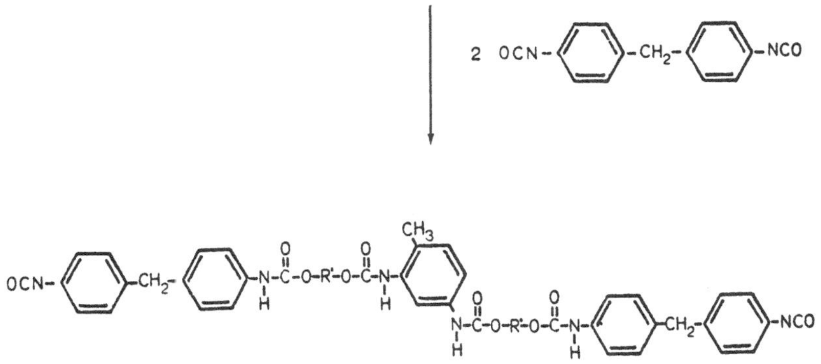

Step 3: "Chain extension" to create segmented polyurethane

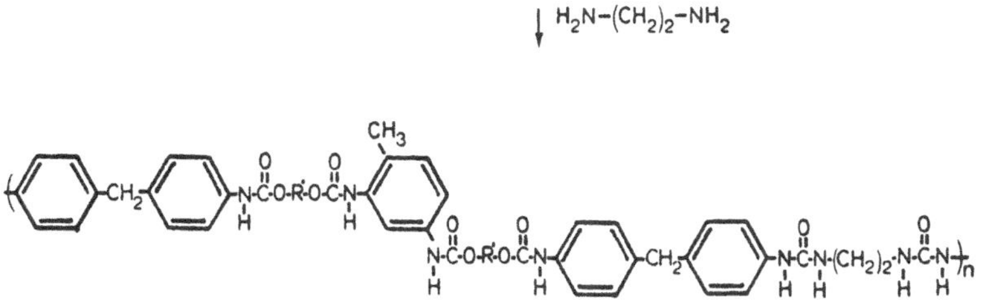

Figure 2. Three step synthesis of polyurethanes - Type II polymers

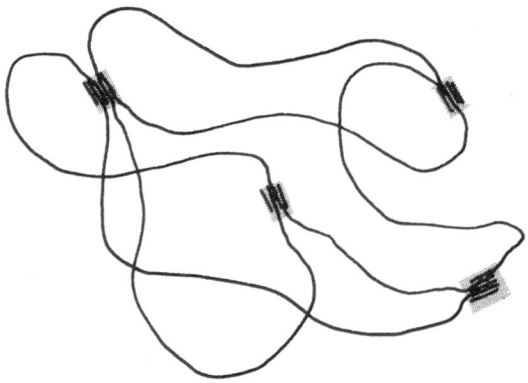

Figure 3. Phase separation in Type I polyurethanes

"hard" segment cluster
"soft" segment
TDI connerctor

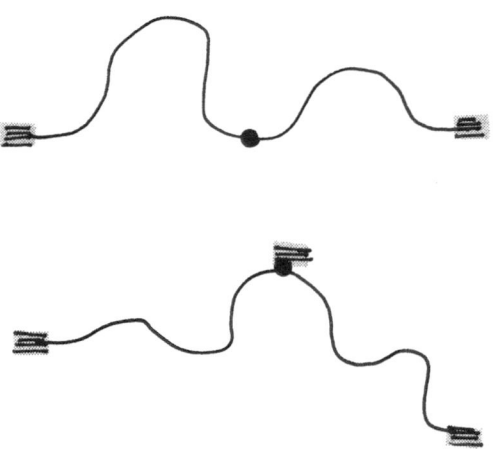

Figure 4. Alternative locations for the TDI
connector in Type II polyurethanes

ESCA examination) and on glass beads used in the Salzman "in vitro" test for platelet retention[1].

Concentration of the polymer, temperature, and evaporation conditions under argon were maintained as closely as possible, specially with casting microscope slides and glass beads, so that the surface morphology in ESCA and platelet retention tests were about the same.

The surface analysed was always the surface exposed to the argon evaporation, not the glass plate side.

BLOOD COMPATIBILITY TESTS

These tests were performed using Salzman Bead Column Test[1] (Fig. 5), where whole human blood, freshly drawn, citrated and thermostated to 37°C is passed from its containing syringe by aliquots of 1 ml through previously coated columns packed with 0.2 mm diameter glass beads. Each column offers an area of 400 cm^2 with a void volume of 0.5 ml.

The platelet count is determined for the blood samples emerging, aliquot by aliquot and is averaged for 5 aliquots and the number of donors. This average divided by the platelet count in the blood in the holding syringe is called the recovery index $\bar{r}$. The platelet retention index $\bar{\rho}$ is defined simply as

$$\bar{\rho} = 1 - \bar{r}$$

thus $\bar{\rho}=0$ corresponding to $\bar{r}=1$ which means an "ideal" surface to which platelets do not adhere.

RESULTS AND DISCUSSION

Table 3 shows the composition of the polyurethanes studied.

The precursor polyethers and "hard" segment analogs[2] were studied by the same methods as the SPU so as to be used as references.

The DSC results confirmed that phase separation was achieved in most SPU synthesised, being fair to good as judged by the appearence of transition temperatures corresponding to the diol and the isocyanate diamine copolymer.

As expected the crystalline melting point of the "soft" segment was depressed by copolymerization as it can be seen in Table 4. All but SPU 24 showed a "soft phase" melting point below the biological test temperature (37°C). The crystallinity of SPU 24 was confirmed by a grossly observable turbidity when equilibrated with isotonic

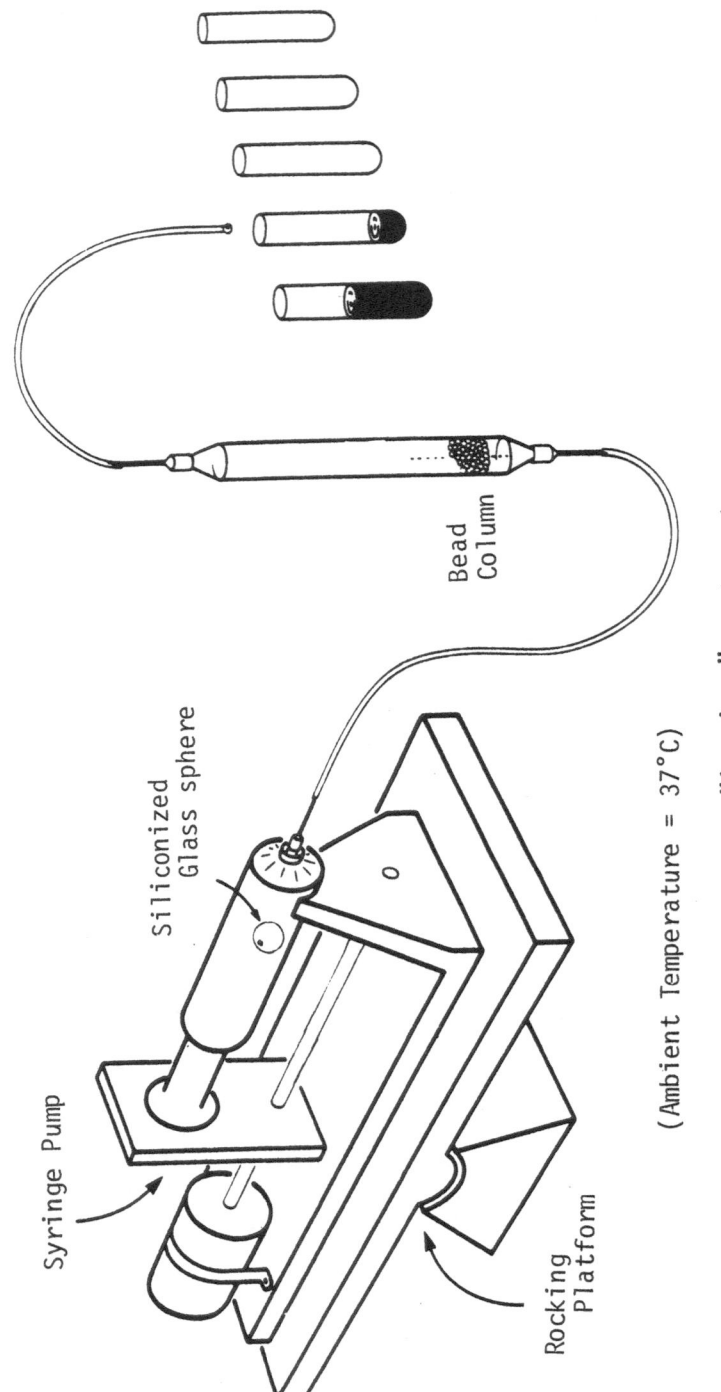

Syringe Pump

Siliconized
Glass sphere

Rocking
Platform

Bead
Column

(Ambient Temperature = 37°C)

Figure 5. "in vitro" test system

Table 3. IDENTIFICATION AND COMPOSITION OF SPU

	Identification				Molar Ratios	
	Type	PE	PE mol wt	HSDI	PE:HSDI:ED	PE:TDI:MDI:ED
24	II	PTMO	2000	MDI		1:0.58:1.o1:0.45
25	I	PTMO	2000	MDI	1:1.39:0.36	
26	I	PTMO	2000	MDI	1:2.25:1.32	
40	II	PEO	1000	MDI		1:0.52:1.08:0.65
401	II	PEO	1000	MDI		1:0.50:0.99:0.50
403	II	PEO	1000	MDI		1:0.51:0.99:0.50
42	I	PPO	2000	TDI	1:2.16:1.05	
44	I	PPO	1200	TDI	1:2.02:0.97	
441	I	PPO	1200	TDI	1:2.08:1.18	
46	I	PPO	1200	TDI	1:2.16:1.17	
48	II	PPO	1200	MDI		1:0.54:0.99:0.49
481	II	PPO	1200	MDI		1:0.52:1.01:0.53
49	II	PPO	400	MDI		1:0.66:0.78:0.49
51	II	PEO	1500	MDI		1:0.56:0.92:0.43
52	II	PEO	8000	MDI		1:0.60:0.84:0.96
53	II	PEO	8000	MDI		1:0.51:0.98:0.50
56	III	PEO	4500	MDI		1:0.49:1.08:0.62
	III	PEO/PPO	1000/1200	TDI	1:2.00:1.05	
	III	PEO/PPO	1000/1200	TDI	1:2.03:1.11	
	IV	PEO/PPO	1000/1200	MDI		1:0.54:1.04:0.57

Table 4. TRANSITION TEMPERATURES OF SEGMENTED POLYURETHANES

PU #	Polyether	Diisocyanate	Ratio	2nd order transitions (°C)	T_m (°C)	T_g (°C)	T_m (°C)	T_{cr} (°C)
49	PPO 400	TDI,MDI	1:0.66:0.78:0.42	189		17		
50	PPO 400	MDI	1:2.00:0.97	180	277	44		
48	PPO 1200	TDI,MDI	1:0.54:0.99:0.49	194.5 229.5 269		-32		
44	PPO 1200	TDI	1:2.02:0.97	233	244	-29		
45	PPO 1200	MDI	1:2.06:1.13	202	284	-33		
42	PPO 2000	TDI	1:2.16:1.05	182 234		-55		
43	PPO 2000	MDI	1:2.21:1.51	185.5 278		-49		
21	PTMO 1000	TDI	1:2.02:0.98	129 149 209	194,244	-72		
24	PTMO 2000	TDI,MDI	1:0.58:1.01:0.45	188 229.5		-71	14.5,37.5	-36.5 (broad)
22	PTMO 2000	TDI	1:2.22:1.20	224	273	-75	3.5	-42 (broad)
31	PTMO 2000	MDI	1:2.03:1.22	233.5	297	-73.5	6.5	-39.5
25	PTMO 2000	MDI	1:1.39:0.36	244		-69	13	
26	PTMO 2000	MDI	1:2.25:1.32		292	-74	2	-45.5 (broad)
39	PEO 600	TDI, MDI	1:0.61:0.86:0.48		274	-15		
32	PEO 600	TDI	1:2.02:1.81	249	253	-15		
35	PEO 600	TDI	1:2.02:1.29	140 249		-20.5		
40	PEO 1000	TDI,MDI	1:0.52:1.08:0.65	degrades at 193°C		-30.5		
36	PEO 1000	TDI	1:1.98:1.33	210		-25.5		

saline at 37°C. All other SPU were transparent in the same conditions.

FTIR did not give any further information about the morphology of phase separation but from the absorption spectra it was possible to determine the concentration of the different forms of nitrogen in the bulk polymer[3] (urethane, urea and amine nitrogen).

ESCA, which showed to be the most promising method in assessing blood compatibility, was used to determine the surface composition of the outermost 30 Å.

Comparing the composition obtained by ESCA with the data obtained from FTIR (Table 5), we see that they agree fairly well for the unreacted diols and "hard" segment analogs, meaning homogeneous polymers. For the SPU there is significantly lower amount (34-40%) of nitrogen in the outermost 30 Å. This decrease in the nitrogen concentration at the surface means that the surface is deficient in "hard" segment. Since the ESCA spectrometer used did not have angular dependent resolution, it was not possible to determine if the "hard" segment present was at the surface or burried somewhere at a depth below 30 Å. This is an important problem since blood only "sees" the surface.

Since the C_{1s} spectra depends on the carbon environment[3], ESCA was also used to determine the relative concentration of the different kinds of carbon. Carbon bound to ether and other forms of carbon could undoubtly be identified, but urethane and urea carbonyl are difficult to find since the molar fractions of carbon in these forms is small compared with the molar fraction of carbon in all other forms.

If we define ϕ as the ratio of the peak height of ethereal carbon (C_{1s} peak chemically shifted 1.5 eV) to all forms of carbon, it will give a measure of carbon at the surface in the form of ethereal bonded carbon, necessarily part of the "soft" segment phase. Thus in a PEO based polyurethane, if all carbon in the scanned layer belonged to PEO, ϕ would be iqual to 1.0.

Ploting $\bar{\rho}$ vs ϕ (Fig. 6) we see that there is a relation between the fraction of ethereal carbon at the surface and blood compatibility.

"Hard" segment analogs are the most reactive surfaces in the study and we would think that as long as "hard" segment domains are present at the surface that will make the surface less compatible. This can be verified with SPU 49 which had a bad phase separation as determined by DSC.

Crystalline SPU (24) is more platelet retaining than the amorphous form of the same substance (26), probably because of differences in adsorption of proteins prior to platelet contact. This may be the reason why the unreacted diols show a higher retention index

Table 5. SURFACE AND BULK COMPOSITION OF SEGMENTED POLYURETHANES "HARD" SEGMENT ANALOGS AND "SOFT" SEGMENT PRECURSORS

Sample	Polyether	Diisocyanate	Surface Composition			Bulk Composition			r*
			C%	O%	N%	C%	O%	N%	
24	PTMO 2000	TDI,MDI	57.2	42.7		79.4	18.2	2.5	0.62
25	PTMO 2000	MDI	73.5	25.2	1.3	79.6	18.3	2.1	0.39
26	PTMO 2000	MDI	73.0	25.5	1.5	78.9	17.3	3.8	0.57
31	PTMO 2000	MDI	74.0	24.0	2.0	79.0	17.5	3.6	0.42
39	PEO 600	TDI,MDI	70.2	27.2	2.6	69.3	24.5	6.2	0.42
40	PEO 1000	TDI,MDI	70.5	27.6	1.9	68.8	26.7	4.6	0.41
47	PPO 400	TDI	71.0	24.9	4.1	69.8	19.8	10.5	0.39
49	PPO 400	TDI,MDI	65.5	30.7	3.8	74.1	19.1	6.8	0.56
48	PPO 1200	TDI,MDI	70.9	27.3	1.8	74.7	21.7	5.3	0.34
44	PPO 1200	TDI	70.2	25.6	4.2	72.4	22.4	5.2	0.81
46	PPO 1200	TDI	70.1	26.9	3.0	72.2	22.2	5.7	0.53
urea		TDI	59.3	13.2	27.5	62.5	12.5	25.0	
urea		MDI	67.7	9.8	22.5	73.8	8.7	17.5	
urethane		TDI	65.4	23.5	11.1	66.7	21.8	11.5	
PTMO 2000			78.3	21.7		80.0	20.0		
PEO 1000			69.5	30.5		66.7	33.3		

*r - ratio: %N surface/%N bulk

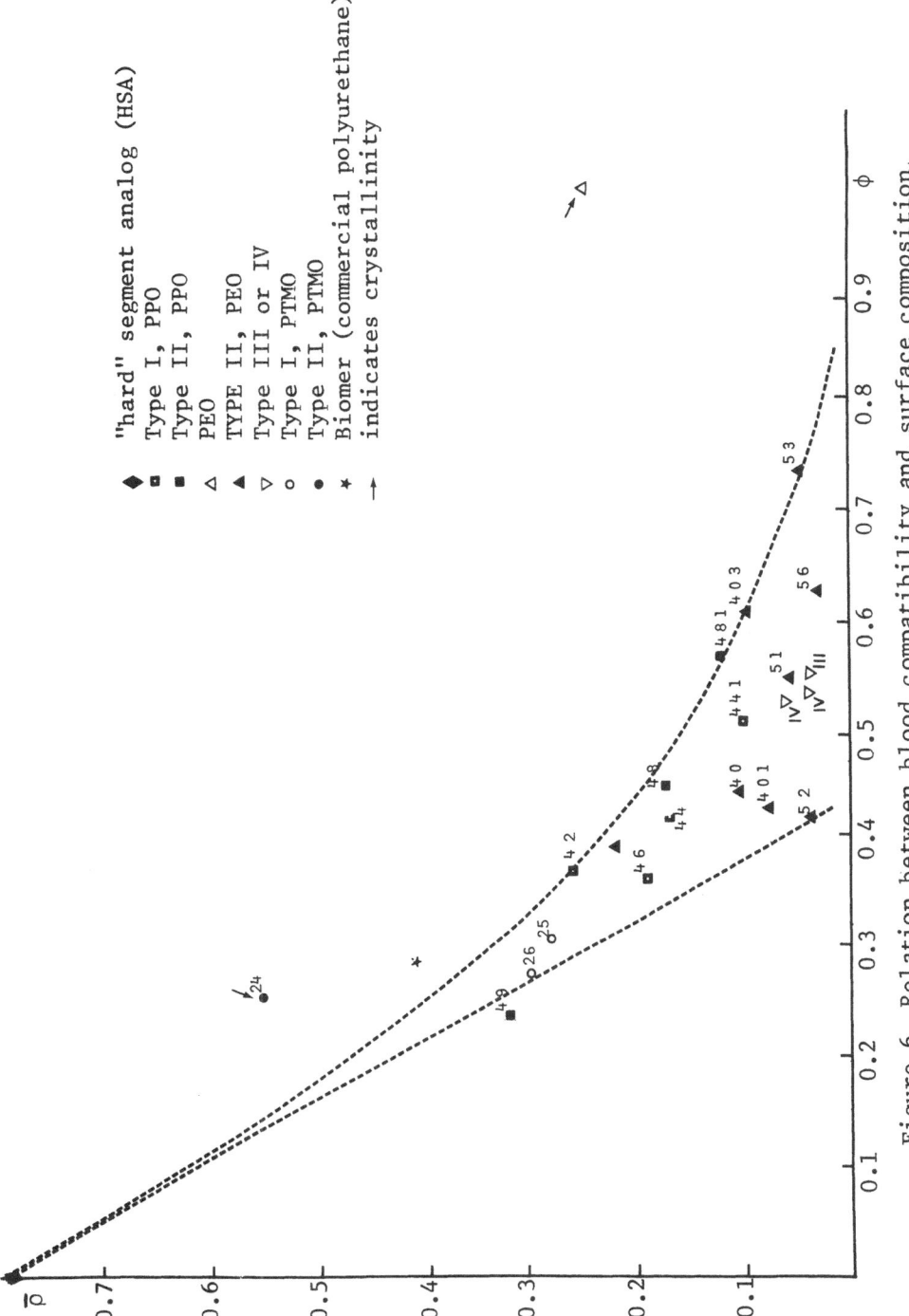

Figure 6. Relation between blood compatibility and surface composition.

than expected on the basis of their ϕ value.

Types III and IV SPU have very low $\bar{\rho}$. It is possible that in these polymers the surface in contact with blood is rich in PEO, or the molecular disorder thereby created is useful.

Parallel tests[4] show that thrombin adsorption is minimal on SPU containing amorphous PEO, greater on the ones with PTMO and very high on "hard" segments analogs.

From this study it can be seen that the good blood compatibility of polyurethanes is no magic of this type of material, but it is due to the "soft" segment part. Our results are in accord with the ones reported by Brash[5,6], Gilding and Reed[7,8] and Furusawa[9]. They do not agree with the results reported by Ratner[10] and Lyman[11]. Ratner's conclusion was that the platelet consumption decreased as the fraction of C_{18} at the surface belonging to C-H (and not C-O-C) increased. For Lyman the fact that when using PPO he obtains a more compatible polymer with PPO 1200 means that there is a certain amount of "hard" segment which is needed at the surface.

From the three ethers tested PEO appears to be the most soft one and it should give non-clotting surfaces whenever rendered into a non dissolving amorphous network.

With this purpose Merrill's research group at MIT is synthesising a new series of SPU with PEO and CHDI (cyclohexane diisocyanate), avoiding in this way the presence of the aromatic ring[12]. By a suitable selection of casting solvents and PEO molecular weight they aim to obtain a SPU capable of segregating basically PEO segments at the surface. The blood compatibility of these surfaces is better than that observed with SPU based on aromatic diisocyanates, suggesting either a better phase separation with CHDI or the fact that aromatic residues at the surface are thrombogenic.

REFERENCES

1. J.N.Lindon, R.Rodvien, D.Brier, R.Greenberg, E.W.Merrill, E.W.Salzman, In vitro assessment of interaction of blood with model surfaces, J.Lab.Clin.Med. 92:904 (1978)
2. S.R.Stopper, Synthesis and characterization of polyurethanes and polyureas in connection with biomedical polymers, M.S. Thesis, MIT (1977)
3. V.Sa da Costa, Study of polyurethanes for biomaterials, Ph.D. Thesis, Dept. of Chemical Engineering, MIT (1979)
4. V.Sa da Costa, D.Brier-Russell, G.Trudel III, D.F.Waugh,

E.W.Salzman, E.W.Merrill, Polyether-polyurethanes surfaces: Thrombin adsorption, platelet adsorption, and ESCA
scanning, J. Colloid Interface Sci. 76:594 (1980)

5. S.J.Whicher, J.L.Brash, Platelet-foreign surface interactions, J.Biomed.Mat.Res. 12:181 (1978)

6. J.L.Brash, S.Uniyal, Dependence of albumin-fibrinigen simple
and competitive adsorption on surface properties of biomaterials, J.Polym.Sci., Polym.Symp. 66:377 (1979)

7. D.K.Gilding, A.M.Reed, Systematic development of polyurethanes for biomedical applications: I.Synthesis, structure
and bulk properties, Trans. 11th Symposium on Biomaterials,
Vol. 3, Society for Biomaterials, San Antonio, Texas

8. A.M.Reed, D.K.Gilding, J.Wilson, M.Johnson, Systematic development of polyurethanes for biomedical applications:
II.Surface and biological properties, Trans. 11th Symposium on Biomaterials, Vol. 3, Society for Biomaterials,
San Antonio, Texas

9. K.Furusawa, Y.Shimura, K.Otobe, K.Atsumi, K.Tsuda, The relation between the properties of the surface and the blood
compatibility of polyurethanes, Konbunshi Ronbunshu
34:309 (1977) |C.A. 87:206455g (1977)|

10. B.D.Ratner, ESCA & SEM studies on polyurethanes for biomedical
applications, ACS Polym.Prep. 21(1):152 (1980)

11. D.J.Lyman, Polymers in medicine. An overview, International
symposium on polymers in Medicine, Porto Cervo, Italy, 1982

12. E.W.Merrill, private communication

SYNTHESIS, STRUCTURAL STUDY AND PRELIMINARY HEMOCOMPATIBILITY TESTS

OF AB AND ABA BLOCK COPOLYMERS WITH POLYVINYL AND POLYPEPTIDE BLOCKS

B. Gallot[+], A. Douy[+], H. Hayany[++] and C. Vigneron[++]

[+]Centre de Biophysique Moléculaire, 45045 Orléans, and
[++]Centre Régional de Transfusion Sanguine, 54500
Vandoeuvre, France

During the last fifteen years a large number of polymers have been examined for biomedical uses and the most popular polymeric materials were probably silicones, polyurethanes and hydrogels. In order to take advantage of our experience about the synthesis and the structure of block copolymers[1,2] we decided, some years ago[3] to synthesize and study more sophisticated polymeric materials owning both good physical properties and a high probability of exhibiting an interesting biocompatibility : we choose block copolymers with polyvinyl and polypeptide blocks. The polyvinyl blocks were selected for their good mechanical properties and the polypeptides for their potential aptitude to improve the compatibility with blood and with different tissues. An additional advantage of the polypeptides was the high variety of their side chains that allows to modify at will the hydrophobicity and the electrical state of the polymer and thus to modify its interactions with blood constituents and with tissues.

In this paper we describe the synthesis and the structure of 5 types of vinyl-peptide block copolymers, namely : polybutadiene-poly(γ-benzyl-L-glutamate) (BG), polybutadiene-poly(ε-carbobenzoxy-L-lysine) (BCK), polystyrene-poly(γ-benzyl-L-glutamate) (SG), polystyrene-poly(ε-carbobenzoxy-L-lysine) (SCK) and poly(γ-benzyl-L-glutamate)-polybutadiene-poly(γ-benzyl-L-glutamate) (GBG), and the results of the hemocompatibility tests performed on some of these polymers.

SYNTHESIS

The synthesis of AB and ABA block copolymers with a polyvinyl and one or two polypeptide blocks is performed in 3 steps. At first the polyvinyl block is obtained by anionic or radical polymerization ; then a primary amine function is introduced at one or at the two ends of the vinyl polymer ; at last the ω aminated polymer PV-NH$_2$ or the α-ω aminated polymer NH$_2$-PV-NH$_2$ is used as a macromolecular initiator for the polymerization of the N-carboxy anhydride (NCA) of the desired amino acid.

Synthesis of Copolymers with a Polyvinyl and a Polypeptide Block

Polymerization of the polyvinyl block, The polybutadiene or polystyrene block is prepared by anionic polymerization in an all glass apparatus, under vacuum, in THF dilute solution (less than 5 %) at low temperature (- 70°C) with cumyl potassium as initiator[4].

Amination of the living end, At the end of the polymerization of the polyvinyl block we carry out a chemical modification of the living end in order to obtain a polymer terminated by a primary amine function. To perform this modification we have followed two different ways : in the first way an intermediary hydroxyl function is involved while in the second way an intermediary carboxyl function is used.

The first way consists in the addition of carbon dioxyde to the THF solution of the living polymer and gives a polymer terminated by a hydroxyl group. Then the addition of phosgen affords a polymer terminated by an oxychloroformyl group. At last the addition of an excess of a primary diamine yields a polymer terminated by a primary amine function. The yield of the amination is good (75 %) and no detectable coupling between macromolecules is observed[5].

The second way consists in addition of carbon dioxyde do the THF solution of the living polymer and gives a polymer terminated by a carboxylic group ; the carboxylation is quantitative[5,6]. Then depending upon the nature of the polyvinyl block two different methods of amination are followed. When the polyvinyl block is a polystyrene one, the addition of sulfonyl chloride to the carboxylated polystyrene gives a polymer terminated by a chloroformyl group, then the addition of a primary diamine yields a polymer terminated by a primary amine function and the yield in aminated polystyrene is higher than 80 %[6]. When the polyvinyl block is a polybutadiene one, the addition to the carboxylated polybutadiene of a primary diamine in presence of a coupling agent (dicyclohexylcarbodiimide = DCCI) gives a polymer terminated by a primary amine function and the yield in aminated polybutadiene is higher than 80 %[5].

Polymerization of the polypeptide block, The polymerization of the NCA of γ-benzyl L glutamate or of ε-carbobenzoxy L lysine initiated by the primary amine function of the aminated polystyrene or polybutadiene is carried out in the absence of moisture, at room temperature in DMF solution in the case of polystyrene[6,7] and in benzene solution in the case of polybutadiene[5].

Copolymers with a Polyvinyl Block and two Polypeptide Blocks

Copolymers with a polyvinyl block and two polypeptide blocks may be obtained by two different ways differing by the type of polymerization used in the synthesis of the polyvinyl block.

In the first way, the polyvinyl block is obtained by anionic polymerization as described before except that the monofunctional initiator (cumylpotassium) is replaced by a bifunctional initiator (dimer dianion of α-methyl-styrene/K)[8]. Then the amination of the two living ends and the synthesis of the polypeptide blocks are performed as described before for AB copolymers.

In the second way an α-ω carboxylated polymer is obtained by radical polymerization of isoprene, butadiene or other monomers[9]. Then the α-ω carboxylated polymer, an α-ω carboxylated polybutadiene (HOOC-PB-COOH) of industrial origin kindly provided by Polyplastic (Rueil-Malmaison, France) in our case, is aminated by addition of ethylene diamine in presence of DCCI in toluene solution[10].

$$HOOC-PB-COOH + H_2N-CH_2-CH_2NH_2 \xrightarrow{DCCI} H_2N-CH_2-CH_2NH-CO-PB-CO-NH-CH_2-CH_2-NH_2$$

and the α-ω aminated polybutadiene is used as a macromolecular bifunctional initiator of the polymerization of the NCA of γ benzyl-L-glutamate or of other amino acids[10].

Characterization of AB and ABA Copolymers

Eventual homopolymers are eliminated and block copolymers are fractionated as described elsewhere[5-7]. Molecular weight of polyvinyl blocks and copolymers are measured by osmometry, viscosimetry and gel permeation chromatography[5-7] and the composition of the copolymers is determined by analysis of the nitrogen content of the copolymer, U.V. spectroscopy and ^{1}H n.m.r. spectroscopy as already described[5-7].

STRUCTURE

We have studied the structure of AB block copolymers polybutadiene-poly(γ-benzyl-L-glutamate) (BG), polybutadiene-poly(ϵ-carbobenzoxy-L-lysine) (BCK), polystyrene-poly(γ-benzyl-L-glutamate) (SG), polystyrene-poly(ϵ-carbobenzoxy-L-lysine)(SCK) with polypeptide compositions between 18 and 84 % and ABA block copolymers poly(γ benzyl L-glutamate)-polybutadiene-poly(γ-benzyl-L-glutamate) (GBG) with polypeptide compositions between 55 and 86 %, by X ray diffraction[11], electronmicroscopy[12], infrared spectroscopy and circular dichroism. We have shown that they exhibit in dioxane concentrated solution (less than about 50 % of solvent) and in the dry state (after evaporation of dioxane at a slow rate) a periodic lamellar structure.

Description of the Structure

The lamellar structure consists of plane, parallel, equidistant sheets ; each elementary sheet of thickness d results from the superposition of two layers ; one of thickness d_A formed by the polyvinyl blocks (polybutadiene or polystyrene) in a more or less random coil conformation ; the other, of thickness d_B formed by the polypeptide chains (polybenzylglutamate or polycarbobenzoxylysine) in an α helix conformation, perpendicular to the interface, assembled in a bidimensional hexagonal array and generally folded (Fig. 1).

The lamellar character of the structure is demonstrated by X-ray diffraction (presence of a set of 3 to 5 sharp lines with Bragg spacings in the ratio 1, 2, 3, 4, 5 in the region of very low angles on X-ray patterns) and is confirmed by electron microscopy in the case of AB and ABA copolymers containing a polybutadiene block. The Fig. 2 gives an example of electron micrographs obtained with ultrathin sections of copolymers stained with osmium tetroxide[8,12] ; the lamellar character of the structure is demonstrated by the presence of parallel stripes alternatively black and white and containing respectively the polybutadiene blocks stained by osmium tetroxide and the polypeptide blocks.

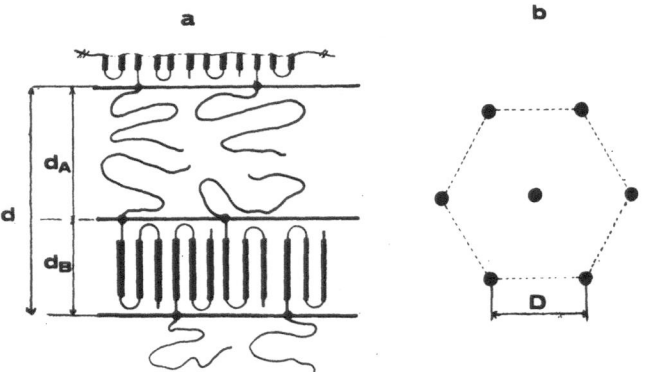

Fig. 1. Schematic representation of the lamellar structure of AB copolymers with a polyvinyl and a polypeptide block[7].
(a) d = intersheet spacing ; d_A = thickness of the layer containing the polyvinyl chains ; d_B = thickness of the layer containing the folded polypeptide helices.
(b) Hexagonal array of the polypeptide helices : D = distance between two helices.

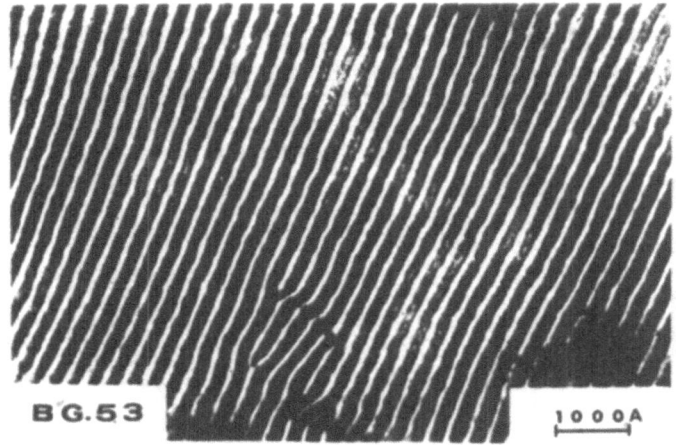

Fig. 2. Electron micrograph of the copolymer polybutadiene-poly(γ-benzyl-L-glutamate BG.53 containing 33 % of polypeptide.

The hexagonal packing of the polypeptide chains is also demonstrated by X ray diffraction (presence of a set of 3 sharp lines with Bragg spacings in the ratio 1, $\sqrt{3}$, $\sqrt{4}$ in the region of low angles on X ray patterns).

The α helix conformation of the polypeptide chains is demonstrated by three techniques : infrared spectroscopy, circular dichroism and X ray diffraction. Infrared spectra exhibit the two characteristic bands of the α helix ; the amide I band at 1655 cm^{-1} and the amide II band at 1545 cm^{-1}. Circular dichroism spectra exhibit the two negative peaks characteristic of the α helix : a broad peak centered at 222 nm and a peak at 209 nm. Low angle X ray patterns provide the parameter D of the hexagonal array of the polypeptide chains ; from D and the molecular chracteristics of the polypeptide block (molecular weight m of the monomer unit and specific volume V_B of the polypeptide block) one obtains the length of the projection on the helix axis of the distance between two monomer units : $h = 2mV_B (\sqrt{3}\ N\ D_o^2)^{-1}$ with N = Avogadro's number ; the value found $h = 1.50 \pm 0.02$ Å is the characteristic value of the α helix.

The folding of the polypeptide chains is deduced from the comparison of their average length $L = hPn$ (Pn = number average degree of polymerization of the polypeptide block) with the thickness d_B of the polypeptide layer. The number of folds ν of the polypeptide chains and the number of times μ that a polypeptide chain crosses the polypeptide layer are given by the formula :

$$\mu = \nu + 1 = L/d_B$$

Influence of the Solvent Concentration

All the five types of copolymer studied BG[3,5], SG[7], BCK[6], SCK[6] and GBK[10] exhibit a similar behaviour versus solvent concentration and the Fig. 3, corresponding to the copolymer SG.14 with a molecular weight of 25,000 for its polystyrene block, containing 47 % of polypeptide and with polybenzyl glutamate chains refolded 2 times, illustrates their behaviour.

When the dioxane concentration increases :
- the total thickness d of a sheet (directly obtained from X ray patterns) increases.
- the thickness d_A of the polyvinyl layer calculated from d[1,5] also increases.

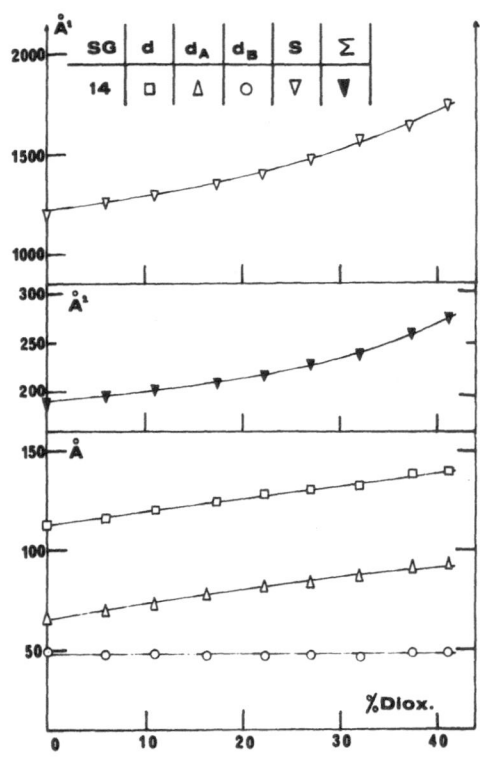

Fig. 3. Variation of the structure parameters of the lamellar structure with dioxane concentration in the case of the copolymer SG.14 containing 47 % of polypeptide. □: d = intersheet spacing : △: d_A = thickness of the polystyrene layer; ○: d_B = thickness of the polypeptide layer ; ▽: S = specific surface ; ▼: Σ = surface per polypeptide helix.

- the thickness d_B of the polypeptide layer, calculated from $d^{1,7}$ remains nearly constant.
- the specific surface S at the interface (calculated from $d^{1,7}$) and the surface Σ ($\Sigma = D^2 \sqrt{3}/2$) occupied by a polypeptide helix both increase, but the ratio S/Σ is independent of the solvent concentration and is equal to 2 μ in agreement with our model of structure.

Factors Governing the Folding of the Polypeptide Chains

The principal factors governing the folding of the polypeptide chains in AB copolymers with a polyvinyl and a polypeptide block and in ABA copolymers with a polyvinyl and 2 polypeptide blocks are the solvent concentration, the molecular weight of the blocks and the nature of the blocks.

The number of folds of the polypeptide block is independent of solvent concentration for the 5 types of copolymers studied BG^5, SG^7, BCK^6, SCK^6 and GBG^{10}.

The number of folds of the polypeptide chains increases with the molecular weight of the polypeptide block[5-7]. For instance, for a set of SG copolymers with a polystyrene molecular weight 25,000, the number of folds of the polypeptide chains increases from 0 to 7 when the molecular weight of the poly(γ-benzyl-L-glutamate) block increases from 11,200 to 131,200 (Table I) and for a set of GBG copolymers with a polybutadiene molecular weight 9,600 the number of folds of the polypeptide block increases from 0 to 1 when the poly (benzyl-L-glutamate) molecular weight increases from 11,700 to 33,500. The number of folds of the polypeptide chains increases with the molecular weight of the polyvinyl block[7].

Table I. Influence of the molecular weight of the polypeptide block on the number of folds of the polypeptide chains in a set of block copolymers SG with a constant molecular of 25,000 for the polystyrene block.

Copolymer	% PG[a]	$\overline{M}$ PG[a]	L (Å)	d_B (Å)	μ	ν
SG.11	31	11200	77	79	0.98	0
SG.12	39	16000	109.5	56	1.96	1
SG.13	41	17400	119	61	1.95	1
SG.14	47	22200	152	49	3.10	2
SG.15	50	25000	171	56	3.05	2
SG.16	58	34500	236.5	78	3.03	2
SG.17	64	44500	304.5	78	3.90	3
SG.18	71	61200	419	82	5.11	4
SG.19	84	131200	899	112	8.03	7

a : PG = poly(γ-benzyl-L-glutamate)

The number of folds of the polypeptide chains depends upon the nature of the polyvinyl block : the polypeptide chains are more folded when they are linked to polystyrene than when they are linked to polybutadiene[7].

The number of folds of the polypeptide chains depends upon the nature of the polypeptide block ; the poly(ε-carbobenzoxy-L-lysine) chains are less folded than the poly(γ-benzyl-L-glutamate) chains[7].

HEMOCOMPATIBILITY TESTS

We have studied two vinyl-peptide block copolymers : a copolymer polystyrene-poly(γ-benzyl-L-glutamate) (SG) and a copolymer polystyrene-poly(L-glutamic acid)[13] (SE) and the corresponding hydrophobic homopolymers : polystyrene (S) and poly(γ-benzyl-L-glutamate) (G). The molecular weight of the polystyrene blocks was 25,000 and the polypeptide content of the copolymers was about 50 %.

Principle of the Study

Samples studied, In the study of polymer hemocompatibility, the first problem to resolve is the choice of the physical and geometrical shape of the sample. We have used to perform our tests copolymer films because they exhibit the following advantages :
    - an easy preparation for any type of starting material,
    - the possibility of studying the film surface before and after blood contact, for instance by scanning electron microscopy,
    - the direct and permanent contact between blood and the material under study at the exclusion of any foreign surface,
    - the possibility of following and controlling a high number of blood parameters,
    - the possibility of performing dynamic studies by submitting the whole system to different movements such as rotation for instance.

Our films were prepared from dilute solution (1 % of polymer) in chloroform. The evaporation of the solvent was carried out at a slow rate at 40°C and the films were directly prepared on the supports used for the various tests. The films were clear, transparent and smooth. Their surface was examined by scanning electron microscopy and was found reproducible from one film to the other.

Tests performed, We studied three types of properties of our polymer films : their hemocompatibility, their influence on the blood coagulation process and their eventual toxicity.

The hemocompatibility of the polymer films was studied at first versus the whole blood and then versus its specific constituents : erythrocytes, leukocytes and platelets.

The influence of the polymer films on the blood coagulation process was studied by thromboelastography.

The eventual toxicity of the polymers was determined by growing living cells on polymer films.

Hemocompatibility Tests

Hemocompatibility tests were performed with polymer films covering the inner surface of 50 ml round bottomed flasks and obtained by slow evaporation, at 40°C, of a 1 % solution of polymer in chloroform using a rotating evaporator.

Citrated fresh blood or its components were put in contact with the films for times between 30 minutes and 4 or 24 or 48 hours. The flasks were thermostated at 37°C and rotated at a constant speed of 20 revolutions / minute that allowed the blood to cover uniformly all the film surface. At regular intervals the amount of hemoglobin liberated by the lysis of the erythrocytes was determined by the benzidin technique, the number of erythrocytes and leukocytes was measured by a Coulter Counter S and the number of platelets was determined with a Technicon Autocounter.

Tests on the whole blood, Citrated fresh blood was used and the behaviour of 4 polymers (a copolymer SG, a copolymer SE, a homopolymer G and a homopolymer S) was compared with the glass behaviour.

Hemoglobin liberated : For all the materials studied, the amount of hemoglobin liberated by the lysis of erythrocytes increases with time, but the homopolymer G and the copolymer SE behave better than the glass while the copolymer SG and the homopolymer S behave worst than the glass. For instance, after 4 hours of contact the films of G and SE release respectively - 8 % and - 5 % hemoglobin than the glass while the films of SG and S release respectively + 6 % and + 12 % of hemoglobin than glass.

Lysis of erythrocytes : For all the materials the number of erythrocytes decreases when the time of contact with blood increases.

The copolymer SG and the glass exhibit a similar behaviour ; the polystyrene S behaves very bad while the copolymer SE and the homopolymer G behave pretty well (Table II).

Lysis of leukocytes : For all the materials the number of leukocytes decreases when the time of contact with blood increases.

The behaviour of the materials improves from glass to copolymer SE, to homopolymer S, to homopolymer G and to copolymer SG (Table II).

Table II. Effect of glass and polymer films on the constituents of citrated fresh blood.[a]

	Glass	S	G	SG	SE
Erythrocytes	0	- 10	+ 9.3	- 2.7	+ 6.8
Lymphocytes	0	+ 0.6	+ 1.3	+ 1.6	+ 0.2
Platelets	0	+ 29	+ 29	+ 27	+ 57

[a]Time of contact between blood and materials 4 hours. Number of blood components expressed in % taking the results obtained with glass as reference. S : polystyrene ; G : poly(γ-benzyl-L-glutamate) ; SG : copolymer polystyrene-poly(γ-benzyl-L-glutamate) ; SE : copolymer polystyrene-poly(glutamic acid).

Lysis of platelets : For all the materials the number of platelets decreases when the time of contact with blood increases.

All the polymers are better than glass ; homopolymers S and G and copolymer SG exhibit a similar behaviour but the copolymer SE is by far the best material (Table II).

Tests on the blood constituents, Suspensions of erythrocytes : For suspensions of different concentrations (5, 10, 20, 30, 40 and 50 %) of washed erythrocytes in physiological water, the amount of hemoglobin liberated by the lysis of the cells was measured after different times of contact with the materials. The classification of the materials is the same for all the concentrations studied. For instance, at a concentration of 40 % that is in the vicinity of a normal hematocrit and after 4 hours of contact, the homopolymer S and the copolymer SG liberate respectively + 9 % and + 7 % of hemoglobin than the glass, while the homopolymer G and the copolymer SE liberate respectively - 14 % and - 13 % of hemoglobin than the glass.

Suspension of leukocytes : For leukocytes concentrates obtained by action of Dextran T.500 on heparinated fresh blood the phagocytosis power and chimiotaction were measured. The following classification was obtained

SG > G > SE   > glass

Plasma rich in platelets (PRP) : For plasma rich in platelets obtained by centrifugation (at 1200 revolutions/minute) of citrated fresh blood, the number of platelets was measured with a Thoma's cell and the aggregability in presence of ADP or collagen, the factor F3P and the time of cephalin kaolin (TCK) were determined.

The copolymer SE was found by far the best material.

Replacement of the glass surface by films of the 4 polymers increases the number of platelets present in the plasma after contact with the materials. After one hour of contact, for instance, the number of platelets was increased by 3 % for the homopolymer S, 18 % for the copolymer SG, 19 % for the homopolymer G and 75 % for the copolymer SE. Furthermore, this classification was confirmed by the determination of the aggregability, the factor F3P and the time of cephalin kaolin.

Thromboelastography

In order to determine the effect of the polymers on the blood coagulation process we used the thromboelastography that allow a full and easy preliminary study of the different phases of the coagulation.

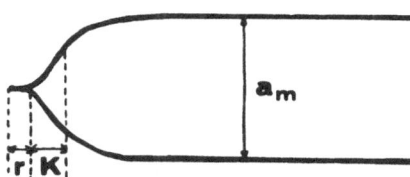

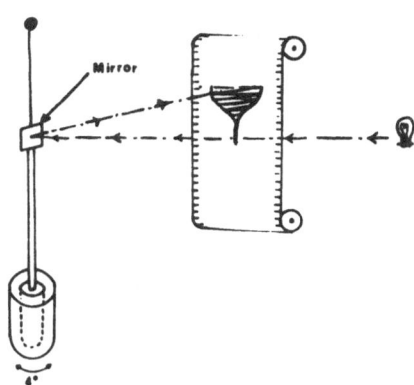

Fig. 4.   Schematic representation of a thromboelastography apparatus, and example of thromboelastogram obtained.

The thromboelastography apparatus is sketched in Fig. 4. It consists in a cylinder covered with a film of the material under study and containing the blood. In the cylinder is immersed a piston suspended by a torsion wire on which is fixed a plane mirror that reflects a light beam on a photo-sensitive paper. The external cylinder swings slowly with an amplitude of 4°. When the blood begins to coagulate the piston swings and provides a diapazon shape curve (Fig. 4).

The principal parameters of a thromboelastogram are :

- the time r of thromboplastinoformation and thrombinoformation, i.e. the time before the beginning of the formation of the clot,
- the time K of the fibrinoformation that represents the dynamic of the beginning of the formation of the clot under the influence of platelets.
- the maximum amplitude a_m that corresponds to the maximum width of the curve.

To characterize a material one uses the index of thrombodynamic potential (ITP) :

$$ITP = \frac{E_m}{K} \quad \text{with} \quad E_m = \frac{100a_m}{100-a_m}$$

and usually the ITP of normal blood is situated between 6 and 12.

Among our polymers the best material is the copolymer SE with an ITP of 7, then come the homopolymer G with 8, the copolymer SG with 10 and the homopolymer S with 11.

Cell Cultures

In order to determine the possible toxicity of the polymers and their compatibility with diversified living cells, four types of cells were grown on polymer films sterilized at 120°C in the autoclave during 2 hours.

Human fibroblasts, Human fibroblasts obtained from fragments of fresh skin grow on all the 4 polymers studied ; but the most favourable polymer is the homopolymer G, then comes the copolymer SE, the glass and the copolymer SG.

Diploid cells, Culture of the diploid cells gave the following classification :

glass > SE > G > SG

Cells from monkey kidney, Culture of monkey Kidney cells gave the following classification :

$$glass > SE = G > SG$$

Cells in unbroken lineage, Culture of cells in unbroken lineage gave the following classification :

$$glass > SE > G >> SG$$

CONCLUDING REMARKS

From the present study it results that the addition of a peptide block to a vinyl polymer improves the hemocompatibility of the polymer. Nevertheless before reaching a definitive conclusion further studies have to be performed. Vinyl-peptide copolymers exhibit a phase separation between the two blocks and the degree of phase separation so the composition of the film surface may depend upon the nature of the solvent used during the preparation of the films. Therefore it will be necessary to study the influence of the nature of the solvent on the hemocompatibility of the films and to correlate it with the degree of phase separation determined by X ray diffraction, with the geometrical state of the film surface determined by electron microscopy and with the composition of the film surface determined by ESCA.

Another interesting result of the present study is that the classification of the polypeptides used in our polymers varies with the nature of the blood constituents involved in the hemocompatibility tests. The best polymer is polybenzyglutamate (G) for erythrocytes and leukocytes but is copolymer polystyrene-polyglutamic acid (SE) for platelets. So a copolymer containing both benzyl glutamate and glutamic acid groups in its polypeptide block will probably exhibit the best hemocompatibility. Such copolymers have already been synthesized and their hemocompatibility will be studied in the near future.

REFERENCES

1. B. Gallot, Preparation and study of block copolymers with ordered structure, Adv. Polym. Sci. 29:87 (1978).
2. B. Gallot, Liquid crystalline structure of block copolymers, in: "Liquid Crystalline Order in Polymers", A. Blumstein,ed. Academic Press, New York (1978) p. 191.
3. B. Perly, A. Douy and B. Gallot, Synthèse de copolymères séquencés polybutadiene-poly(L-glutamate de benzyle) et étude structurale de leurs mésophases, C.R. Acad. Sci. Paris C279:1109 (1974).
4. A. Douy, G. Jouan and B. Gallot, Synthèse par polymérisation anionique de copolymères biséquencés polybutadiene-poly(vinyl-2-naphtalene). Etude de leur structure en solution concentrée avant et après polymérisation du solvant, Makromol. Chem. 177:2945 (1976).
5. B. Perly, A. Douy and B. Gallot, Block copolymers polybutadiene-

poly(benzyl-L-glutamate) and polybutadiene-poly(N^5-hydroxypro-pylglutamine). Preparation and structural study by X-ray and electron microscopy, Makromol. Chem. 177:2569 (1976).

6. J.P. Billot, A. Douy and B. Gallot, Preparation, fractionation and structure of block copolymers polystyrene-poly(carbobenzo-xy-L-lysine) and polybutadiene-poly(carbobenzoxy-L-lysine), Makromol. Chem. 178:1641 (1977).

7. A. Douy and B. Gallot, Block copolymers with a polyvinyl and a polypeptide block : factors governing the folding of the poly-peptide chains, Polymer 23:1039 (1982).

8. A. Douy and B. Gallot, Polybutadiene-polystyrene-polybutadiene block copolymers. Study of organized structures by X-ray dif-fraction, electron microscopy and differential scanning calo-rimetry, Makromol. Chem. 156:81 (1972).

9. S.F. Reed, Telechelic diene prepolymers : 2 carboxyl-terminated polydienes, J. Polym. Sci. A.1, 9:2147 (1971).

10. A. Douy and B. Gallot, unpublished results.

11. A. Douy, R. Mayer, J. Rossi and B. Gallot, Structure of liquid crystalline phases from amorphous block copolymers, Mol. Cryst. Liq. Cryst. 7:103 (1969).

12. A. Douy and B. Gallot, Study of liquid crystalline structures of polystyrene-polybutadiene block copolymers by small angle X ray scattering and electron microscopy, Mol. Cryst. Liq. Cryst. 14:191 (1971).

13. B. Gallot and A. Douy, Copolymères peptidiques amphipatiques, leur obtention et leur application comme émulsifiants, French Patent n° 82 15 975 (1982).

INTERACTIONS OF SOME PLASMATIC PROTEINS WITH ANTICOAGULANT POLY-
STYRENE RESINS: MECHANISM OF CATALYTIC ACTIVITY TOWARDS THE THROMBIN-
ANTITHROMBIN REACTION

Christine Fougnot[*], Marcel Jozefowicz[*] and
Robert D. Rosenberg[**]
[*]Laboratoire de Recherches sur les Macromolécules, C.S.P.
 Université Paris-Nord, 93430 Villetaneuse, France
[**]Harvard Medical School, MIT, Boston, Mass, USA

INTRODUCTION

Polymeric materials having antithrombogenic activity are very
important and their development is expected in the field of artificial organs such as the artificial vessel or the devices for extracorporeal circulation. In previous papers[1,2] we described that the
binding of sulfonate and amino acid sulfamide groups onto crosslinked polystyrene endows these materials with antithrombic activity
which requires the presence of a plasma cofactor, antithrombin III[1].
These insoluble materials operate as catalysts of the inactivation
of thrombin by its inhibitor as does soluble heparin[3]. The catalytic
effect of this mucopolysaccharide was demonstrated to require the
formation of complexes between heparin and either antithrombin III
or thrombin or both[4,5,6].

We now report results concerning the interactions between these
two proteins and two of these polystyrene derivatives : sulfonated
polystyrene (PSSO3) and sulfonate-glutamic acid sulfonamide polystyrene (PSSO2Glu). The adsorption of each protein was studied either
in purified system or in plasma and the binding constants were
determined. Moreover, the catalysed formation of thrombin-antithrombin III complex was examined and compared with the antithrombic
activity measured through thrombin clotting times of plasma preincubated with these polymers.

MATERIALS AND METHODS

Preparation of Polystyrene Derivatives

Sulfonated polystyrene (PSSO3) was prepared by quantitative

261

hydrolysis of chlorosulfonated polystyrene. Sulfonate-glutamic acid
sulfonamide polystyrene (PSSO$_2$Glu) was prepared by reacting glutamic
acid sodium salt with chlorosulfonated polystyrene as previously
described[1].

The chemical composition of these resins was determined by
elemental analysis PSSO$_3$: 4.2 meq/g of –SO$_3$ Na group   ; PSSO$_2$Glu :
1.02 meq/g of –SO$_3$Na and 2.23 meq/g of –SO$_2$Glu group.

The beads were crushed in order to increase their specific
surface[1] and the final particle average size, determined by quantitative microscopy was 25µm.

Proteins and Chemicals

Human thrombin and human antithrombin III were isolated in physically homogeneous form by methods previously reported from
R.D. Rosenberg Laboratory[7].

Polybrene was supplied by the Aldrich Chemical Co.

Measurements of Proteins Concentrations

Thrombin concentrations were measured using amidolytic activity
of the enzyme towards a chromogenic substrate (S2238).

Antithrombin III (AT) concentrations were measured using the
thrombin-antithrombin inhibition reaction[7] or a specific radioimmunoassay.
Thrombin-antithrombin complex (TAT) concentrations were measured
using a specific radioimmunoassay[8].

Antithrombin III Adsorption

The polymer suspension (in 0.15 M NaCl in 0.01 M Tris-HCl,
pH 7.5) was mixed (v/v) in a polystyrene tube with either a solution
of purified AT or fresh platelet poor plasma (PPP) diluted (v/v)
with a solution of purified AT. After 15 min. at room temperature,
the mixture was centrifuged and the AT concentration was measured
in the supernatant. The adsorption on the polymer surface was calculated by difference between a control performed with buffer instead
of polymer suspension and the protein concentration found in the
supernatant.

Thrombin Adsorption

The adsorption was performed under the conditions of AT
adsorption. Purified thrombin was modified – by alkylphosphorylation
of the active serine site – before it was mixed with PPP. After the
centrifugation, the supernatant was removed and the polymer was

washed twice with Tris-buffer in order to eliminate all the non-
adsorbed protein. Then a polybrene solution (50 mg/ml in tris-
buffer) was added up to the initial volume. The mixture was mixed
for 15 min at room temperature and then centrifuged. The amount of
thrombin adsorbed on the polymer surface was deduced from the protease concentration found in the last supernatant.

Thrombin Clotting Times

100 µl of polymer suspension (various contents) were incubated
with 200 µl of PPP during 15 min. at 37°C. Then 100 µl of thrombin
solution (10 µg/ml) were added and then the clotting time was
measured. Control times were performed with buffer instead of polymer
suspension.

Catalysis of the Thrombin-Antithrombin Complex (TAT) Formation

50 µl of polymer suspension (various concentrations) were
incubated in a polystyrene tube with 100 µl of either a solution
of purified AT (150 µg/ml) or fresh platelet poor plasma for 15 min.
at room temperature. Then 10 µl of a solution of purified thrombin
(30 µg/ml) were added and the mixture was vortexed for exactly 10
seconds. The reaction was stopped by addition of 50 µl of a hirudin
solution (25 u/ml). After centrifugation, the TAT level was measured
in the supernatant.

The level of non-catalysed reaction was determined by replacing
the polymer suspension by the same volume of buffer. The maximum
of TAT which can be generated under the experimental conditions was
determined with buffer but, in this case, the reaction between
thrombin and AT was allowed to occur for at least 15 min. before
the addition of hirudin. All the results obtained in the presence
of polymer were expressed in percent of this maximum value.

RESULTS

Adsorption Studies

Antithrombin III (AT) adsorption was studied using different
concentrations of both materials and a range of AT initial concentrations varying from 0 to 1.5 mg/ml in purified system or 0 to
1.0 mg/ml in diluted plasma system. In the second case, the mixture
of PPP with the concentrated purified AT was made prior to use and
was the same for all the polymer concentrations. After incubation
between the insoluble material and the AT solution, the suspension
was centrifuged and the concentration of AT in the supernatant was
compared to a control performed under the same conditions without
polymer.

The Fig. 1 shows the amounts of AT adsorbed on 100 mg of

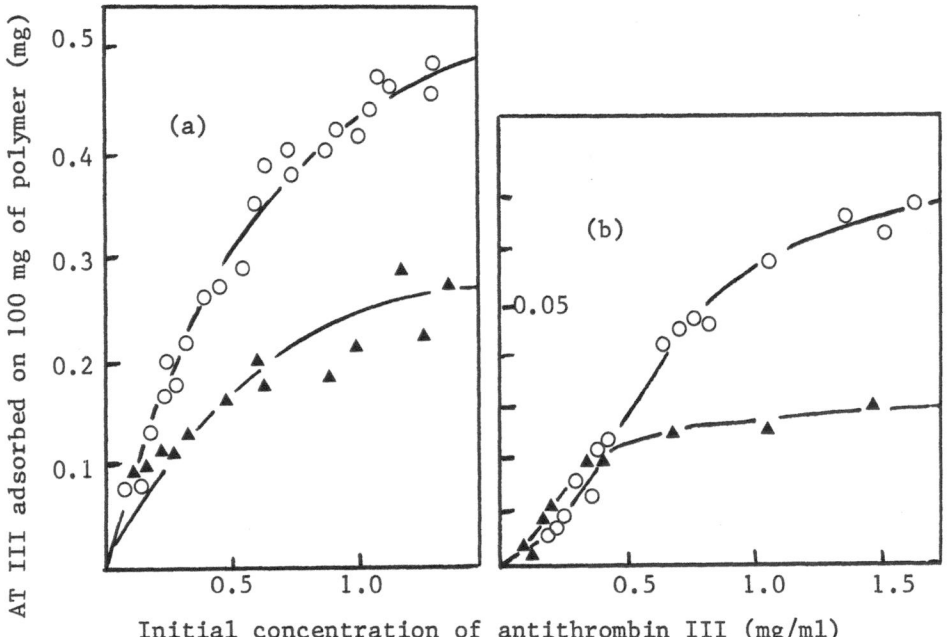

Fig. 1. Antithrombin III adsorption on PSSO₂Glu O or PSSO₃ ▲
(a) purified system -(b) plasma system.

insoluble material ; Fig.1a depicts the results obtained when AT
is the only protein present in the mixture ; Fig. 1b shows the AT
adsorption observed when the other plasma proteins may compete with
the antiprotease.

The saturation, which is virtually proportional to the amount
of polymer, is observed with approximately the same initial AT
concentration for both materials ; but the amount of AT adsorbed
on $PSSO_2Glu$ is about twice the amount of AT adsorbed on the same
surface area of $PSSO_3$ although the total substituent content is
higher in $PSSO_3$ than in $PSSO_2Glu$[9]. Furthermore the saturations
determined in plasma system are approximately 8 times lower than
the corresponding values obtained in purified system. This decrease
was demonstrated to be essentially due to albumin which competes
with AT for the adsorption on the polymer surface[10].

Thrombin adsorption was first measured with purified thrombin.
In order to study its interaction with the polymers in plasma the
thrombin molecule was blocked with a diisopropylphosphoryl group
on the active serine site (DIP-thrombin). The adsorption of this

modified thrombin was studied either in purified system or in
diluted plasma system.

    The amounts of thrombin adsorbed on 10 mg of insoluble material
are reported in Fig. 2. In all cases, they were determined by
desorption from the surface[9].

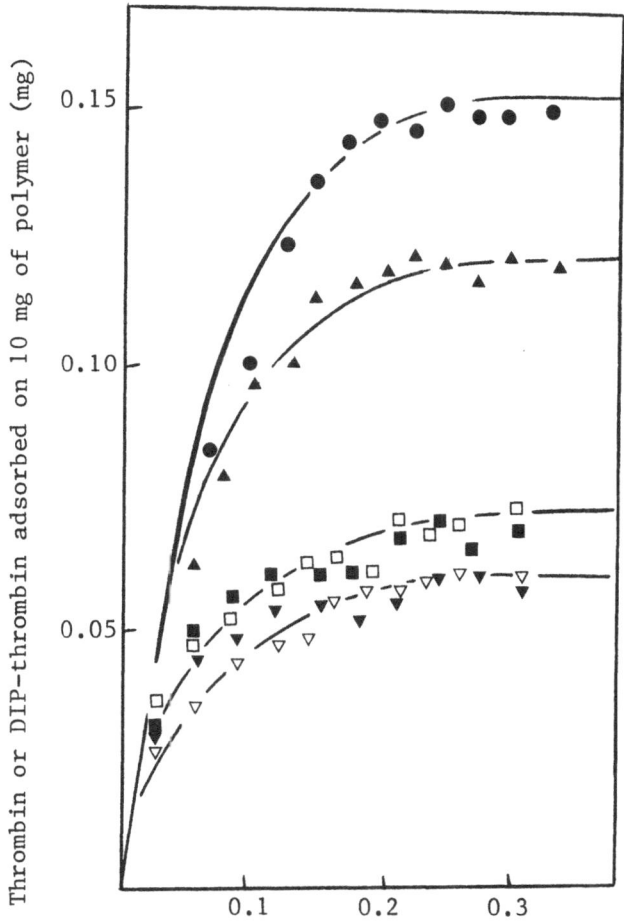

Fig. 2 : Thrombin adsorption on PSSO$_2$Glu ● ■ □ or PSSO$_3$ ▲ ▼ ▽
         Normal thrombin in purified system ● ▲ . DIP-thrombin in
         purified system ■ ▼ or in plasma system □ ▽ .

    The thrombin adsorption was studied using a range of protease
concentrations varying from 0 to 0.35 mg/ml. The curves show that

at low thrombin concentration, all the protein is adsorbed on the
polymer. The quantity adsorbed at saturation is slightly higher for
$PSSO_2Glu$ than for $PSSO_3$ and the ratio between the two plateau values
is identical to the ratio of the negative charges located on the
surface. When thrombin is modified, its adsorption is approximately
divided by a factor 2. The two thrombin molecules (modified and non-
modified were demonstrated to interact with the same polymer sites[10]
and then DIP-thrombin can be used to study the interaction between
the protease and the surface in diluted plasma. In contrast with the
AT adsorption, the presence of the other plasmatic proteins does not
affect the thrombin adsorption (Fig. 2)

The adsorption of the thrombin-antithrombin (TAT) complex on
the polymer surface was also examined in plasma. The study was
conducted as the AT adsorption was and the results are reported in
Fig. 3 for normal plasma and defibrinated plasma. No significant
differences can be seen between the two sets of experiments which

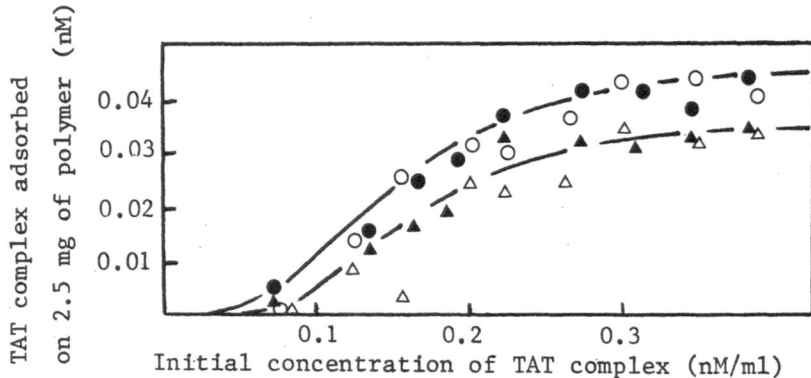

Fig. 3 : Thrombin-antithrombin complex adsorption on $PSSO_2Glu$ ● ○
or $PSSO_3$ ▲ △ . Normal plasma ● ▲ . Defibrinated plasma ○ △.

evidences that fibrinogen, contrary to albumin, does not compete
with the TAT complex for the adsorption on the polymer surface.

All the adsorption curves were analysed with respect to the
most common adsorption isotherms used in chemisorption[9]. The AT,
thrombin and DIP-thrombin adsorptions seem to correspond respectively to a Tempkin isotherm (AT) or a Langmiur isotherm (thrombin).
Then the equilibrium between the surface, the protein in solution
and the adsorbed protein can be caracterised with an affinity
constant. These values are reported in Table I.

Table 1. Affinities $(M/1)^{-1}$ of different proteins for
         polystyrene derivatives : $PSSO_3$ - $PSSO_2Glu$

	$PSSO_3$	$PSSO_2Glu$
Thrombin	$1.0.10^7$	$1.8.10^7$
DIP-thrombin	$2.9.10^6$	$3.3.10^6$
TAT complex	$2.5.10^6$	$2.8.10^6$
AT III	$3.2.10^5$	$3.2.10^5$
Albumin	5.10^4	5.10^4

The affinity of albumin can be evaluated from the results of
AT adsorption obtained in plasma system. This calculation assumes
that the adsorption of the proteins on the surface is reversible.
Then, in the case of albumin, only an apparent affinity can be
evaluated.

Finally a rough evalutation of the TAT affinity was also done
assuming that TAT and albumin compete for the adsorption on the
surface in plasma system.

All these values are of the same order of magnitude for both
materials and the affinities of the surface for these proteins
increase according to the following sequence : $k_{Alb} < k_{AT} < k_{TAT} < k_T$.

Inactivation of thrombin in plasma

The antithrombic activity of the polymers was first compared
through the thrombin clotting times of platelet poor plasma (PPP)
incubated with various amounts of suspended resin. The results were
translated into the amount of inactivated thrombin as previously
described[1] and are reported in Fig. 4. These curves show that the
inactivation of the same thrombin activity in plasma requires
approximately 3 times more $PSSO_3$ than $PSSO_2Glu$.

In previous papers[1,3] the antithrombic activity of these
polymers was demonstrated to involve one (or several) plasma cofactor as for instance antithrombin III. The catalytic effect of the
polymers on the thrombin-AT reaction was evaluated by comparison
of the TAT amounts generated during 10 seconds in presence or
absence of polymer. In each case, the polymer suspension - or
buffer - was first incubated with fresh platelet poor plasma. Then
thrombin was introduced and after 10 seconds the reaction was
blocked by addition of hirudin. The Fig. 5 shows a typical variation
of TAT generation as a function of the amount of polymer present in
the test. The TAT levels were measured in the supernatant bbtained
by centrifugation of the final suspension and the results are
expressed in percent of the maximum of TAT complex which can be

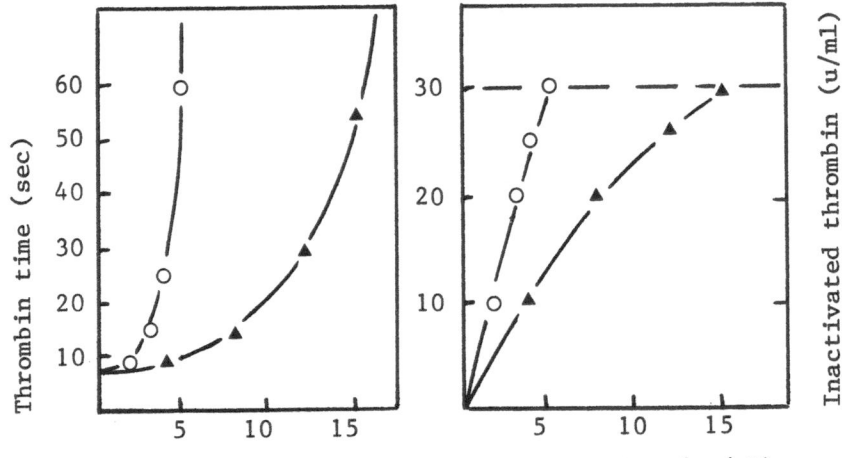

Fig. 4. Antithrombic activity measured in plasma. PSSO$_2$Glu O —
PSSO$_3$ ▲ .

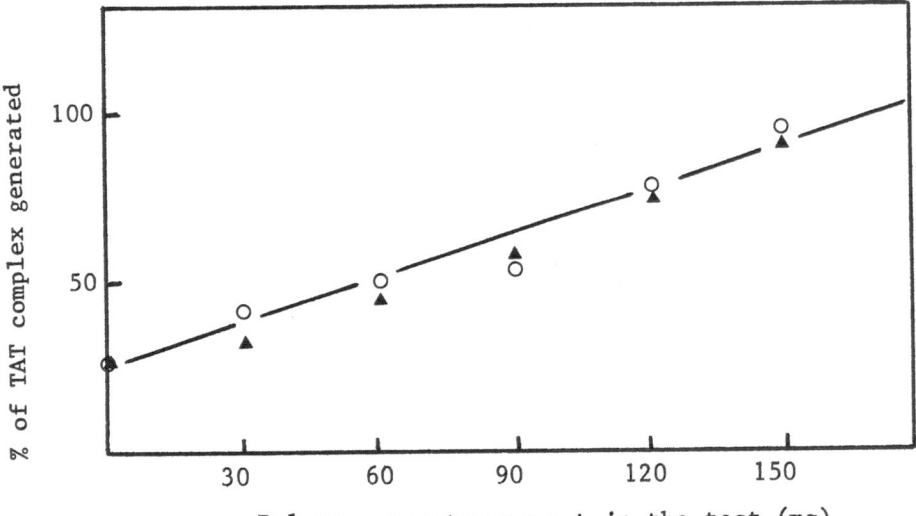

Fig. 5. Thrombin-antithrombin complex generation in plasma —
PSSO$_2$Glu O – PSSO$_3$ ▲ .

generated under the experimental conditions[11]. These results

clearly evidence that these insoluble polymers catalyse the inhibition of thrombin by antithrombin III as soluble heparin. However,
the catalytic effect evaluated through the generation of TAT complex
is quantitatively identical for both polymers in contrast with the
total inhibition of thrombin in plasma determined through the
thrombin clotting times.

DISCUSSION

   The different results of adsorption show that the characteristics of the interaction surface-protein depends on the protein.

   Antithrombin competes with albumin for the adsorption on the
polymer sites but cannot be desorbed by addition of a polycationic
compound. Moreover the theoretical isotherm which fits the experimental results is the Tempkin isotherm which means that the
variation of heat of adsorption with coverage can be interpreted
in terms of surface heterogeneity. These results suggest that AT
may interact simultaneously with several polymer sites including
or not the negatively charged functions.

   In contrast, thrombin does not compete with albumin but is
quantitatively desorbed by polybrene. Moreover the quantities of
thrombin adsorbed at saturation on both polymers are in the same
ratio as the total negative charges borne by the polymers backbones.
These results suggest that the protease interacts with the sulfonate
and carboxyl functions which are nearly equivalent with respect to
this interaction. This conclusion is consistent with the fact that
thrombin adsorption obeys the Langmuir isotherm which means that
the polymer surface is homogeneous and the heat of adsorption is
independent of the surface coverage.

   The comparison of the polymer affinities for the different
proteins and the TAT complex might suggest a possible mechanism
for the catalysis of the TAT generation. When incubated with the
polymer, thrombin and AT adsorb on the surface and a fast reaction
between the adsorbed species occurs. The TAT complex generated on
the insoluble material may be desorbed either by thrombin which
has a higher affinity than TAT for the polymer or by albumin which
has a lower affinity but is in much larger concentration than TAT
complex in plasma. Therefore a catalytic effect can be observed.

   The efficiency of $PSSO_3$ and $PSSO_2Glu$ to accelerate the TAT
complex generation in plasma is about the same. In contrast, when
thrombin clotting times are measured on plasma preincubated with
these polymers $PSSO_2Glu$ has approximately 3 times the activity
of $PSSO_3$. This difference between the total inhibition of thrombin
and its neutralization by antithrombin suggests that some polymers
of this series - as for instance $PSSO_2Glu$ - may also accelerate

another inhibition reaction of thrombin in plasma.

REFERENCES

1. C. Fougnot, J. Jozefonvicz, M. Samama and L. Bara, New heparin-like insoluble materials : Part I, Ann. Biomed. Eng. 7 : 429 (1979).
2. C. Fougnot, M. Jozefowicz, M. Samama and L. Bara, New-heparin-like insoluble materials : Part II, Ann. Biomed. Eng. 7 : 441 (1979).
3. C. Fougnot, M. Jozefowicz, M. Samama and L. Bara, Interactions of anticoagulant insoluble modified polystyrene resins with plasmatic proteins, Thromb. Res., in press.
4. R.D. Rosenberg and P.S. Damus, The purification and mechanism of action of human antithrombin-heparin cofactor, J. Biol. Chem., 248 : 6490 (1973).
5. R. Machovich and P. Aranyi, Effect of heparin on thrombin inactivation by antithrombin III, Biochem. J., 173 : 869 (1978).
6. E. Holmer, G. Söderström and L.O. Andersson, Studies on the mechanism of the rate - enhancing effect of heparin on the thrombin - antithrombin III reaction, Eur. J. Biochem., 93 : 1 (1979).
7. L.H. Lam, J.E. Silbert and R.D. Rosenberg, The separation of active and inactive forms of heparin, Biochem. Biophys. Res. Com., 69 : 570 (1976).
8. H.K. Lau and R.D. Rosenberg, The inactivation and characterization of a specific antibody population directed against the thrombin-antithrombin complex, J. Biol. Chem., 255 : 5885 (1980).
9. C. Fougnot, R.D. Rosenberg and M. Jozefowicz, Affinity of puri-fied thrombin or antithrombin III for two insoluble anticoagu-lant polystyrene derivatives, Biomat., submitted.
10. C. Fougnot, M. Jozefowicz and R.D. Rosenberg, Adsorption of thrombin or antithrombin III on two insoluble anticoagulant polystyrene derivatives. Competition with the other plasmatic proteins, Biomat., submitted.
11. C. Fougnot, M. Jozefowicz and R.D. Rosenberg, Catalysis of the generation of thrombin-antithrombin complex by insoluble anti-coagulant polystyrene derivatives, Biomat., submitted.

CHARACTERIZATION AND ALBUMIN ADSORPTION ON SURFACE OXIDIZED POLYETHYLENE FILMS

Adam Baszkin & Marie Martine Boissonnade

Physico-Chimie des Surfaces et des Membranes, Equipe
de Recherche du CNRS associée à l'Université Paris V,
UER Biomédicale, 45 rue des Saints-Pères, 75270 Paris
cedex 06, France

INTRODUCTION

One of the initial events occuring as blood comes in contact
with a polymer is the adsorption of a protein layer at the blood-
material interface [1-6]. This layer modifies the original surface
and has an important influence on subsequent phenomena such as
platelet adhesion and blood coagulation [1-7].

Studies on the adhesion of platelets to polymer surfaces precoated with a variety of proteins have indicated that albuminated
surfaces appear to prevent platelet adhesion and confer non-throm-
bogenicity to the base polymer. In contrast surfaces precoated with
fibrinogen or γ-globulin show increased platelet adhesion and
release reactions leading to thrombosis [8,9]. The selectivity of
protein adsorption appears to be dictated by the chemical and physical structure of the polymer surface [1,10]. It has also been shown
by means of the electron microscopy examination of 2-hydroxyethyl
methacrylate (HEMA) and styrene block copolymers with adsorbed
plasma proteins, that albumin selectively adsorbs on hydrophilic
domains composed of HEMA and is absent in hydrophobic domains where
γ-globulin is adsorbed [11]. This finding is important because a
normal vascular endothelium, which is a perfect non-thrombogenic
material, was reported to have separated microphase structure composed of hydrophilic and hydrophobic microdomains. The purpose of
the present paper is to report on albumin adsorption on chemically
oxidized polyethylene films. The most significant aspect of the
system studied by us is a relatively simple method of transformation,
by oxidation, of the non-polar hydrophobic polyethylene surface into
a polar one. Another advantage of the system is that chemical

271

treatment does not modify the bulk properties of the polymer
influencing only the short-range interactions at the solid-liquid
interface. Also, the amount of groups supplied to the surface of
the polymer can be varied and quantitatively determined.

EXPERIMENTAL

Oxidation

Low density polyethylene film (PE), known under the commercial
name "Polyane" and produced by "La Cellophane" was used in this
work. Its thickness was 100 μm and its density was 0.929 g/cm^3. It
melts between 109-112°C. The PE samples were extracted with acetone,
dried under reduced pressure and oxidized by immersion in the mixture of sulfuric acid (sq.gr. 1.84 pure grade) and potassium
chlorate (reagent grade). The oxidation was performed at room temperature and the percentage of potassium chlorate in the mixture
for each set of experiments was varied. Thus, different degrees of
oxidation could be obtained. After oxidation PE samples are rinsed
several times with triple distilled water and dried in a desicator
under reduced pressure.

The IR and ESR [Mn(II) probes] spectroscopies showed the presence of carbonyl (C = 0) and olefinic bonds (— C = C —) in the
surface layer of the oxidized PE [12]. The presence of these carbonyl
groups accounts for adsorption of calcium (Ca^{2+}) ions on the surfaces of oxidized PE. They increase the wettability and self
adhesion of the oxidized PE samples [13].

Adsorption of calcium ions (Ca^{2+}) on oxidized PE surfaces

To determine the surface density of polar sites created on
PE surfaces during oxidation a radiotracer method elaborated in
laboratory was used [14]. The adsorption of ^{45}Ca^{2+} ions on oxidized
PE/solution interface was performed from an aqueous solution of
calcium chloride. Plotting the concentration of Ca^{2+} ions in the
solution against the measured radioactivity of adsorbed ^{45}Ca^{2+} ions,
it was possible, for each degree of oxidation, to establish an
adsorption isotherm. The measured radioactivity is then converted
into the surface density of polar sites (sites/cm^2). The obtained
isotherms were Langmuir Type I isotherms with the plateau value
attained at ≃ 3.10^{-5} M CaCl$_2$ concentration in solution.

Cleaning of polymer samples for protein adsorption measurements

Polymer films were rinsed several times with absolute ethanol
and extracted during 12 hours with acetone. Then, they were dried
and immersed in distilled water. The leaching out of different
polymer constituents decrease the surface tension of water. Water
was changed several times until its constant surface tension value

(72.1 nM/m) was found. Such surfaces were considered clean.

Adsorption of bovine serum albumin (BSA) on PE samples

Bovine serum albumin (BSA) used in this study was obtained from Sigma. This material, crystallized and lyophilized, was essentially globulin free and was used without further purification.

Labelling of BSA with ^{125}I was performed according to the method previously described [15]. Levels of iodination were less than 1 atom of iodine per molecule of protein. BSA adsorption on PE films was studied from NaH_2PO_4 - NaOH buffer solution at pH = 7.5 with $[Na^+] = 5.10^{-2}$ M. In the experiments of adsorption at different pH, potassium phosphate buffer and potassium hydrogen phtalate buffer with a constant $[K^+] = 9.4.10^{-4}$ M, were used.

In adsorption experiments at pH = 7.5, where the ionic strength has been varied by addition of NaCl, the constant pH was maintained adding NaOH to the solutions.

The adsorption experiments were carried out according to the following two distinct methods :
   a) The circular PE films were hung by a stainless wire in a glass chamber filled with phosphate buffer solution and degassed under vacuum with constant stirring. Then while polymer films were still submerged in the buffer solution, 25 ml of a new buffer solution, containing a given amount of unlabelled protein and about 20 µl of labelled ^{125}I albumin was added. The protein solution was gently mixed 5 minutes with a magnetic stirrer to ensure homogeneity. After 5 minutes a 1 ml sample of the solution was taken and placed in a γ-counter vial. The adsorption experiment was continued for 3 hr to ensure a plateau value of protein adsorption [15]. Each film was then rinsed with 200 ml of the buffer solution (flow rate 0.5 l/min) and placed in a γ-counter vial for activity measurements. The amount of albumin in the initial 1 ml of solution is known and its radioactivity in counts per time unity can then be used to calculate the amount of protein adsorbed on a PE sample.
   b) Protein denaturation may occur at the air-solution interface [16,17]. In order to avoid this interface to contact the PE surface during its withdrawal (Langmuir-Blodgett transfer) from the solution for rinsing, as in procedure (a), two main modifications to the experimental set-up were introduced.

The concentrated protein solution was supplied to the adsorption vessel by a central tube situated at the bottom of the vessel, well below the liquid-air interface. After the protein adsorption process was finished, the same tube has been used to introduce the buffer solution in order to rinse the samples.

To evacuate the protein solution after adsorption a lateral
tube connected with a pump was used. This tube is located in the
upper part of the apparatus (about 1 cm from the solution/air interface). When rinsing solution, entering with a constant flow by a
central tube, mounted to the level where the lateral tube was fixed,
the protein solution started to be evacuated. Rinsing of the samples
was stopped when radioactivity of solution in the vessel fell to
zero (see Fig. 1).

The experiments performed according to both procedures gave
similar results within of an experimental error ± 10%.

Contact angle measurements

Contact angles on unoxidized and oxidized PE films were measured by the method of Andrade et al [18]. The films were immersed in
triple distilled water for 24 hr prior to the contact angle

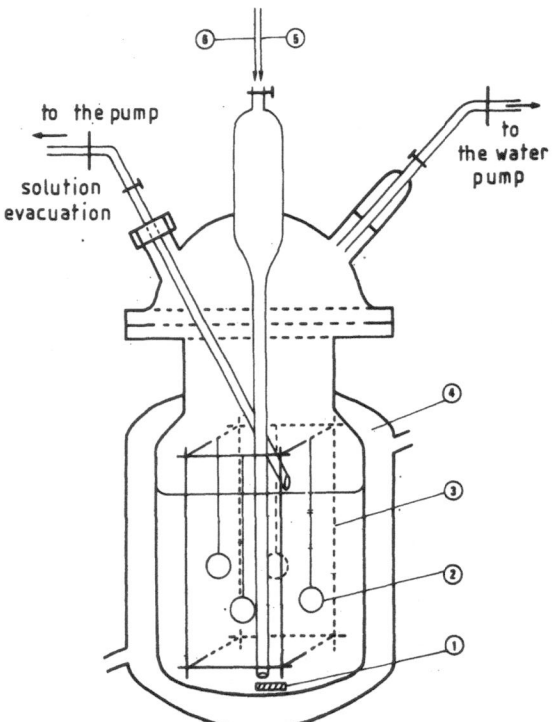

Fig. 1.   Experimental set-up for protein adsorption. (1) magnetic
          stirrer ; (2) sample holder ; (3) samples of polymer ;
          (4) thermostated double wall glass chamber ; (5) and (6)
          entrance of concentrated protein solution or entrance of
          rinsing solution.

measurements in order to ensure their complete hydration. The hydrated films were put into a chamber filled with octane saturated water and the octane contact angles on each polymer surface were measured. With this technique one can directly calculate the non-dispersive interaction at the polymer/water interface (I_{SW}).

$$I_{SW} = \gamma'_{WV} - \gamma_{OV} - \gamma_{OW} \cos \Theta \qquad [1]$$

where Θ is the octane under water contact angle on a polymer, γ_{OW} the octane/water interfacial tension (50.5 mN/m), γ_{OV} the octane surface tension (21.6 mN/m) and γ'_{WV} the surface tension of octane saturated water (72.1 mN/m). With all these values we have :

$$I_{SW} = 50.5 \ (1 - \cos \Theta). \qquad [2]$$

Surface tension measurements at the air/protein solution interface

The Wilhelmy plate technique was used. The decrease of the surface tension with time due to the adsorption of protein molecules at the air/protein solution interface was recorded. The values after 40 minutes were taken. To set up a new pH we used an automatic micropump which introduced 10^{-1} M NaOH to the measuring cell equipped with a magnetic stirrer. The same surface tension values for each pH were obtained with albumin initially dissolved in 10^{-3}M NaOH and titrated with 10^{-1} M HCl. All albumin solutions were freshly made each day.

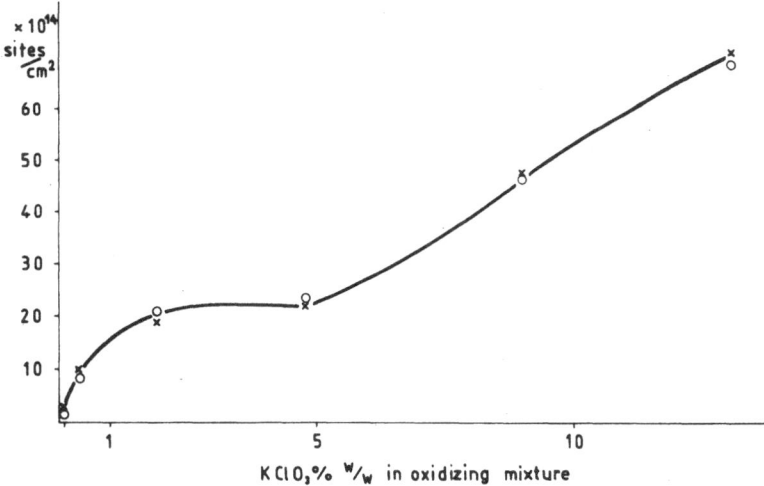

Fig. 2.   Surface density of polar groups versus KClO₃ concentration in oxidizing mixture.

RESULTS AND DISCUSSION

Characterization of polymer surfaces

     The isotherm of calcium adsorption on PE samples is shown in
Fig. 2. When polyethylene films were immersed in sulfuric acid
which did not contain any potassium chlorate, polar groups were not
formed and calcium ions did not adsorb at the solution/polymer interface. It is clear from Fig.2 that the increase of the $KClO_3$ concentration in the oxidizing mixture increases the surface density
(sites/cm^2) of polar groups. It should be noted also that a plateau
value was obtained when the $KClO_3$ concentration in the oxidizing
mixture was between 2 and 5% (w/w). For these concentrations the
surface density of polar sites was 20.10^{14} sites/cm^2. It is interesting to compare the calcium ion adsorption isotherm with the
contact angle data on oxidized PE surfaces.

     Data in Table 1 indicate a rapid decrease of the hydrophobi-
city of PE surfaces for the treatments where the $KClO_3$ concentration in the oxidizing mixture did not exceed 2%. Within 5-13% of
$KClO_3$ in the oxidizing mixture, neither the contact angle nor the
polar (donor-acceptor) interactions, I_{SW}, at the polymer-water
interface, change.

     Stronger chemical action of the oxidizing mixture (with a
$KClO_3$ percentage > 5%) introduces polar sites in a thicker surface
region of the polymer. These groups distributed in a three dimensional zone of the polymer are accessible to Ca^{2+} ion adsorption
(Fig. 2) but do not influence the value of the contact angle
(Table 1).

     Indeed as it has previously been shown by us [13] and by others
19,20, the wettability measurements are mainly sensitive to the
short range interactions and they give information only on the
outermost layer of a polymer surface.

     Table 1.   Octane under water contact angles and non-
               dispersive energies I_{SW} on PE/water interfaces

$KClO_3$ (%)	$\theta°$	$\cos \theta$	I_{SW} (mN/m)
0	31	0.8572	7.23
0.1	37	0.7986	10.19
1.98	50	0.6427	18.07
4.76	54	0.5878	20.86
9.09	57	0.5446	23.04
13.0	57	0.5446	23.04

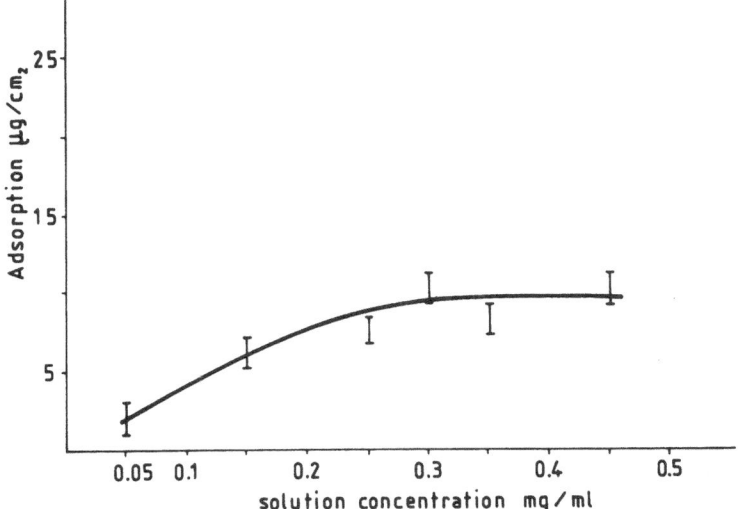

Fig. 3.   BSA adsorption on oxidized polyethylene ($\delta = 20 \cdot 10^{14}$ sites/cm^2) vs solution concentration ; pH = 7.5.

It may be considered, therefore, that the maximum density of polar groups on the oxidized PE surface is approximately equal to $20 \cdot 10^{14}$ sites/cm^2.

Protein adsorption

The results are presented in the form of adsorption isotherms by plotting the adscrbed amount of BSA against the BSA concentration in solutions.

Fig. 3 shows that the adsorption is dependent on albumin concentration in solution and that the plateau value was attained for the albumin concentration in solution equal to 0.35 mg/ml. Some other authors [21,23] reported plateau values of adsorption at this concentration.

The effect of oxidation of PE surfaces on protein adsorption is shown in Fig. 4. It can be noticed that the surface concentration of albumin increases with the KClO$_3$ percentage in the oxidizing mixture and that this increase is essentially pronounced in the 0-2% range of KClO$_3$ in the mixture (see the enlarged portion of the curve in the left corner of Fig. 4).

It has been suggested [22,27] that the main driving force for protein adsorption at the solid/liquid interface is due to hydrophobic bonding between protein aliphatic residues and a solid substrate. In this case the break down of organised water structure

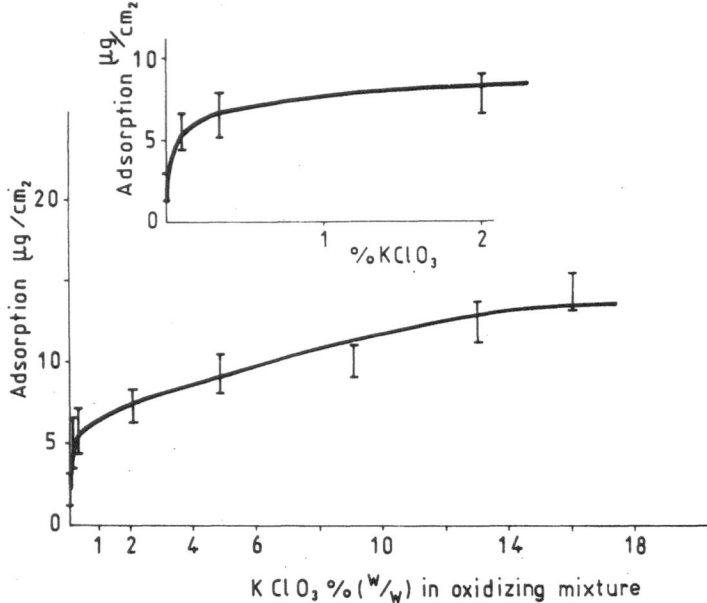

Fig. 4.   BSA adsorption as a function of polyethylene oxidation.
          The initial part of the curve is represented in magni-
          fication in the upper corner of the Figure ; pH = 7.5 ;
          BSA concentration = 0.35 mg/ml.

by the non-polar solid surface has to take place. Recent data of
Bagghi and Birnbaum [27] on IqG adsorption on hydrophobic latex
indicate that charge interactions between protein molecules and
latex particles have virtually no influence on adsorbed amounts.
Mizutani [25] has demonstrated that adsorption of proteins to silicone
coated glass surfaces was not affected by urea solutions of high
concentrations which normally dissociate hydrogen bonds.

     Our results show that the short range forces originating from
the surface layer of the oxidized polyethylene (where polar, hydrophilic sites were created) play an important role in albumin
adsorption. When these forces, as represented by I_{SW} (Table 1), do
not change any more with polymer oxidation, protein adsorption
reaches a plateau value. Specific behavior of surface oxidized
polyethylene samples can also be derived from albumin thin wetting
film studies on these substrates [28]. Proust, using previously
developed technique for measuring dry spot formation on contact
lenses [29], showed that albumin thin wetting films are completely
unstable on unoxidized polyethylene samples. The stability of these
films increases with the amount of polar sites at the polymer
surface. For the samples with the surface density > 10^{15} sites/cm^2
the film stability is higher than on a hydrophilic cleaned glass

surface . It seems therefore that the stability of these wetting
films is more reinforced by the presence of strong short range
forces on oxidized PE surface than by those originating from the
hydrophobic or hydrophilic solids.

The adsorbed amounts of albumin on oxidized and unoxidized PE
surfaces are much higher than one can accomodate in a flat monolayer (0.25 - 0.90 $\mu g/cm^2$ depending on side-on or end-on configuration) [15] of a non-denaturated protein.

Many authors found that proteins adsorb in quantities exceeding
monolayer packing [4,9,15,24,30-34]. Protein purity, conformational
change, cleaning of polymer, surface roughness, heterogeneity of
polymers and interfacial coagulation were quoted as factors influencing the adsorbed amounts. Quite recently Chan and Brash [35] suggesting that conformational change in adsorbed proteins is not a
general phenomenon and each protein-surface combination should be
individually studied.

We have previously shown by means of SEM micrographs [12], that
our polyethylene has an uneven surface and that during oxidation
etching of polymer (mainly amorphous regions) takes place. This
would mean that the roughness factor of pitted surfaces may be
considerably high and that the amount of protein adsorbed per unit
of true area for PE samples would in reality be less by an unknown
factor than those shown in Fig. 3 and Fig. 4.

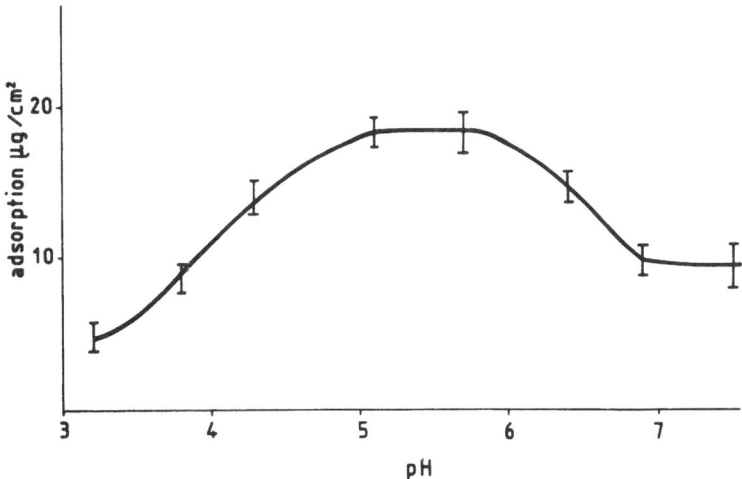

Fig. 5.   pH dependence of BSA adsorption on oxidized polyethylene
          film (δ = 2.10^{15} sites/cm^2). [Na^+] = $9.4.10^{-4}$ M ; BSA
          concentration in solution = 0.35 mg/ml.

pH dependence of albumin adsorption on oxidized PE surface is shown in Fig. 5. It can be noticed that maximum adsorption is found in the region of the isoelectric point (pH = 4.9) and that both at low and high pH the adsorbed amounts of albumin are smaller. Curves of similar shape were found on different synthetic charged polymer lattices [22,26] and on hydrophilic silica [36]. On hydrophobic, specially treated, silica [24] and on hydrophobic polyethylene [37] the extent of adsorption was reported to be independent of pH.

On hydrophilic surfaces and also on oxidized polyethylene films (Fig. 5) the maximum adsorption close to the isoelectric point may be explained by reduction of electrostatical lateral repulsions between adsorbing molecules. Other workers [22,26] show that ionic strenght increase in the substrate solution weakens the effect of these repulsions so that more molecules can adsorb at low than at high pH.

In Table 2 is illustrated the effect of NaCl in the substrate solution on albumin adsorption on unoxidized PE. Both the rinsing buffer and the substrate solution have the same pH = 7.5 and identical concentration of NaCl. It can be seen, in agreement with the observations made on silicone coated glass surfaces [25], that on unoxidized PE samples high salt concentrations had a positive effect on the amount of protein adsorbed.

Although adsorption at solid-liquid and liquid-air interfaces differs in nature (non-polar residues of proteins may penetrate into air phase in the latter case), it was interesting to see how the pH influences air-solution interfacial tension of BSA solutions. This pH dependence of interfacial tension of albumin solutions for two different salt concentrations (NaCl) is illustrated in Fig. 6. Slight lowering of the interfacial tension with increasing bulk concentration can be noticed.

The decrease of the interfacial tension with time was attributed to the adsorption of new protein molecules at the surface and also to their expansion and rearrangement in the adsorbed state [16]. It was also suggested that this rearrangement is controlled by the

Table 2. Effect of NaCl concentration on the amount of BSA adsorbed on unoxidized PE ($\mu g/cm^2$) ; pH = 7.5 ; BSA concentration in substrate solution = 0.35 mg/ml ; adsorption time = 3hr.

0	0.15 M	0.5 M	1 M	2 M
1.95	2.05	2.25	3.25	3.60

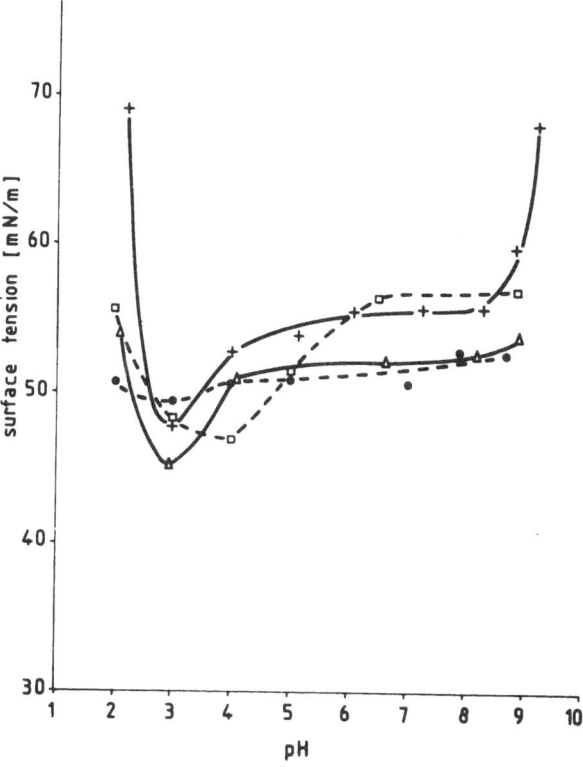

Fig. 6. pH dependence of surface tensions of BSA solution. (+)
BSA solution 0.1 mg/ml ; (☐) BSA solution 0.1 mg/ml with
NaCl 5.10^{-2} M ; (Δ) BSA solution 0.25 mg/ml ; (●) BSA
solution 0.35 mg/ml with NaCl 5.10^{-2} M. Temperature = 23°C

water molecules in the interfacial region which may hydrate a small
number of internal polar groups of globular proteins, and salt bonds
and bring about their rearrangement or unfolding, particularly at
low surface occupancy [39]. Graham and Phillips [16] assumed that a low
bulk concentration favors monolayer formation containing a mixture
of unfoiled and native molecules, while a high concentration leads
to multilayer thick film formation. This further adsorption to form
thicker protein films does not affect the interfacial tension
significantly.

The adsorption of BSA is insensitive to pH in the neigh-
bourhood of the i.e. point indicating that the apolar air phase
influences adsorption in the same way as hydrophobic solid
substrates (where the main driving force of adsorption are hydrophobic interactions).

Pronounced minima of interfacial tension at pH = 3 (Fig. 6) may reflect the isomerization reaction of albumin and the presence of a Fast isomeric form (F). It can be seen (Fig. 6), in agreement with other authors [16,38], that N-F transition is sensitive to the salt concentration in the bulk solution. Once again similar influence of the apolar air phase, in our case, and of a hydrophobic polyethylene [37], on albumin adsorption at this pH is demonstrated.

CONCLUSION

Hydrophilic/hydrophobic force balance of polymers may be varied by incorporation of a hydrophilic polymer in a bulk of a hydrophobic polymer (grafting of a hydrophilic monomer would be such a case). Also different block copolymers and polymer alloys are prepared with the aim to ensure the desired properties of the polymer. All these procedures alter the nature of the polymer, changing the magnitude of its long and short range forces.

The surface oxidized polyethylene is a specific surface where the long range hydrophobic forces do not change with oxidation and where the short range forces are very intensive [14]. More, polar groups at this surface are upright oriented to the plane, what is not necessarily true for hydrophilic polymer for which the distribution and orientation of functional groups may be aleatory. Furthermore, even for low surface densities of polar sites (well below the surface saturation) albumin adsorption is considerably enhanced, as if new points of anchorage for protein were introduced at the surface.

It follows, therefore, that such a surface has a specific effect upon charge distribution on the albumin molecule causing its unfolding and multilayer formation.

ACKNOWLEDGMENTS

The authors wish to thank Dr. L. Ter-Minassian-Saraga for stimulating discussions and suggestions concerning the experimental part of this work.

REFERENCES

1. D.J. Lyman, W.M. Muir and I.J. Lee, The effect of chemical structure and surface properties of polymers on the coagulation of blood. Surface free energy effects, Trans. Amer. Soc. Artif. Int. Organs 11:301 (1965).
2. L. Vroman, A.L. Adams, M. Klings and G. Fisher, Fibrinogen, Globulins, Albumin and Plasma at Interfaces in "Applied Chemistry at Protein Interfaces", E. Baier Ed., Advances in Chemistry Series ACS Washington D.C. (1975).

3. D.J. Lyman, Structural order and blood compatibility of poly-
    meric prostheses in "Structural Order in Polymers" IUPAC,
    F. Ciardelli and P. Giusti Eds., Pergamon Press, Oxford and
    N.Y. (1981).
4. S.W. Kim and R.G. Lee, Adsorption of blood proteins onto
    polymer surfaces in "Applied Chemistry of Protein Interfaces",
    E. Baier Ed., Advances in Chemistry Series ACS Washington D.C.,
    (1975).
5. R.M. Gendreau, S. Winters, R.I. Leininger, D. Fink, C.R. Hassler
    and R.J. Jackobsen, Fourier transform infrared spectroscopy
    of protein adsorption from whole blood. Ex vivo dog studies,
    Applied Spectroscopy 35:353 (1981).
6. B.D. Ratner, D. Shuttleworth, T.A. Horbett and H.R. Thomas,
    The organization of adsorbed protein films. An ESCA method
    for analysis, First World Biomaters. Congress, Baden near
    Vienna, Austria, Book of Abstracts, p.431 (1980).
7. T. Okano, S. Nishiyama, I. Shinohara, T. Akaike, N. Sakurai,
    K. Kataoka and T. Tsuruta, Role of microphase separated
    structure on the interfacial interaction of polymer with
    blood, ACS Polymer Preprints 20:571 (1979).
8. D.J. Lyman, K. Knutson, B. McNeill and K. Shibatani, The effects
    of chemical structure and surface properties of synthetic
    polymers on the coagulation of blood. The relation between
    polymer morphology and protein adsorption, Trans. Amer. Soc.
    Artif. Int. Organs 21:49 (1975).
9. S.W. Kim and E.S. Lee, The role of adsorbed proteins in platelet
    adhesion onto polymer surfaces, J. Polymer Sci. Polymer Symp.
    66:429 (1979).
10. A. Baszkin and D.J. Lyman, Adsorption/desorption studies of
    plasma proteins on polymer surfaces in "Advances in Bio-
    materials", Vol.3, G.D. Winter, D.F. Gibbons and H. Plenk Eds.,
    J. Wiley and Sons, Chichester UK (1982).
11. T. Okono, S. Nishiyama, I. Shinohara, T. Akaike and Y. Sakurai,
    Interaction between plasma proteins and microphase separated
    structure of copolymers, Polymer J. 10:223 (1978).
12. B. Catoire, P. Bouriot, A. Baszkin, L. Ter-Minassian-Saraga and
    M.M. Boissonnade, Polymer interface analysis by ESR spectra
    of Mn(II), J. Colloid Interface Sci. 79:143 (1981).
13. A. Baszkin and L. Ter-Minassian-Saraga, Effect of surface pola-
    rity on self-adhesion of polymers, Polymer 19:1083 (1978).
14. A. Baszkin, M. Deyme, M. Nishino and L. Ter-Minassian-Saraga,
    Surface chemistry and wettability of modified polyethylene,
    J. Colloid Polymer Sci. 61:97 (1976).
15. A. Baszkin and D.J. Lyman, The interaction of plasma proteins
    with polymers. Relationship between polymer surface energy
    and protein adsorption/desorption, J. Biomed. Mater. Res.
    14:393 (1980).
16. D.E. Graham and M.C. Phillips, Proteins at liquid interfaces.
    Molecular structures of adsorbed films, J. Colloid Interface
    Sci. 70:403 (1979).

17. F. MacRitchie and A.E. Alexander, Kinetics of adsorption of proteins at interfaces, J. Colloid Sci. 18:453 (1963).
18. J.D. Andrade, R.N. King, D.E. Gregonis and D.L. Coleman, Surface characterization of poly(hydroxyethyl methacrylate) and related polymers. Contact angle methods in water, J. Polymer Sci. Polymer Symp. 66:313 (1979).
19. D.M. Brewis and D. Briggs, Adhesion to polyethylene and poly-propylene, Polymer 22:7 (1981).
20. H. Schonhorn, Surface modification of polymers for adhesive bonding in "Polymer Surfaces", D.T. Clark and W.J. Feast Eds., J. Wiley and Sons, Chichester UK (1978).
21. J.L. Brash and V.J. Davidson, Adsorption on glass and poly-ethylene from solutions of fibrinogen and albumin, Thrombosis Res. 9:249 (1976).
22. W. Norde and J. Lyklema, The adsorption of human plasma albumin and bovine pancreas ribonuclease at negatively charged poly-styrene surfaces. Adsorption isotherms, effect of charge, ionic strength and temperature, J. Colloid Interface Sci. 66:257 (1978).
23. V. Hlady and H. Fueredi-Milhofer, Adsorption of human serum albumin on precipitated hydroxyapatite, J. Colloid Interface Sci. 69:460 (1979).
24. F. MacRitchie, The adsorption of proteins at the solid/liquid interface, J. Colloid Interface Sci. 38:484 (1972).
25. T. Mizutani, Adsorption of proteins on silicone coated glass surfaces, J. Colloid Interface Sci. 82:162 (1981).
26. T. Suzawa and T. Murakami, Adsorption of bovine serum albumin on synthetic polymer lattices, J. Colloid Interface Sci. 78:266 (1980).
27. P. Bagchi and S.M. Birnbaum, Effect of pH on the adsorption of immunoglobulin G on anionic poly(vinyltoluene) model latex particles, J. Colloid Interface Sci. 83:460 (1981).
28. G. Wajs, L. Ter-Minassian-Saraga and P. Payrau, Copolymères silicone-poly(vinylpyrrolidone). Utilisation en optique de contact, Rapport DGRST, Génie Biologique et Médical, 78.7.159 (1980).
29. A. Baszkin, M.M. Boissonnade, J.E. Proust, S. Tchaliovska, L. Ter-Minassian-Saraga and G. Wajs, Silicone grafted with poly(vinylpyrrolidone) for contact lenses. Surface properties and stability of thin tear films, J. Bioengineering 2:527 (1978).
30. B. Jansen and G. Ellinghorst, Radiation initiated grafting of hydrophilic and reactive monomers on polyetherurethane for biomedical application, Radiat. Phys. Chem. 18:1195 (1981).
31. J.D. Andrade, Interfacial phenomena and biomaterials, Med. Instrumentation 7:110 (1973).
32. C.J. Van Oss, D.R. Absolom, A.W. Neumann and W. Zingg, Deter-mination of the surface tension of proteins. Surface tension of native serum proteins in aqueous media, Biochim. Biophys. Acta 670:64 (1981).

33. W. Lemm and V. Unger, Adsorption of blood proteins on different polymer surface in vitro in "Evaluation of Biomaterials", G.D. Winter, J.L. Leray and K. de Groot Eds., J. Wiley and Sons (1980).

34. Y. Sakurai, T. Akaike, K. Kataoka and T. Okano, Interfacial phenomena in biomaterials chemistry in "Biomedical Polymers", E.P. Goldberg and A. Nakajima Eds., Academic Press, N.Y. (1980).

35. B.M.C. Chan and J.L. Brash, Conformational change in fibrinogen desorbed from glass surfaces, J. Colloid Interface Sci. 84:263 (1981).

36. B.W. Morrissey and R.R. Stromberg, The conformation of adsorbed blood proteins by infrared bound fraction measurements, J. Colloid Interface Sci. 46:152 (1974).

37. E. Brynda, M. Houska, Z. Pokorna, N.A. Cepalova, Y.V. Moiseev and J. Kalal, Irreversible adsorption of human serum albumin onto polyethylene films, J. Bioengineering 2:411 (1978).

38. E. Tornberg, The application of the drop volume technique to measurements of the adsorption of proteins at interfaces, J. Colloid Interface Sci. 64:391 (1978).

39. L. Ter-Minassian-Saraga, Protein denaturation on adsorption and water activity at interfaces. An analysis and suggestion, J. Colloid Interface Sci. 80:393 (1981).

RADIATION INDUCED MODIFICATION OF POLYETHERURETHANE TUBES WITH HEMA AND ACRYLAMIDE

Bernd Jansen

Institute for Physical Chemistry
University of Cologne
Luxemburgerstr. 116, 5000 Köln 41, FRG

INTRODUCTION

In a previous work we described the modification of polyetherurethane films with hydrophilic monomers via radiation grafting in order to improve the thromboresistance of the polymer surface.[1,2] Following the "preswelling technique", the trunk polymer films were swollen with monomer (2-hydroxyethyl methacrylate, 2,3-dihydroxypropyl methacrylate, acrylamide) for certain lengths of time and irradiated afterwards in the absence of surrounding monomer. Using this method, graft copolymers with different penetration depths of the graft component in the trunk polymer could be obtained, leading to different surface properties of the grafted films. The investigation of the mechanical behaviour of the grafted products showed that mechanical strength decreases with increasing grafting yield, but that low grafted films (grafting yield < 5%) nearly have the same mechanical properties as the trunk polymer.[3] Furthermore in this study it was tried to correlate surface parameters as contact angle and interfacial free energy γ_{sw} between polymer and water with the results of protein adsorption to the film surfaces.[2,3] We found that for grafted films with a grafting yield below 5% the protein adsorption decreases as the interfacial free energy γ_{sw} decreases, as it was postulated by J.D. ANDRADE in his "minimum interfacial free energy hypothesis".[4] Films with higher grafting yields do not follow this hypothesis, due to an increase in surface area caused by the grafting process. These high grafted

samples preferentially do adsorb albumin, the extent of
the albumin adsorption being a function of the grafting
yield. Recent results of platelet adhesion measurements
as performed with the SPFE (Stagnation Point Flow Experiment)-test[5] revealed, that platelets have a low tendency
to adhere to HEMA grafted polyetherurethane films with
low grafting yield and low interfacial free energy.[6]

Because of these encouraging preliminary results
we tried to apply the "preswelling grafting technique"
to the modification of tubes, in order to have materials
which can be tested in animal experiments. As tube
material a polyetherurethane with excellent mechanical
properties and hydrolysis resistance was chosen. To
preserve the good mechanical properties of the original
trunk polymer, it was tried to modify only the internal
surface of the tubes and to yield a high degree of
modification at low grafting yields.

MATERIALS AND METHODS

Trunk Polymer

As trunk polymer a polyetherurethane semicarbazide
("Rö 325", Fa. Technochemie, Dossenheim, FRG), composed
of 4,4'-diisocyanatodiphenylmethane, poly(oxytetra-
methylene) and isophthalic acid dihydrazide was used.
In Tab. 1 some properties of this polymer are listed.

Table 1. Properties of Rö 325-polyetherurethane[7]

tensile strength:	$\sim 8.2 \cdot 10^7 \ N \cdot m^{-2}$
elongation at break:	$\sim 512\%$
hydrolysis resistance:	after 8 weeks storage in H_2O at 343 K no change of mechanical properties

Graft Monomers

Highly purified 2-hydroxyethyl methacrylate (HEMA),
supplied by Fa. Röhm & Haas, Darmstadt, FRG, and recrystallized acrylamide (AAm), supplied by Fa. Stock-
hausen, Krefeld, FRG, were used as graft monomers. In the
case of grafting with AAm, a 10 molar solution of AAm in
methanol was used.

Preparation of Polyetherurethane Tubes

Polyetherurethane tubes (I.D. 4 mm and 6 mm) were prepared from a 15% polymer solution in DMF by the dipping method. Glass rods with adequate diameters were cleaned with chromic sulfuric acid, acetone and distilled water. After drying they were dipped into the polymer solution, and the solution was allowed to run freely from the rods in air for $\sim$ 10 minutes. Afterwards the glass rods were dried in an oven for 1.5 h at 343 K. For the second layer the glass rods were turned before dipping. By repeating this procedure for 10-12 times the desired wall thickness of $\sim$ 0.6 mm could be achieved. After extracting residual solvent with distilled water at 313 K for 48 h the tubes were removed from the glass rods by simply swelling them in an EtOH/water mixture over night. For the surface grafting the "glass-side" of the tubes was used.

Grafting Procedure

Internal grafting of tubes (4 mm and 6 mm I.D.) was performed by pumping the monomer (or monomer solution) through the tubes for different lengths of time and irradiating the preswollen tubes in a ^{60}Co-γ-source with a dose rate of 1 J$\cdot$kg$\cdot$s^{-1} ($\hat{=}$ 3.6$\cdot$10^5 rad$\cdot$h^{-1}). After extraction with an EtOH/water mixture and pure water, the grafting yield was determined gravimetrically.

Water Sorption Measurements

The grafted tubes were stored in distilled water at 298 K for 24 h; superficial water on the inner surface was removed with filter paper. The water swollen tubes were weighed and their water content calculated as follows:

$$\frac{W - W_o}{W_o} \times 100 = \% \text{ water content}$$

W = weight of water swollen tube
W_o = weight of dry tube

Contact Angle Measurements

The octane-in-water contact angle was measured using the captive-bubble method described by HAMILTON.[8,9] The measurements were carried out with a contact angle gonio-meter of the Fa. Lorentzen and Wettres (Stockholm),

fitted with a special cell (Ramé-Hart, N.J., USA). To
achieve a plane surface of the samples before measurement, the tubes were cut with a sharp knife in the
longitudinal direction and then glued to a PMMA-plate
with a cyanoacrylate adhesive. All samples were stored in
tridistilled water for 24 h before the contact angle was
determined.

Sterilization of Grafted Tubes

The grafted tubes were sealed in polyethylene bags
and irradiated in a ^{60}Co-γ-source with a total dose of
$\sim 3 \cdot 10^4$ J·kg^{-1} ($\triangleq$ 3 Mrad) at a dose rate of
~ 1.1 J·kg^{-1}·s^{-1}. After incubation of 0.5 cm long tube
segments in BHI (Brain Heart Infusion)-bouillon for 24 h
at 310 K no growth of bacteria was observable.

RESULTS

First attempts to graft the internal surface of
polyetherurethane tubes by filling the tubes (which were
clamped at one end) with monomer (HEMA or AAm solution),
to remove the monomer after a certain time and afterwards
to irradiate the preswollen tubes, proved to be not very
successful: tubes grafted in this way exhibited an irreg-
ular graft distribution at the inner surface, as contact
angle measurements revealed. To achieve a more uniform
swelling of the tubes, the monomer (or monomer solution)
was pumped through the tubes with the aid of a roller
pump. By keeping the monomer flow constant, definite
swelling degrees could be obtained (Fig. 1). The grafting
of tubes with HEMA by the above method led to similar
time conversion curves as already found for the grafting
of polyetherurethane films with HEMA[2,3]: in the very
first beginning of the reaction there is a kind of induction period, followed by a steep increase of the grafting
yield before the saturation value is reached (Fig. 2).
A more detailed discussion of this grafting behaviour is
given elsewhere.[3] The dependence of the grafting yield
on the preswelling time in HEMA monomer at a constant
irradiation dose is shown in Fig. 3. By simply varying
the preswelling time, grafting yields up to 70 mg HEMA/cm^2
can be obtained.

Qualitatively the same results are found for the
grafting with acrylamide (Fig. 4). As a solvent for the
solid monomer, methanol was used. In comparison with
HEMA-grafting, only low grafting yields up to 4 mg
AAm/cm^2 were yielded, due to a poorer swelling of the
AAm solution into the polyetherurethane.

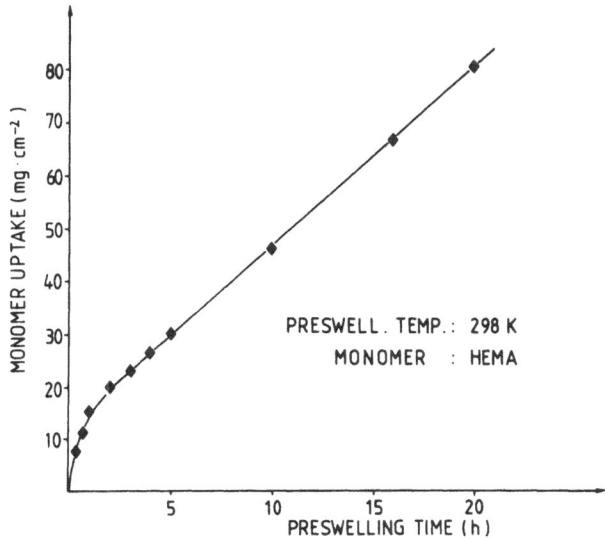

Fig. 1   Preswelling of a Rö 325-polyetherurethane
tube with 2-hydroxyethyl methacrylate (HEMA)

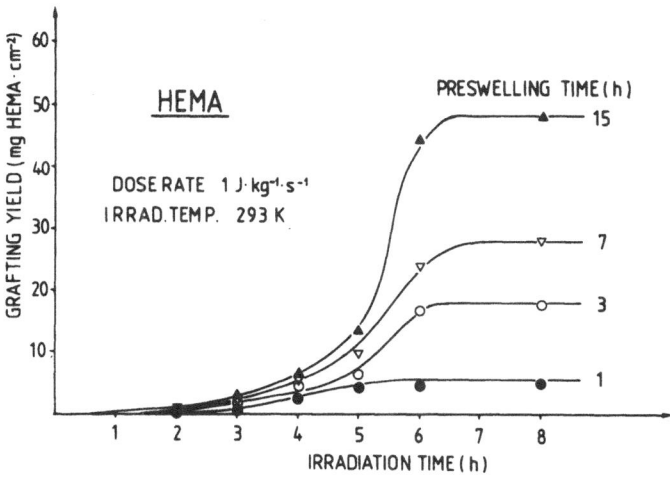

Fig. 2   Time conversion curves of the grafting of
Rö 325-polyetherurethane tubes with HEMA

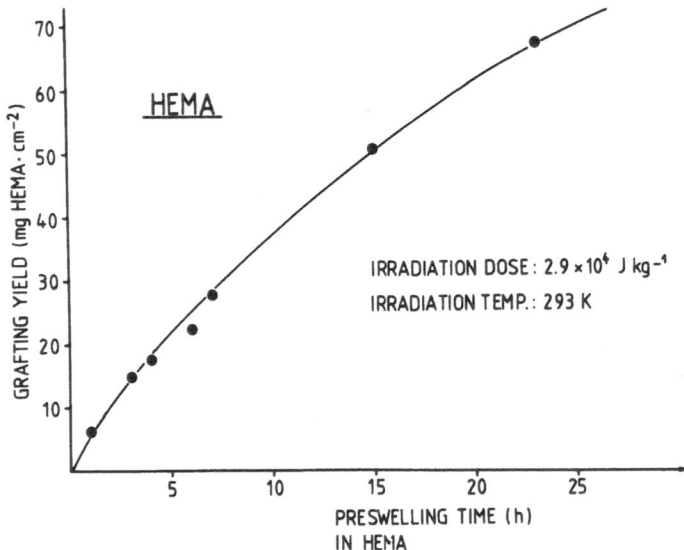

Fig. 3   Grafting of Rö 325-polyetherurethane tubes
          with HEMA; dependence of the grafting yield
          on the preswelling time in monomer

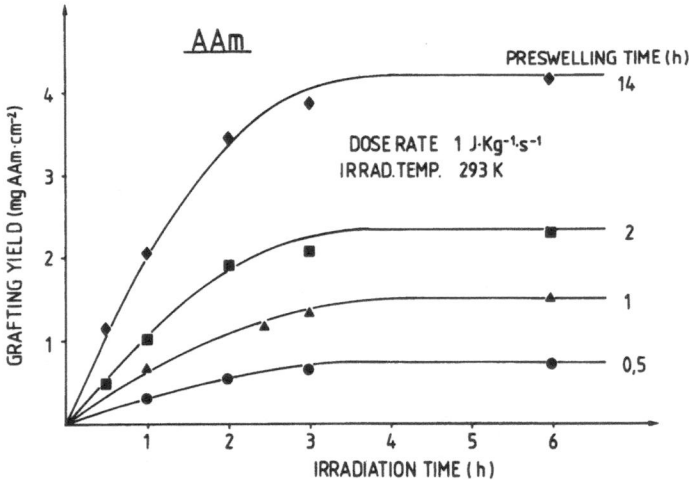

Fig. 4   Time conversion curves of the grafting of
          Rö 325-polyetherurethane tubes with
          acrylamide

The water uptake of such modified tubes increases linearly with increasing grafting yield for both systems (Fig. 5). To investigate whether the surface had become more hydrophilic after grafting, the octane-under-water contact angles of the inner surfaces of the tubes were measured (Figs. 6 and 7). Independent of the grafting yield, the contact angle of HEMA-grafted tubes lies in the region of ~ 140°, the contact angle of AAm-grafted tubes is about ~ 130° (the unmodified polyetherurethane tube has an octane-under-water contact angle of ~ 106°). Obviously a constant surface monomer content is reached when the monomer is pumped with a constant flow rate through the tubes, leading to an uniform surface grafting after irradiation with a dose sufficient large to convert all monomer.

For some grafted tubes with low grafting yields (HEMA: 5 mg / cm^2; AAm: 1.5 mg/cm^2) the interfacial free energy γ_{sw} against water was determined according to a method described in reference (2). In both cases, γ_{sw} is below 20 $\mu N \cdot cm^{-1}$. As we know from our preliminary studies, surfaces with such low interfacial free energy do adsorb only a small amount of protein, and moreover platelets should have a little tendency to adhere to such a surface.

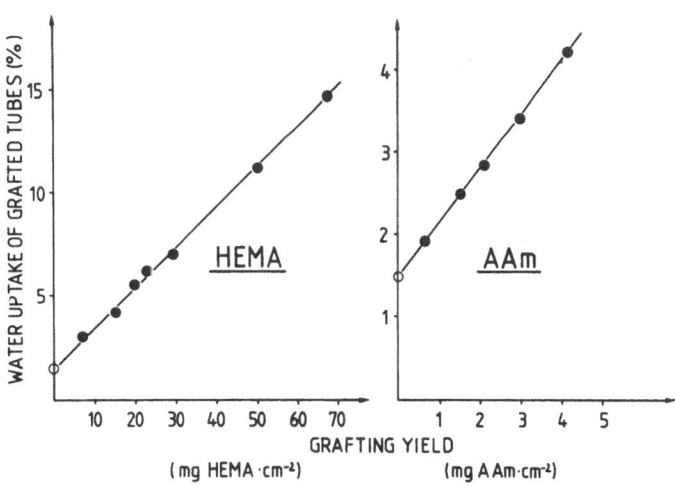

Fig. 5   Water uptake (%) of HEMA- and AAm-grafted Rö 325-polyetherurethane tubes

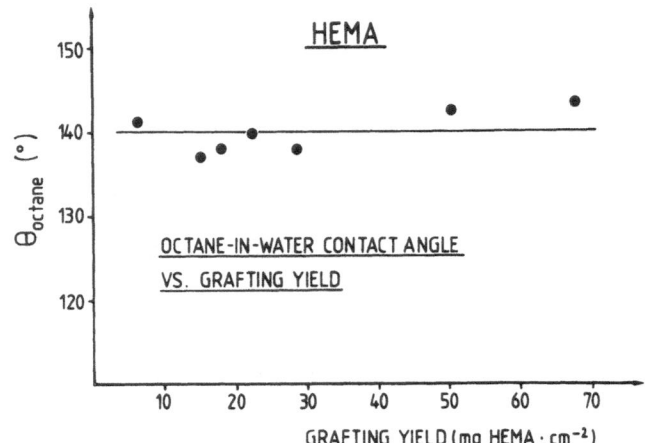

Fig. 6   Octane-under-water contact angles of HEMA-
         grafted Rö 325-polyetherurethane tubes of
         various grafting yields

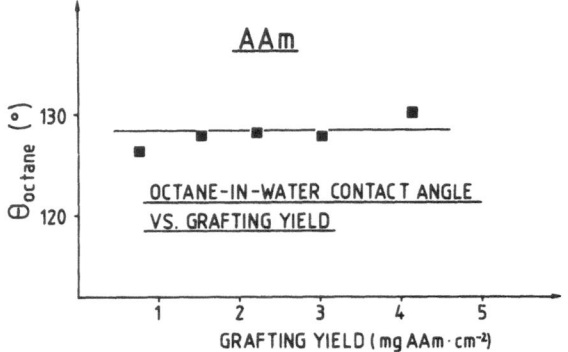

Fig. 7   Octane-under-water contact angles of AAm-
         grafted Rö 325-polyetherurethane tubes of
         various grafting yields

    Summarizing, the application of the 'preswelling
grafting technique' to the hydrophilic modification of
polyetherurethane tubes leads to grafted products with
a well defined hydrogel content at the surface. By
keeping the grafting yield low, the mechanical properties

of the trunk polymer can be largely saved, whilst a relatively high surface hydrophilicity is achieved. Tubes modified in this manner can be easily sterilized with ^{60}Co-γ-radiation, without remarkable loss of mechanical strength. Our current work is concerned with a more detailed investigation of mechanical parameters, especially of the compliance, of the grafted tubes, and with the evaluation of these tubes in implantation experiments.

REFERENCES

1. B. Jansen, G. Ellinghorst, J. Polym. Sci. Polym. Symp. 66, 465 (1979).
2. B. Jansen, G. Ellinghorst, Radiat. Phys. Chem. 18 (5-6), 1195 (1981).
3. B. Jansen, G. Ellinghorst, submitted for publication in J. Biomed. Mater. Res.
4. J.D. Andrade, Med. Instrumentation 7, 11o (1973).
5. H. Petschek, D. Adams, A. R. Kantrowitz, Trans. Amer. Soc. Artif. Int. Org. Vol. XIV, 256 (1968).
6. B. Jansen, Paper presented at the International Conference 'Biomedical Polymers', 12-15 July 1982, Durham, GB.
7. H. D. Stenzenberger, D. O. Hummel, Angew. Makromol. Chem. 82, 103 (1979).
8. W. C. Hamilton, J. Colloid. Interf. Sci. 40, 219 (1972).
9. W. C. Hamilton, J. Colloid. Interf. Sci. 47, 672 (1974).

This work was supported by the Fritz Thyssen Stiftung, Cologne, FRG.

IMPROVEMENT OF THE BIOCOMPATIBILITY OF POLYMERS THROUGH SURFACE MODIFICATION

Herbert Bauser and Horst Chmiel

Fraunhofer-Institut für Grenzflächen- und Bioverfahrenstechnik

Stuttgart, Germany

1. Introduction

Biomaterials are by definition materials that assume the functions of tissue in natural organs or organ parts. They must therefore imitate the properties of such tissue as well as possible. For example, a vascular prosthesis must exhibit a tension-expansion curve highly similar to that of a natural blood vessel, as well as a smooth inner surface which corresponds to the endothelial covering. In other words, a biomaterial must be made to act as much as possible like the natural tissue in its biological environment - in the case of vascular prostheses in the environment of blood, tissue, and interstitial fluid -, it must withstand biodegradation and prove to be biocompatible (Table 1).

Table 1.   Requirements for Biomaterials

Functional feasibility

The function of the organ or tissue must be guaranteed, even for long-term use

Biostability

Biological environment must not impair the functioning of the biomaterial

Biocompatibility

Biomaterial must not disturb the biological system

Steribizability

Sterilisation procedure must not impair the functioning of the biomaterial

In addition to the general necessity for clinically used materials to be sterilizable, the requirements placed on biomaterials can then be divided into two categories: 1) those involving physical properties (for example, material strength), thus applying to bulk material, and 2) those requirements which involve interaction between the biomaterial and the biological system, thus applying to interface or surface properties. It then follows that the first set of requirements can be met by chosing or developing a suitable bulk material, the second set by the proper modification of surfaces. The production of polymers with surface layers suitable for contact with blood was the goal of the research program being presented here.

2. Blood compatibility and protein adsorption

Biocompatible materials in contact with blood must meet the conditions listed in Table 2. These conditions are partially interdependent. Thus, the requirement of antithrombogenicity rests on the condition that neither thrombocytes, clotting factors nor inhibitors may be damaged. Since failure to meet this rather complex requirement can have fatal consequences in only seconds, it is the main – and often only – requirement considered, during the development of biomaterials and one that is truly difficult to fulfill.

Table 2.   Conditions for blood compatibility

Biomaterial must not induce

- generation of thrombi or emboli,
- impairment of cell function or destruction of cells,
- alteration of plasma proteins,
- toxic or allergic reactions,
- immuno reactions,
- development of cancers,
- electrolyte depletion

But not only thrombotic occlusions and stenoses are to be avoided. Even the preliminary forms of undesirable clotting processes such as large and irreversible deposits of protein (especially fibrinogen) can lead to considerable disturbances in sundry cases, in particular in membrane function. We call your attention here to the phenomenon described in detail by Baier /1/, Lyman /2/, Nyilas /3/ and others that the protein layer which forms in a very short time on the biomaterial is of decisive importance for the clotting process. Hence, good blood compatibility requires for the proteins: No conformational changes or denaturization on contact with the polymer surface, and, consequently, reversible adsorption. In general, this means that interaction between the protein and the surface of the biomaterial should be as slight as possible.

The various side groups of proteins - ionized and non-ionized polar as well as non-polar groups - make it possible for an appropriate foreign membrane to capture and retain proteins. This takes place by means of dispersion forces, through electrostatic and entropic forces (hydrophobic interactions) or hydrogen bonds. Plain concepts for the minimization of these interactive forces, such as a negatively charged surface (e.g. anionic groups), hydrogel coating or avoidance of hydrogen-bond forming sites have been suggested, and these concepts have given a useful impetus to biomaterial research, but have often led to dead ends, too. They ignore numerous factors that may play a role in the interaction between proteins and foreign surfaces. For example, the adsorption of small counter ions (e.g. prior to protein adsorption /4/) can reverse the polarity of a surface and transform the intended electrostatic repellance into attraction. Other factors are

- contribution to entropic forces by conformational changes of protein /3/,
- inhomogenous surface (e.g. microphase structure) /5/,
- Redox processes (electron transfer) /6/.

These reservations do not, however, change the validity of the basis idea, which says that in general minimal interaction between proteins and the biomaterial surface should be the goal. Convincing proof of this interpretation is the high blood compatibility of isotropic carbon (or LTI carbon), in which a very slight interaction with proteins has been shown empirically /7/. The excellent performance of isotropic carbon suggests to deposit such protein-compatible material as a blood-compatible thin layer on polymers /8/, and the purpose of our research was, among other things, to realize this idea /9/. We also tried coating polymers with strongly hydrophilic layers, that is, layers similar to hydrogel.

3.   Test methods

Because of the significance of interaction with proteins, a protein adsorption/desorption measurement was carried out as a first test /10/. A protein solution (2 g/l fibrinogen) with a constant velocity of 0.5 ml/sec flows through a measuring cell which, depending on the form of the sample, is constructed as either a plate or a cylinder condenser (Fig. 1). By changing the cell's capacitance and dielectric load, the build-up or dissolving of an adsorbate layer on sample surfaces can be determined. Calibration of the measuring equipment with bovine fibrinogen solution and checking it with an ESCA measurement yielded a detection limit of 3×10^{11} fibrinogen molecules per cm^2, that is, roughly one third of a monolayer /10/. Adsorption and - after the replacement of the fibrinogen solution with water or saline (after about 6 h) - desorption are measured as functions of time (sect. 4). The desorption behaviour with respect to time is especially informative, since it provides

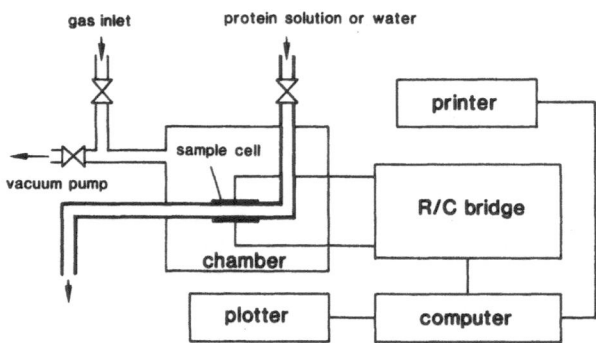

Fig. 1   Set-up for protein adsorption/desorption measurement

data on the irreversibly adsorbed fraction and thereby on the degree of interaction of protein with the surface. Hence, the amount of adsorbate remaining
after 6 h of desorption (irreversibly adsorbed fibrinogen) is considered to be
one way of measuring the thrombogenicity of the surface. This method has the
advantage over in vitro tests with blood that it can be more easily standardized, since it is not dependent on the blood donor. However, this fibrinogen
adsorption/desorption test has to be correlated with in vitro tests, and first results comparing this method with a coagulation parameter test confirm its relevance to blood compatibility. Meanwhile, testing of vascular prostheses in an
in vivo experiment was begun. The prosthesis is implanted in the carotic artery of a dog (a beagle) and removed after 4 or 8 weeks. The statistical material
is here still insufficient for final conclusions, but some of the results will be
reported at the end of this paper.

4.   Coating processes

       We used two different materials for the coating of polymer biomaterials:
Polyacrylonitrile (PAN) and isotropic carbon. Polyacrylo-

$$-CH-CH_2-$$
$$|$$
$$CN$$

nitrile layers are generated by plasma polymerization. As shown schematically in Fig. 2 a glow discharge is maintained in a mixture of inert gas and monomer vapor by means of radio frequency power fed into the reactor.

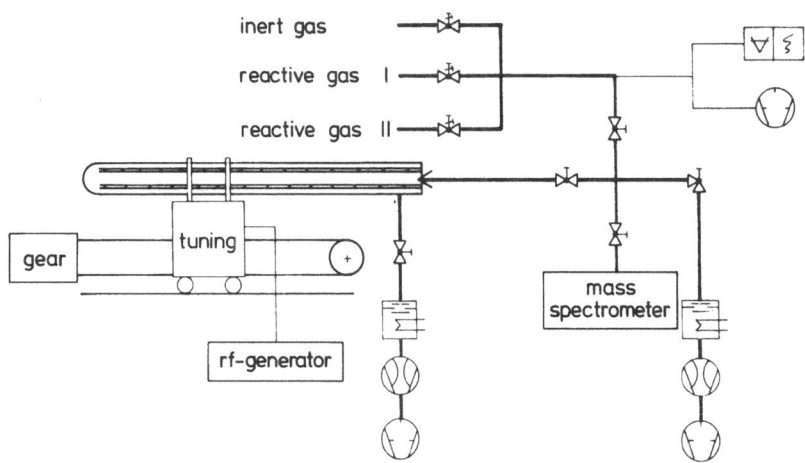

Fig. 2   Set-up for plasma polymerization and deposition in a tube

The monomer is activated in the discharge (at a pressure of 0.1 mbar). From the activated monomer a grafted PAN layer forms on the polymer substrate. One advantage of this process is the possibility of coating the inner surface of the tubing. The effect of a PAN coating with polyetherurethane tubing on fibrinogen desorption is shown in Table 3.

Table 3.   Desorption of fibrinogen from surface modified polyurethane tubings, after approx. 6 h adsorption. O means below detection limit $(3 \cdot 10^{11}$ molecules/cm^2)

Desorption time (min)	0	100	200	300	400	500
PUR (Pellethane)	285	190	190	190	190	190
PUR treated in Ar-plasma	120	130	130	120	28	0
PUR coated with PAN	545	0	0	0	0	0

With an initial value of several 10^{14} fibrinogen molecules per cm^2 (obtained after 6 h of adsorption) the standardized rinse process with water (also 0.5 ml/sec, at a shear rate of approx. 10 sec^{-1}) caused only a relatively small

reduction of the adsorbate thickness for uncoated PUR, whereas a rapid desorption took place for PAN-coated PUR. After only 100 min the amount of
adsorbate had fallen below the detection limit. This represents a reduction of
the adsorbed fibrinogen by at least a factor of 1000 .

As will be seen below, we obtained similar and sometimes even better results
with carbon layers. It is well known that so-called isotropic carbon is especially compatible with blood /7/. It is deposited as pyrolytic carbon in a flui-
dized bed of a hydrocarbon-noble gas mixture at relatively low pyrolysis temperatures between 1200 and 1500 C (LT = low temperature). Since this process
can only be used with a heat-stable substrate material with low thermal expansion coefficients, we (and others /8/) used vacuum-coating processes for the
deposition of similar coatings onto polymers. In our work we used two processes: with one, the results of which will be reported in sect. 5, the carbon is
separated out of a hydrocarbon-noble gas mixture in the set-up shown in Fig.2.
However, not heat, as in pyrolysis, but rather a glow discharge is used for the
decomposition.

In addition to plasma deposition, we also used cathode sputtering. In order to
optimize this process, we coated PUR films with a smooth surface. Under various process conditions we were able to deposit layers onto the films that were
at least as smooth as the polished surfaces of LTI carbon (the surface smooth-
ness had proved to be an important condition for good blood compatibility of
LTI-carbon).This is demonstrated in Fig. 3.

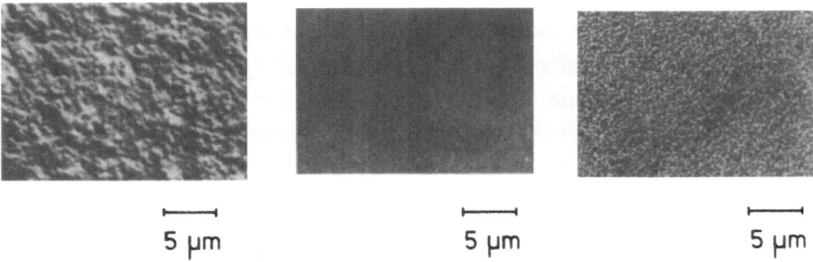

5 µm   5 µm   5 µm

Fig.3   Scanning electron micrographs comparing the surface of
             sputtered carbon layers with that of LTI carbon /11/

According to evidence proved by electron diffraction micrographs, the layer's
crystalline structure also corresponds largely to that of isotropic carbon. An
important problem is, of course, the low flexibility of isotropic carbon
(Table 4). The slight elongation at break of LTI carbon seems at first to
make the use of such a material on an elastic substance impossible. This problem becomes somewhat less serious in view of the fact that in very thin layers
(thicknesses of the order of 10 nm are sufficient) bulk properties do not completely develop, and, hence the elongation at break is somewhat higher
with vacuum deposited carbon layers /7/. This is, however, not enough.

Therefore, in addition to solid sheets, we have developed layers with a granular structure in which expansion spaces between the single granules are
formed during stretching, giving the layer a quasiflexibility. Whereas   layers

Table 4.   Mechanical properties of LTI carbon /7/ and polyurethane
             elastomer /12/. Figures related to dilatability

	LTI carbon	PUR elastomer	
Young's modulus	17000 to 28000	700	N/mm^2
Elongation at break	1.6 to 2.1	400 to 500	%

already show cracks due to usual manipulations and develop new cracks under
tension, granular layers remain intact even under an elongation of 6 %. The
microexpansion joints can be kept to approx. 0.1 μm or less in width, and the
granules can be placed in a scale-like manner, so that the microexpansion
joints are small compared to the diameter of the blood cells. Due to the scale
effect, the flowing blood does not directly contact the polymer surface. Several examples of fibrinogen adsorption-desorption curves are shown in Fig. 4.

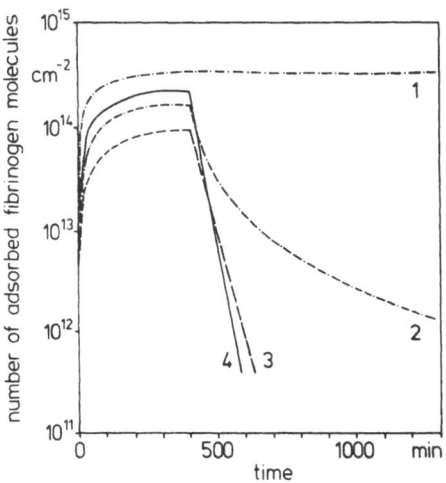

Fig. 4   Fibrinogen adsorption and desorption curves for an uncoated poly-
           urethane film (1) and for carbon layers

The desorption process that begins after about 400 min of adsorption can
barely be recognized in the semi-logarithmic representation for the PUR substrate (1). By comparison, the amount of adsorbate for the carbon layers

(3 and 4) has dropped within a short amount of time below the detection limit, i.e. by at least three powers of ten. Curve 2 shows that, depending on the choice of the process parameters, carbon layers can also be produced which do not have such slight interaction with fibrinogen as do those in 3 and 4. It should be pointed out that not only the material determines the protein inter- action or the antithrombogenicity, but also the surface produced, of which the actual structure and trace impurity content depends on the deposition conditions. These relationships have not yet been researched systematically, but reproduce- able interdependencies among several process parameters have been found. Studies on LIT carbon by Bokros and co-workers /11/ have shown that, i.e., the incorporation of oxygen into the carbon layer reduces blood compatibility, and trace amounts of oxygen may - as one example - depend on process cond- itions.

On some of the carbon layers competitive adsorption of albumin and fibrinogens was studied by Dr. Lemm (Berlin) using the radioactive tracer method in a batch experiment. As can be seen in Table 5 an initial high fibrinogen adsorp- tion (after 1 h) declines in the course of time (24 h), while the albumin adsorp- tion increases. The displacement of fibrinogen by albumin is a good sign, since a coating of albumin causes a higher antithrombogenicity /2/.

Table 5   Competitive adsorption of albumin and fibrinogen. Figures are given in $\mu g/cm^2$.

Sample	Adsorb. albumin $\mu g/cm^2$		Adsorb. fibrinogen $\mu g/cm^2$	
	1 hour	24 hours	1 hour	24 hours
CELH 12/24	0	20.0	14.7	2.9
CELH	11.8	16.1	14.3	5.3
CELH 12/33	8.7	28.4	17.1	5.4
CELH 33/34	9.0	23.4	28.0	5.9

5.   Internal coating of vascular prostheses with carbon

     Since it is now possible to make the rigid carbon layer quasi-flexible, the next step is to attempt to introduce this new carbon layer into the inside of tubes, particularly in vascular prostheses. There are now so-called carbon- ized or graphitized vascular prostheses available with relatively thick black carbon layers. However, in view of the favourable preliminary test results of our layers, it would also certainly be useful to get some experience with our very thin (semi-transparent) layers in vascular prostheses, since - as has been shown - carbon surfaces can be quite different from one another, depending on process conditions.

With cathode sputtering, a direct coating of the inside of a tube is not possible. It has been shown, however, that a sputtered layer can be transformed onto the

inside of a tube /9/, although the optimization and evaluation of these prostheses have not yet been completed.
A process with which ready-made tubes can be coated subsequently is that of
plasma deposition (sect. 4). We have coated PTFE tubing and several PTFE
prostheses using this method and obtained, depending on the conditions of the
coating, a reduction (and in one case even an increase) of the irreversibly adsorbed amount of fibrinogen as compared to uncoated prostheses (Fig. 5).

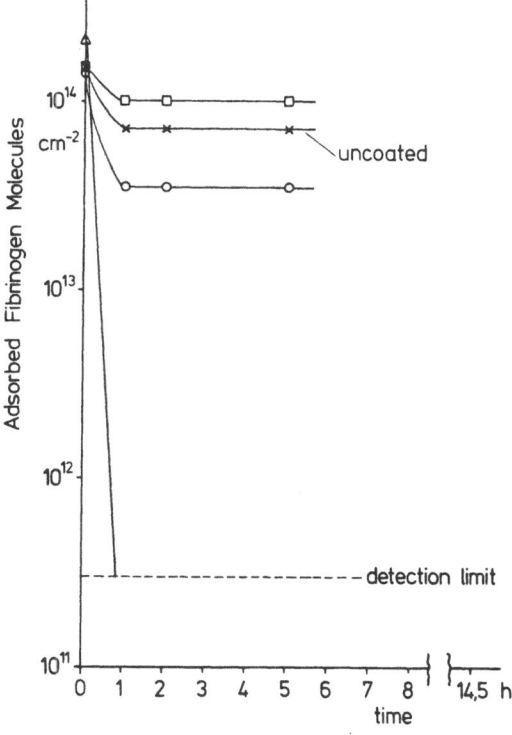

Fig. 5   Desorption of fibrinogen at PTFE prostheses, coated with thin
            carbon layers under different deposition conditions

For a first exploratory in vivo test only 5 prostheses were tried, each of which
was coated under different conditions. Thus, no statistically conclusive state-
ments are yet possible. So far, 1 prosthesis was still patent upon removal after
4 weeks. The others were occluded after 1 week. But among the prostheses
which were still patent at explantation, interesting differences were noted
between the coated prostheses and the uncoated prostheses, which, according
to the opinion of the surgeons, possibly indicates good long-term properties. As
demonstrated in Fig. 6, the uncoated prostheses had a thick, loosely constructed protein layer with numerous entrapped cells, whereas the carbon-coated
prostheses exhibited a thin, dense protein layer with no cellular materials.
Visually a shiny grayish-white layer over more than 90 % of the inner surface

was observed, and less than 10 % of the surface was covered with small micro-thrombi. On the uncoated prostheses, about 40 % of the surface was covered with deposited thrombi.

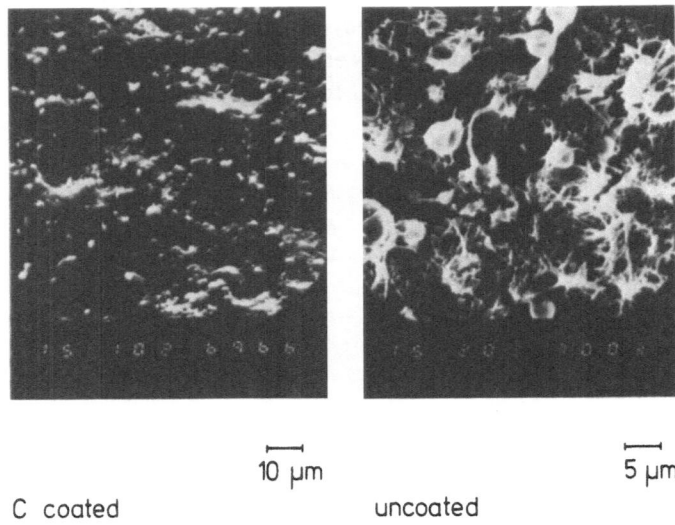

10 μm   5 μm

C coated   uncoated

Fig. 6   Scanning electron micrographs of the inside of a carbon coated (left) and an uncoated (right) PTFE vascular prosthesis (central sections both) after 4 weeks in carotis position of beagles. Both prostheses were patent on explantation

6.   Protein adsorption by coated and uncoated membranes

Protein adsorption not only plays a role within the bounds of a stan-dardized test, but can also have practical significance for the performance of membranes /13/. Fig. 7 shows the influence of a PAN or carbon layer on the feed-solution side of a 0,5 μm Nuclepore[R] porous membrane on the time depen-dence of its transmembrane flow (or ultra filtration rate). A bovine serum flows through the membrane and the decrease in the transmembrane flow is a result of fouling caused by protein adsorption. This can be derived from the ratio of the pore radius to the size of the protein molecule, and is confirmed by scanning electron microscopy. If membranes of this type were used for therapeutic plasma separation – a procedure which lasts 1 to 2 h – a higher average filtration rate could be achieved by appropriately coating the membrane (and thereby possibly reducing treatment time). There is still clogging of the pores since the pores have not been coated, and the proteins can penetrate into the pores. A similar and even more pronounced improvement was observed with a Cuprophan[R] mem-brane where whey was used as a protein solution (Fig. 8). Since the proteins in this example cannot penetrate into the pores, an apparently permanent improve-ment of the filtration was achieved by coating the feed solution side of the membrane with carbon.

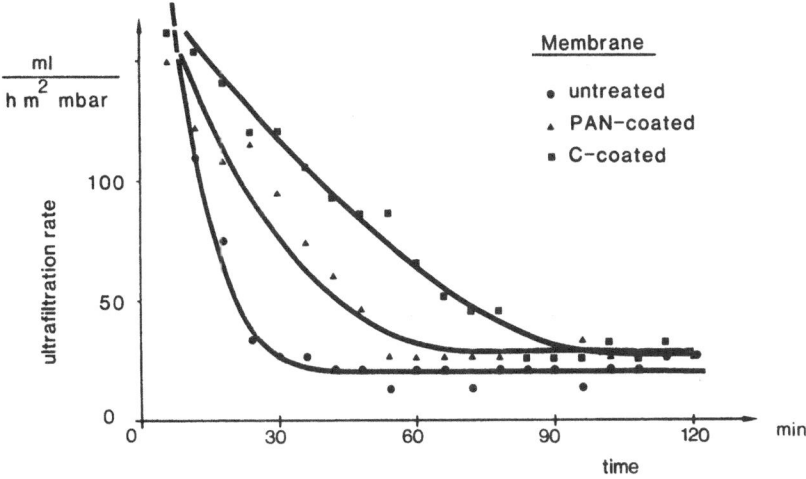

Fig. 7   Filtration of bovine serum with Nuclepore[R] membrane
          (pore size 0, 4 um)

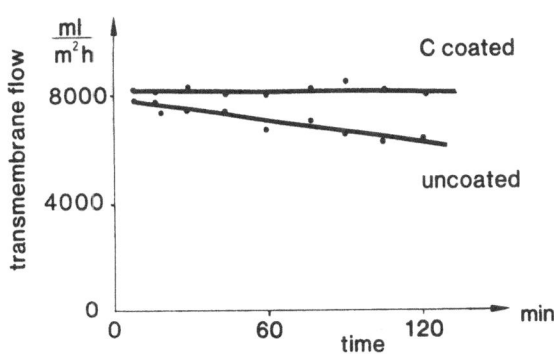

Fig. 8   Filtration of whey with Cuprophan[R] HDF membrane. Transmembrane
          flow vs. time for uncoated and carbon coated membranes

7.   Conclusion

    Based on the relationship between protein adsorption and thromboge-
nicity, a method was reported for fibrinogen adsorption/desorption measurements as a prescreening method. Processes for the coating of polymers with polyacrylonitrile and with carbon were discussed, and results on layers on flat films,
the inside of tubes and vascular prostheses, and on membranes were described.

Exploratory in vivo experiments show the influence of coating on the adsorbed protein layer in the prosthesis and the deposit of microthrombi. From this the basic significance of interaction between proteins and the biomaterial, as well as the significance of coatings as surface modifications were stressed.

Acknowledgements

We would like to express thanks to our co-workers, especially Mr. K. Birkner for the sputtering work, Dr. G. Hellwig for protein adsorption studies, Dr. B. Schindler for developing plasma deposition, and Mr. N. Stroh and Mr. M. Timmermann for filtration experiments. We are grateful to Dr. Lemm, Berlin, for competitive adsorption measurements.

REFERENCES

1.   R.E. Baier, The organization of blood components near interfaces
      Am. New York Acad. Sci., 283 (1977) 17

2.   S.W. Kim and D.J. Lyman, The interface of polymers with blood
      Appl. Polym. Symp. 22 (1973) 289

3.   E. Nyilas, T.H. Chin and G.A. Herzinger, Thermodynamics of native
      protein/foreign surface interactions, Trans. Amer. Soc. Artif. Intern.
      Organs 20 (1974) 480

4.   L. Vroman, A.L. Adams and M. Klings, Interactions among human
      blood at interface, Fed. Proc. 30 (1971) 1494

5.   T. Okano, M. Shimuda, L. Shinohara, K. Kataoka, T. Akaike
      and Y. Sakurai, The role of a microphase-separated structure in the
      interaction between polymer and platelet, 1st World Biomed. Congr.,
      Book of Abstracts, Baden (1980), 2.5.6

6.   M. Schaldach, R. Thull, P. Baurschmidt and R. Blaser, Elektro-optische
      Untersuchungen zum Koagulationsmechanismus des Systems Fibrinogen-
      Fibrin, Ber. Bunsen-Ges. 77 (1973) 794

7.   J.C. Bokros, Carbon biomedical devices, Carbon 15 (1977) 355

8.   H.S. Borovetz, D.D. Mateer, R.L. Hardesty and A.D. Haubold,
      Oxygen permeability of carbon-surfaced microporous membranes,
      J. Biomed. Mat. Res. 14 (1980) 145

9.   H. Bauser, K. Birkner, H. Chmiel, G. Hellwig and B. Schindler,
      Entwicklung blutverträglicher Kohlenstoffschichten auf Kunststoffen
      Biomed. Techn. 26 (1981) 116

10.   G. Hellwig, Messung der Proteinadsorption an Kunststoffoberflächen
       mittels dielektrischer und elektronenspektroskopischer Methoden,
       Chem.-Ing.-Technik 51 (1979) 530

11.   J.C. Bokros, L.D. LaGrange and F.J. Schoen, Control of structure
       of carbon for use in bioengineering, Chem. Phys. Carbon 9 (1972) 103

12.   H. Domininghaus, Die Kunststoffe und ihre Eigenschaften, VDI-Verlag
       Düsseldorf (1976)

13.   H. Bauser, H. Chmiel, N. Stroh and E. Walitza, Interfacial effects
       with microfiltration membranes, J. Membrane Sci. (1982) in press

PHYSICAL CHARACTERIZATION OF POLY(2-HYDROXYETHYLMETHACRYLATE)
GELS : EFFECT OF THE DILUENT CONTENT ON THE MECHANICAL AND
TRANSPORT PROPERTIES

C. Carfagna, C. Migliaresi, L. Nicolais and A. Sacerdoti

Polymer Engineering Laboratory
University of Naples
80125 Naples, Italy

ABSTRACT

Crosslinked poly(2-hydroxyethylmethacrylate), PHEMA, has been
prepared by radical polymerization in presence of different amounts
of water and water-diacetine solutions. While homogeneous samples
can be prepared at all water-diacetine content, amounts of water
higher than 40% causes the formation of macroporous opaque sponges.
Mechanical properties of water swollen samples have been measured
and related to the crosslink formation during the polymerization
process. From water diffusion kinetics in sorption and desorption
experiments at 37°C the diffusion coefficients have been calculated.
Measurements of the swelling at equilibrium in water-diacetine mix-
res indicate the presence of "cosolvency" phenomenon for the samples
studied.

INTRODUCTION

The use of polymeric materials for biomedical applications has
become more and more important in the last years.

Poly(2-hydroxyethylmethacrylate) gels (PHEMA) and its composites have been widely proposed for several uses as prosthetic materials for the good biocompatibility and/or hemocompatibility[1] which
seems to be connected to the water structure inside the gel[2].

In previous papers[3,4,5] it has been shown that different struc-

311

tures and permeabilities can be achieved by modifying the amount of additives used during the polymerization.

EXPERIMENTAL

The hydrogel used in this study is the poly(2-hydroxyethylmethacrylate), PHEMA, developed by Wichterle and Lim in 1961[6].

The commercial monomer was obtained from Rohm and Haas, Inc., and crosslinked at 37°C for 1 hour with 0.5% by wt. of ethylenedimethacrylate by using 0.1% by wt. of ammonium persulphate and sodium methabisulphite as redox initiator, in presence of different amounts of water and water-diacetine mixtures as indicated in Table 1. The polymerization was carried out between two glass plates separated by a silicon rubber gasket and two polyester sheets in order to avoid adhesion between polymer and glass.

Once the reaction was completed, the samples were washed and immersed in distilled water in order to eliminate the diacetine and any unreacted material, and to swell up to the equilibrium value.
Mechanical tests were performed on the samples, cut in the dumb-bell shape (ASTM 638), immersed in water at 37°C by using an Instron Universal Testing Machine mod. 1112.
Water sorption and desorption kinetics were determined with a gravimetric technique. The kinetics of desorption was performed by putting the samples in a dessiccator over silica gel at a temperature of 37°C.

Table 1. Mechanical properties of PHEMA gels prepared with different amounts of water and water-diacetine

Sample	Initial water content (wt.%)	Initial diacetine content (wt.%)	E (Kg/cm^2)	σ_b (Kg/cm^2)	ε_b
1	40	--	4.0	1.6	0.61
2	50	--	3.5	2.2	1.11
3	60	--	2.4	1.7	1.25
4	70	--	1.5	0.8	0.81
5	15	15	5.2	2.6	0.75
6	20	20	4.0	2.4	1.07
7	25	25	3.1	1.9	1.07
8	30	30	2.5	1.4	0.88

The loss of water was recorded until no significative change in weight was observed. In order to evaluate the complete water desorption, the samples were put in a dry oven at 100°C. No significative differences were observed between the two asymptotic values of weight loss.

The equilibrium swelling degree in water-diacetine solutions at different compositions and at T = 37°C was finally measured.

RESULTS AND DISCUSSION

Typical stress-strain curves for different initial water and water-diacetine contents are reported in Figures 1 and 2, respectively. In both cases the presence of additives reduces the elastic modulus E, and the strength at break σ_b, slightly affecting the elongation at break ε_b. These mechanical parameters are reported in Table 1. The strong dependence of the elastic modulus (E) on the diluent content in the reaction batch may be attributed to the number of intramolecular rings. In fact, previous analysis, performed on different systems, revealed that the number of intramolecular bonds increased linearly with increasing dilution in the mixture[7]. The reduction of the crosslinking degree with increasing diluent content might be caused by the extension of the cyclization reaction.

In Figure 3 the Young moduli (E) are reported as function of the diluent concentration in the reaction mixture for all the samples studied. The data are well fitted by a straight line which gives a value of E = 8.25 Kg/cm^2 when the crosslink reaction is carried out without diluent. This value well compares with previously reported data[8]. The extrapolation at zero elastic modulus reveals that with a diluent concentration of approximately 80% the crosslinking reaction does not occur. It is interesting to notice that the samples crosslinked in presence of 60% of water and water-diacetine solutions show the same value of elastic modulus in spite of the different water content (see Table 2). This, as will be discussed later, is due to the fact that in the case of the sample number 3 the water in excess of 40% is concentrated in microvoids and therefore does not affect very much the value of the elastic modulus. Also the water transport analysis performed on the studied samples revealed a progressive reduction of the network density by increasing the diluent content. In fact from the data of Table 2 an increase in equilibrium water uptake is always observed increasing the additive concentration for both the series of specimens.

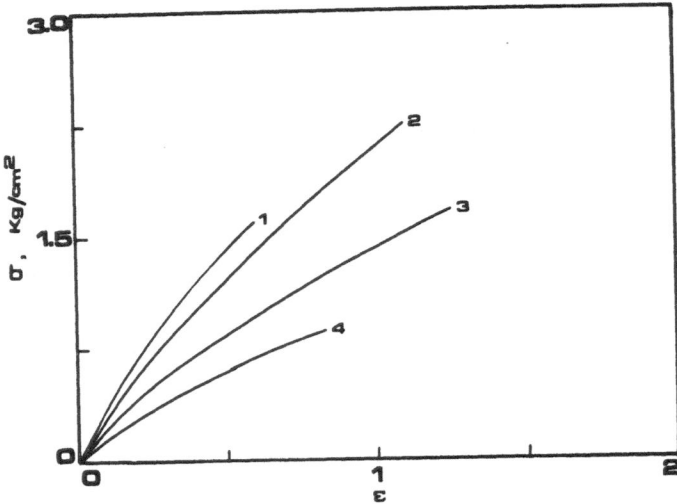

Figure 1. Stress–strain curves for PHEMA gels prepared with initial
water content of: (1) 40%, (2) 50%, (3) 60%, (4) 70%.

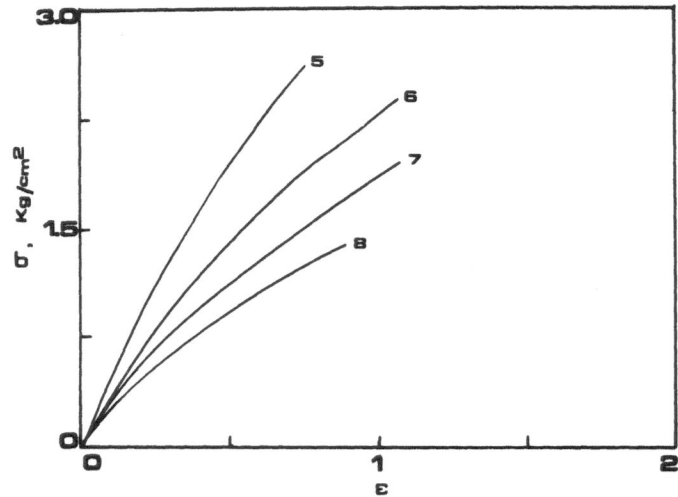

Figure 2. Stress–strain curves for PHEMA gels prepared with initial
diacetine/water content of: (5) 30%, (6) 40%, (7) 50%,
(8) 60%.

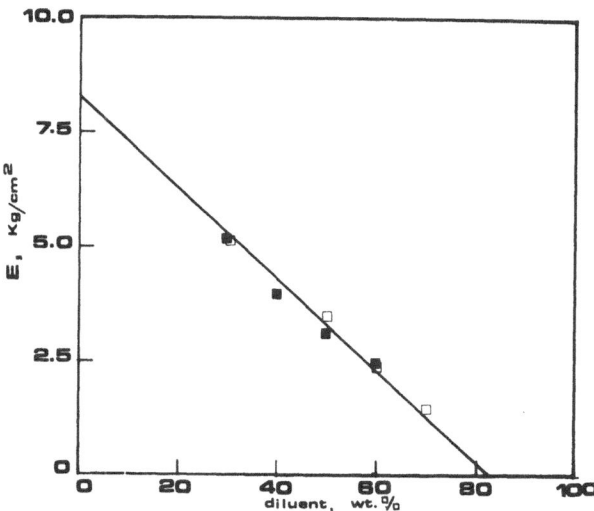

Figure 3. Elastic modulus (E) vs. diluent content for samples prepared from water (□) and water/diacetine mixtures (■).

Table 2. Transport properties of PHEMA gels prepared with different
initial contents of water and water/diacetine

Sample	Initial water content(wt.%)	Initial diacetine content(wt.%)	Eq. water (wt.%)	$D_S x 10^7$ cm^2/s	$D_d x 10^7$ cm^2/s
1	40	–	41.1	3.3	2.9
2	50	–	44.2	2.4	2.8
3	60	–	56.5	1.2	1.7
4	70	–	66.1	0.5	1.2
5	15	15	40.5	2.6	2.2
6	20	20	41.2	2.8	3.1
7	25	25	41.9	3.0	3.0
9	30	30	42.4	2.1	1.9

The introduction of diacetine in the reaction batch reduces the
swelling and the water gain. While the sample prepared with an initial water content of 60% shows an equilibrium water uptake of 56.5%,
the equivalent sample containing 30% of water and 30% of diacetine
shows an equilibrium water uptake of 42.4% by weight. The shrinkage

which results in this latter case can be usefully used for impro-
ving the physical adhesion between fiber and polymer in the case of
fiber reinforced PHEMA composites. In fact, one of the limiting factors in reinforcing swollen polymers is the lack of adhesion caused
by the swelling of the matrix. In contrast, once the matrix shrinks
during the crosslinking reaction, a state of compressive stress aro-
und the fillers[9] enhances the interfacial adhesion between the two
phases and consequently the resulting mechanical properties.

The Figures 4 and 5 report the water content kinetics for the
samples prepared with water/diacetine mixtures and water respectively. Following Crank[10] the diffusion of a penetrant in a swelling
matrix can be analyzed similarly to that in a non-swelling system
by assuming a proper measure of the penetration depth. The solution
of the differential equation for diffusion gives[10,11], for sorption
and desorption in the initial stage:

$$(1) \qquad \frac{W_w}{W_{aw}} = \frac{4}{\sqrt{\pi}} \cdot \sqrt{\frac{D \cdot t}{L_d^2}}$$

where W_w is the water content, W_{aw} is the asympthotic or initial
water content for sorption or desorption respectively, D is the diffusion coefficient and L_d is the thickness of the dry sample. From
Figures 4 and 5, using a linear regression technique the derivative:

$$(2) \qquad \frac{d\,W_w}{d(\sqrt{t})} = p$$

can be calculated and then the values of:

$$(3) \qquad D = \frac{\pi\,p^2}{16\,W_{aw}^2} \cdot L_d^2$$

are computed and reported in Table 2.

The analysis of the kinetics of water transport reveals that,
while for the samples prepared with water and diacetine (Fig. 4)
the diffusion coefficients show no significant variations with the
diluent concentration, for the specimens 3 and 4 (Fig. 5), prepared
only with water, a reduction of diffusion coefficient is observed.
In these latter cases the diffusion process is well represented by

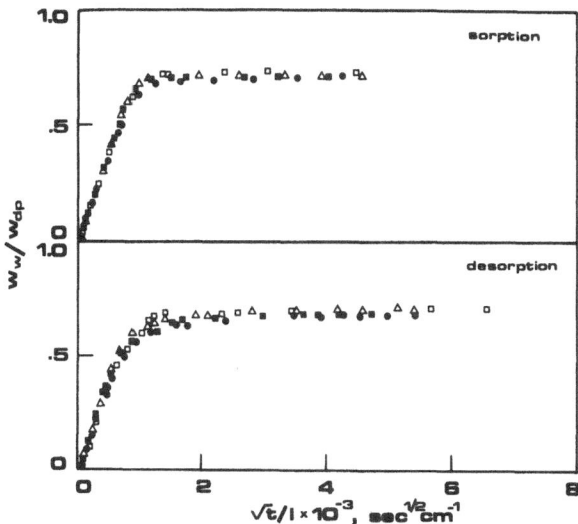

Figure 4. Water content (W_w) referred to the weight of the dry polymer (W_{dp}) versus $\sqrt{t}/l_d$, were "t" is the time and "l_d" is
the thickness of the dry sample, of samples prepared with
amounts of water/diacetine mixtures equal to: (●) 30%,
(■) 40%, (△) 50%, (□) 60%.

a dual mode sorption. In other words, the absorbed water is in part
diluted in the matrix according to a random mechanism, as proposed
by Flory[12], and in part aggregated as clusters in previously existing
voids. For all the samples it can be assumed that the plasticization
water content is of the order of 40%; if a water density equal to
unity is hypothized for the water present in the pores, a void fraction of 16% and 26% may be computed for the specimens prepared with
an initial water content of 60% and 70% respectively (samples 3
and 4). These "spongy" samples were not transparent as those with
an equilibrium water uptake of the order of 40%, but opaque so re-
vealing a phase separation and the presence of liquid water in the
pores. The intensity of the opacity was increasing with the initial
water content.

Moreover, the analysis of the plots relative to the water up-
takes vs. the square root of time (Fig. 5) reveals anomalies for
the spongy sample prepared with an initial water content of 70%.
In fact, while for the clear specimens the transport phenomenon is
successfully described by the Fick's law, in the case of the spongy
samples the water uptake increases also after a long immersion in

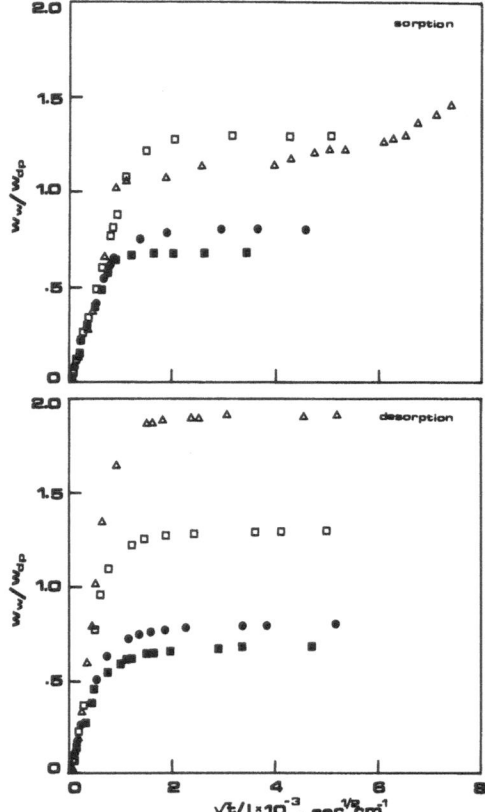

Figure 5. Sorption and desorption curves for the samples prepared
with amounts of water of: (■) 40%, (●) 50%, (□) 60%,
(△) 70%.

water as a consequence of a coupled diffusion-relaxation sorption
mechanism[13].

    The water transport analysis was extended to the dry samples
immersed in water/diacetine solutions of different compositions, in
order to evaluate, if present, cosolvency phenomena. It is well known
that the solubility and the degree of swelling of a polymer in a
mixture of two liquids can be larger than those in each liquid se-
parately[14,15]. This phenomenon, which has been referred to as cosolvency, is related to the free energy of mixing in the ternary system.
Our analysis performed on a very large number of samples revealed
a maximum swelling for the ratio water/diacetine of 40/60 (Figures
6 and 7). These results are similar to those obtained by Refojo et
al.[16] on different systems.

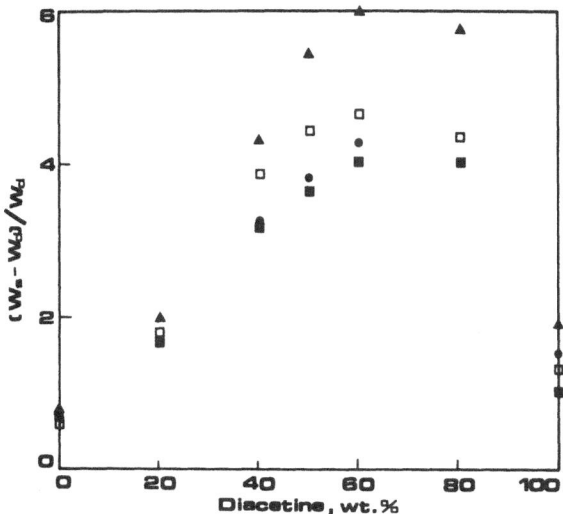

Figure 6. Weight gain (W_s-W_d) referred to the weight of the dry sample (W_d) versus the diacetine weight content in the water/
diacetine solution for the samples prepared with initial
water/diacetine contents of: (□) 30%, (●) 40%, (■) 50%,
(▲) 60%.

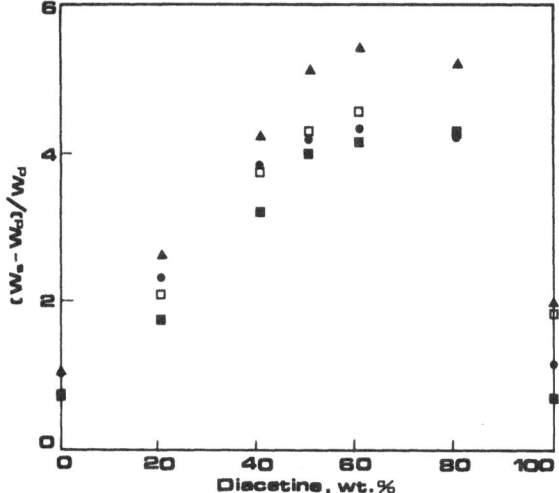

Figure 7. Cosolvency phenomenon for the samples prepared with initial
water contents of: (■) 40%, (□) 50%, (▲) 60%, (●) 70%.

CONCLUSIONS

The results reported in this work showed that mechanical and transport properties of Poly-2-hydroxyethylmethacrylate are strongly affected by the composition of the reaction mixture, and in particular by the ratio water/diacetine. It has been shown that it is possible, using different types and amounts of additives, to obtain a wide range of physical properties of swollen PHEMA as required for various biomedical applications. Moreover, the spongy samples could be of interest in the design of biomedical implants whose morphology is characterized by porous structure with very high water contents.

Work partially supported by CNR Special Project "Chimica Fine Secondaria"

REFERENCES

1. B.D. Ratner,A.S. Hoffman, Am. Chem. Soc. Symp. Ser., 31:1 (1976)
2. S.D. Bruck, J. Biomed. Mater. Res., 7:387 (1973)
3. R.A. Haldon, B.E. Lee, Br. Polym. J., 4:491 (1975)
4. L. Sprincl, J. Kopecek, L. Lim, Calcif. Tissue Res., 13:63 (1973)
5. K. Kliment, M. Stol, M. Raab, J. Stohl, J. Biomed. Mater. Res., 2:473 (1968)
6. O. Wichterle, D. Lim, Nature, 185:117 (1960)
7. J. Hasa, J. Janacek, J. Polym. Sci., Part C, 16:317 (1967)
8. C. Migliaresi, L. Nicodemo, L. Nicolais, P. Passerini, J. Biomed. Mater. Res., 15:307 (1981)
9. L. Nicolais, R.A. Mashelkar, J. Appl. Polym. Sci., 20:561 (1976)
10. J. Crank, "The Mathematics of Diffusion", Clarendon Press, Oxford (1975)
11. R.B. Bird, W.E. Stewart, E.N. Lightfoot, "Transport Phenomena", J. Wiley Sons Inc., New York (1960)
12. P.J. Flory, "Principles of Polymer Chemistry", Cornell University Press, Ithaca (1960)
13. D.J. Enscore, H.B. Hopfemberg, V.T. Stannett, A.R. Berens, Polymer, 18:1105 (1977)
14. A. Beerbower, J.R. Diekey, Asle Trans., 12:1 (1969)
15. Z. Rigbi, Polymer, 19:1229 (1978)
16. M.F. Refojo, F.L. Leong, J. Biomed. Mater. Res., 15:497 (1981)

SECTION III
MEDICAL AND SURGICAL APPLICATIONS OF POLYMERS

THE DESIGN OF A SMALL DIAMETER ARTERIAL REPLACEMENT

D. Annis, T.V. How and R.M. Clarke

Bioengineering & Medical Physics Unit

University of Liverpool, P.O.Box 147, Liverpool, UK

The concept of a tubular conduit of a synthetic material to be
used to replace a length of diseased or damaged artery became a
reality only in 1952 when Voorhees (1) thought it possible that
a tube of textile fabric might succeed where solid walled tubes
had failed, the woven structure of the fabric forming a skeleton
upon which blood might clot and become entrapped within the fabric
ultimately to be transformed into a tube of living collagenous scar
tissue reinforced by the fibres of the woven fabric within it. It
is upon that principle that the first arterial grafts were constructed and it has remained the guiding principle to this day.

The first grafts to be used in man were of fabric of the newly
developed 'Vinyon N', a polymeric material which could be drawn
into fibres whose strength was based upon amide linkages similar
to those that hold together natural polymers. Voorhees' grafts
were seamed tubes constructed of two layers of flat woven material.
In 1954 he was able to report the successful use of these grafts
in 18 patients in the treatment of aortic and popliteal aneurysms.
However, during the following years it became increasingly apparent
that, while the principle was sound, the nylon material he used
lost strength and was not able to withstand the repeated pulsing of
blood at arterial blood pressure. The quest began for a stronger
fibre and for a technique of seamless tubular weaving.

Poly(ethylene terephthalate) (Dacron) proved to be the suitable
material and even to this day remains the most used polymer for the
purpose. The textile industrial already had techniques of seamless
tubular weaving and these were readily applied to the manufacture
of arterial grafts. However, the essential need for a strong,

rigid yarn meant that the woven tubes were inflexible, they kinked
when an attempt was made to place them in the naturally curved
lines of blood vessels. Again, it was to the textile industry
that this problem was directed, and the problem solved by the
application of the process of circumferential heat crimping.

Since that time textile technology has provided the many minor
modifications of design that followed. These have included
changes in the yarn and in the pattern of weaving. The tightness
of the weave determines the degree of porosity and the degree to
which blood passes through the interstices of the graft prior to
its clotting. This has been optimized. Practice, too, over the
years has established the criteria of quality control both for the
manufacture of the polyester itself and for the production of the
graft in its final form. The occasional failure in use today can
usually be ascribed to a failure of the quality control of manufacture and not of design.

Grafts of fabric construction have proved remarkably effective in
the 25 years of their use and indeed the emergence of vascular
surgery has largely depended upon their invention. However, the
limitations of their usefulness have also been increasingly appreciated. Grafts of woven fabric construction usually fail when used
for the replacement of arteries smaller than 7mm internal diameter.
They fail either because of early thrombosis within hours of
implantation or because of later failure due to the development of
an hyperplastic intima at or near the union between the synthetic
graft and the natural artery. Progressive narrowing ultimately
leads to diminished flow and final thrombosis.

A feature of the crimped fabric graft which it can be assumed will
predispose to thrombus formation in the graft is the grossly uneven inner surface due to crimping. This must give rise to turbulence of blood at the surface, the centripetal extension of the
turbulence being relatively greater in the tubes of smaller bore.
Turbulence and the increased shear of blood within the lumen of
the vessel are well-known causes of intravascular thrombosis.
Generally, fabric grafts larger than 7mm internal diameter will
tolerate this disturbance of blood flow but smaller grafts of the
same construction will not.

The invention of a porous flexible arterial graft without the need
for crimping and having a grossly smooth inner lining has been the
only major advance in the construction of arterial grafts during
the last twenty-five years. It is a tube of expanded microfibrous
PTFE (Gore-tex). With grafts of this construction it has been
possible to replace arteries of 6mm internal diameter with only
moderate success but they are successful only rarely when used

for the replacement of smaller arteries. As with the fabric
grafts, Gore-tex, too, was not designed for the construction of
artery grafts. It was designed as a windproof and waterproof
fabric. It was the application of this unrelated technology to
the production of grafts that brought this advance in vascular
surgery.

There is an urgent clinical need for the design of and the development of a synthetic arterial replacement whose internal diameter
ranges between 3mm and 7mm. With others we have worked on the
commonly held but as yet unproven hypothesis that small grafts
are likely to succeed only if they match the physical characteristics of the natural artery. So far the resemblance between the
synthetic artery and the natural one has been wholly anatomical
and, in that gross sense, an expanded PTFE graft is a closer model
of the natural artery than is the crimped graft. However, the
rheological features of Gore-tex are widely different. Rheological features of arterial prostheses have not been a consideration
in the design of any grafts in current use today.

It has been suggested by many authors that failure of small bore
inelastic prostheses may be the result, in part, of a mismatch
in the mechanical properties between the synthetic artery and the
natural artery to which it is attached. We hold to the commonly
held, but as yet unproven, hypothesis that a successful small
diameter arterial prosthesis should be designed to closely match
the mechanical properties of the natural vessel which it replaces.

The mechanical stresses and haemodynamic changes that are brought
about by the implantation of an inelastic graft into a naturally
elastic vessel can, in general, be anticipated by the application
of the laws of haemodynamics. At the point of union between a
natural and a circumferentially rigid synthetic vessel there will
be, with each pulse, an ever changing discrepancy in the diameter
of the rigid prosthesis and the elastic natural vessel to which it
is attached. The creation of sudden narrowing or divergence of
flow boundaries is known to produce flow disturbance, the severity
of which depends upon the relative change in diameter and the
Reynolds number. This is well documented in steady flow conditions.
However, in pulsatile arterial blood flow the highly transitional
nature of the flow, the presence of secondary flow and flow separation may produce extremely complex flow patterns. It is probable
that secondary flow and flow separation will develop at the anastomosis and that high stresses in the wall of the natural artery will
occur close to the anastomosis between the rigidly held graft and
the compliant artery. These stresses would affect both the wall of
the artery itself and the endothelial surface near the site of
union. It is more than likely that these factors have a part to
play in the formation of thrombus and of an hyperplastic intima, the
common causes of failure of small bore grafts.

Woven and knitted fabric grafts in present-day use and the more
recently used expanded PTFE grafts (Gore-tex) have an apparent
longitudinal compliance either because of the crimping or because
of the internal microfibrous construction of the material. Unlike
the natural artery the synthetic prosthesis has insignificant
radial and circumferential compliance, being constructed of highly
rigid polymers (PET:PTFE). The interposition of a circumferentially rigid tube in a blood vessel must have a profound effect not
only upon the flow within and near the graft but upon flow downstream from the graft. With each heart beat a pulse is transmitted down blood vessels at approximately 5 metres per second.
For example, pulse wave velocity is increased in the rigid walled
graft, (absolute rigidity would accelerate it to the speed of
sound in blood) leading to an increase in wave reflection and
kinetic energy losses.

The three dimensional compliance of natural arteries varies with
their anatomical site. Most arteries are orthotropic structures.
That is, the elastic modulus is constant throughout any one of
the three orthogonal directions. However, the elastic modulus in
each of these directions is different. The rheology of blood
vessels has been extensively studied and thorough analyses have
been developed, notably by Patel et al.   (2) . From them have
come complex mathematical models that describe the behaviour of
the walls of natural vessels during pulsatile flow, in which are
described the simultaneous three-dimensional changes of the wall
of the vessel in relation to the haemodynamics of flow. There
are, therefore, mathematical expressions of the behaviour of a
natural vessel from which it might be possible to design a
matching prosthesis.

Design of small diameter vascular grafts requires more than a
mechanical replica of the vessel. Firstly, it requires the design
of an elastomeric polymer with an elastic modulus closely related
to that of the natural vessel. The polymer must be tissue
compatible and blood compatible.

Secondly, it requires a method of fabricating the polymer to produce a material which is porous, in which the interstices are not
so large as to allow of the free flow of blood through them when
in contact with blood at normal blood pressure.

Thirdly, in producing a cylindrical tube of the material, we
suggest that it is desirable to so regulate the construction as
to reproduce the anisotropy of the natural arteries.

In our own work we have gone some way towards these objectives.
During the last five years we have developed a process of electro-

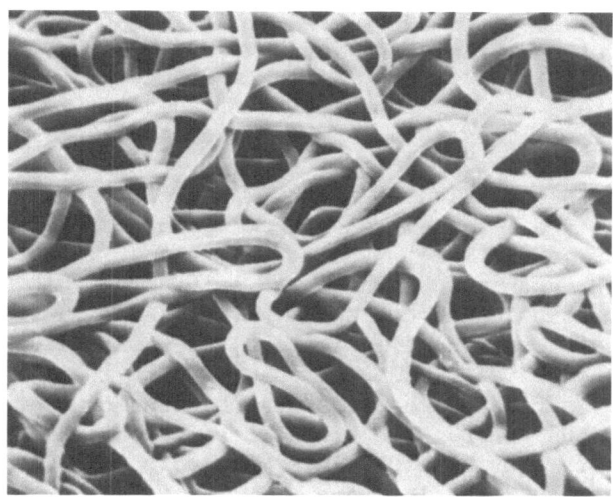

Figure 1.   SEM of the surface of the graft. The diameter
            of the fibres ranges between 1 and 1.5 microns

static spinning by which we have been able to produce anisotropic
cylindrical tubes whose walls are porous and microfibrous, being
composed of a three-dimensional mesh of polyetherurethane fibres
each of which has a diameter of 1 to 1.5 microns (Figure 1). We
have described our work elsewhere (3), but it may suffice to say
that they have excellent handling properties and satisfy the
surgeon in his need to make a firm union at the junction between
the natural and the synthetic artery. A firm but delicate attachment results from the ingrowth of connective tissue into the interstices of the surface of the graft, securing the graft at its
junction with the natural artery and allowing of the development
of a fine surface neoadventitia and, according to the animal
species, a variable extent of new intima on the blood surface.
Thus in the dog a completely new intima formed throughout the
length of a 13cm long, 6mm bore, synthetic artery. More recently
we have had fair success with implantation of 4mm bore grafts as
interposition vessels in the common cartoid artery in dogs.

There are a number of important design features of our graft.
Firstly, the fibres are of approximately 1μm diameter. This
ensures a closer, more delicate attachment during the healing
stage. In the process of attachment motile macrophages recognize
the fine fibres as a size that they can easily engulf. The
arrival of fibroblasts allows of a deposition of collagen fibre

so close to the polyurethane fibres of the graft that on electron
microscopy there is no gap visible between those fibres and the
ingrowing connective tissue at a resolution of 20Å. A second
important feature is that penetration of these living connective
tissue cells is limited to a depth of about 20-30µm, beyond which
the entanglement of polyurethane fibres prevents further penetration of living cells. The usual thickness of the wall of the 4mm
graft is about 500 µm, leaving the great bulk of the thickness of
the wall of our graft free of collagen. After implantation compliance remains unchanged without the stiffening that would follow
the deeper penetration of collagenous fibrous tissue. In consequence our grafts remain pulsatile even two years after implantation. A third design feature of our graft is its grossly smooth
but open inner surface. This ensures the minimum disturbance of
blood through the graft. The inner microfibrous surface becomes
impregnated with fibrin which, in some species, may be replaced by
a new intima and endothelium.

We have developed a technique for predicting the mechanical
behaviour of our graft based on static data obtained experimentally
from the graft, with which it is possible to make direct comparisons between it and the rheological data from natural vessels (2).
Although the graft that we make has circumferential compliance comparable to that of natural arteries, the longitudinal compliance is
greater than some of the peripheral limb vessels that we would wish
to replace. We are working on the problem of modifying the anisotropy of our graft more closely to match that of natural arteries,
and we have shown that this is possible, at least in 10mm diameter
grafts.

We believe that the development of a successful synthetic graft to
replace the smaller arteries of the body will be achieved only
when there is a fuller understanding of the rheology of natural
arteries and of the haemodynamics of blood flow within them. With
this information it should be possible truly to design a successful synthetic replacement with matching rheological and haemodynamic charcteristics.

References

(1) A.B. Voorhees, A. Jaretski, A.H. Blakemore, The use of tubes
constructed from Vinyon 'N' cloth in bridging arterial defects.
Ann. Surg. 135:332 (1952).
(2) D.J. Patel, R.N. Vaishnav, Rheology of large blood vessels, in:
"Cardiovascular Fluid Dynamics", D.H. Bergel ed., Academic Press,
London & N.Y., Vol. 2, 1972.
(3) D. Annis, A. Bornat, R.O. Edwards, A. Higham, B. Loveday, J.
Wilson, An elastomeric vascular prosthesis. Trans. Am. Soc. Artif.
Int. Organs 24:209 (1978).

THE DEVELOPMENT OF SMALL DIAMETER VASCULAR PROSTHESES

Donald J. Lyman
Department of Materials Science and Engineering
University of Utah
Salt Lake City, Utah 84112, USA

We have been investigating the effect of chemical structure and surface properties of synthetic polymers on the coagulation of blood. The ultimate goal of these studies is the development of satisfactory prostheses to bypass obstructed arteries of small diameter (less than 6 mm I.D.) for use in aortocoronary surgery and to replace damaged veins. The surgeon currently must use autogenous tissue such as the saphenous veins in these situations because of the lack of a good synthetic polymer prosthesis. Although the results of using autogenous saphenous veins are good, this method has certain disadvantages. Some patients do not have satisfactory veins either because of prior removal, disease, or because the vein is of inadequate size. Even with patients having satisfactory veins, one has to contend with the problem of limited quantity and a significant increase in operating time. Therefore, the development of a successful small diameter synthetic polymer prosthesis is of utmost importance to both the patient and the surgeon.

Since the current prosthesis materials (Dacron and Teflon) are thrombogenic, our approach to solving the small diameter vascular prosthesis problem was to develop a blood compatible polymer. Much of these studies have been summarized in Reference 1. One of the polymers that we prepared which appeared to have a nonthrombogenic surface was a block copolyether-urethane-urea material. However, when this urethane material was fabricated into small diameter solid wall tubes and implanted in dogs, occlusion occurred within several days. Prostheses recovered from these experiments gave evidence that the thrombus was originating at the anastomosis between the artery and the prosthesis. We believed that a mismatch in compliance between the polymer and

the pulsating artery was causing sufficient mechanical and hemodynamic trauma to stimulate thrombosis and intimal hypertrophy.[2]

While thin-walled tubes were as compliant as the natural
artery, they could not be sutured without tearing. To obtain
both compliance and sutureability, we developed a process to
fabricate tubes whose walls were filled with voids. This lower
wall density compared to the air-dried films resulted in an increased elasticity of the prosthesis. Fourier transform infrared
surface studies and dynamic modulus-temperature studies of the
bulk materials indicated that with precise control of fabricating
conditions, one could retain the surface and bulk morphology of
the air-dried film. Implantation of these compliant urethane
prostheses in dogs gave support to our hypotheses on blood coagulation and compliance since those prostheses in which the elastic
index best matched the dog femoral arteries had the best patency
rate (82%) at one month.[3] Those prostheses which were either
more elastic or less elastic were less successful. Some of these
compliant grafts were left in the experimental animal and were
still patent at over 24 months implantation time. For comparison,
a series of polytetrafluoroethylene grafts implanted under identical conditions were not patent (0%) at one month. Because of
the improvement in patency, these compliant polyurethane grafts
have now been approved by the U.S. Food and Drug Administration
for human implantation.

An alternate approach to increase compliance of urethane
block copolymers is to vary the structure of the soft polyether
block to lower the initial modulus. In our analysis of structure-
property relationship of the block copolyurethanes, it would
appear that a polyether segment based on a

$$\text{+OCH}_2\text{CH}_2\text{CH}_2\text{+}_x \quad \text{or} \quad \text{+OCH}_2\text{CH}_2\overset{\overset{\displaystyle CH_3}{|}}{\text{CH}}\text{+}_x \quad \text{or} \quad \text{+OCH}_2\overset{\overset{\displaystyle CH_3}{|}}{\text{CH}}\text{-CH}_2\text{+}_x$$

repeat segment would give us materials having the desired mechanical properties. Initial synthesis and incorporation of a polytri-
methylene glycol into a block copolyether urethane did show a reduction of the elastic modulus (see Table I).[4] Thus, one could
fabricate a more compliant graft which shows improved suturing
properties. Of interest, though, is whether the new block copolyurethane also show the needed nonthrombogenic surface properties. This is important since our earlier studies have shown
that changes in polyether diol molecular weight and polyether
diol repeat structure can affect protein adsorption and platelet
adhesion. Biological studies on these materials are currently
underway.

TABLE I

MECHANICAL PROPERTIES OF BLOCK COPOLYETHER-URETHANE-UREAS

	Tensile Strength (Psi)	Elastic Force, 100% Elongation (Psi)	Elastic Modulus (Psi)
PEUU Based on Polypropylene Glycol 1025	3564	1515	2758
PEUU Based on Polytrimethylene Glycol (Mol. Wt. 1400)	2122	550	1272

In summary, our studies have shown that it is necessary to incorporate the proper chemical, physical and mechanical properties into a polymer implant if a workable small diameter vascular prosthesis is to be achieved. To do this, one must couple basic research on the polymer surface using a variety of techniques with the more applied implant studies where the variables of mechanical properties show their influence.

REFERENCE

1. D. J. Lyman, Structural order and blood compatibility of polymer prostheses, in: "Structural Order in Polymers," F. Ciardelli and P. Giusti, eds., Pergamon Press Ltd., Oxford (1981).

2. D. J. Lyman, F. J. Fazzio, H. Voorhees, G. Robinson and D. Albo, Jr., Compliance as a factor in effecting patency rates of a new copolyurethane vascular graft, J. Biomed. Maters. Res. 12:337 (1978).

3. K. B. Seifert, D. Albo, Jr., H. Knowlton, and D. J. Lyman, The effect of the elasticity of prosthetic wall on the patency of small diameter arterial prostheses, Surgical Forum 30:206 (1979).

4. D. J. Lyman, Manuscript in preparation.

TESTING DEVICES FOR THE EVALUATION OF PROSTHETIC HEART VALVES

G. Nardi', S. Cicconardi', R. Pietrabissa" , P. Dario'",
M. Lazzoni" and P. Giusti"

'Ist. di Macchine, Fac. Ingegneria, Univ. Pisa, Italy
" Ist. di Chimica, Fac. Ingegneria, Univ. Pisa, Italy
'"Centro 'E. Piaggio', Fac. Ingegneria, Univ. Pisa, Italy

INTRODUCTION

The ever more increasing use in cardiovascular surgery of valvular prostheses, calls for their characterization and evaluation
both in vitro and in vivo. Such a need is felt especially in
studying and realizing new artificial valves. In fact, as the
perfect prosthesis has not yet been designed, many laboratories are
looking for a solution to this problem. The devices able to test
in a more and more exact way the performances of each valve, are
therefore very important[1-5]. To this purpose the pulsatile flow and
the steady flow devices have been developed. In the first class
are the mock circulatory system, for functional evaluations, and
fatigue test devices; some apparatus able to evaluate the variables
of interest, as the pressure drop or the steady retrograde flow,
belong to the second class.

Tests have usually a comparative meaning and, for this reason,
it is useful to test a standard valve, the natural one for instance.
The fluid used in such tests has a great importance: blood would be
perfect of course, but usually a 30% glycerine and 70% water solution is used because its viscosity and density characteristics are
very similar to those of blood. For studying the chemical and biological blood-prosthesis interactions, in vivo tests using animals

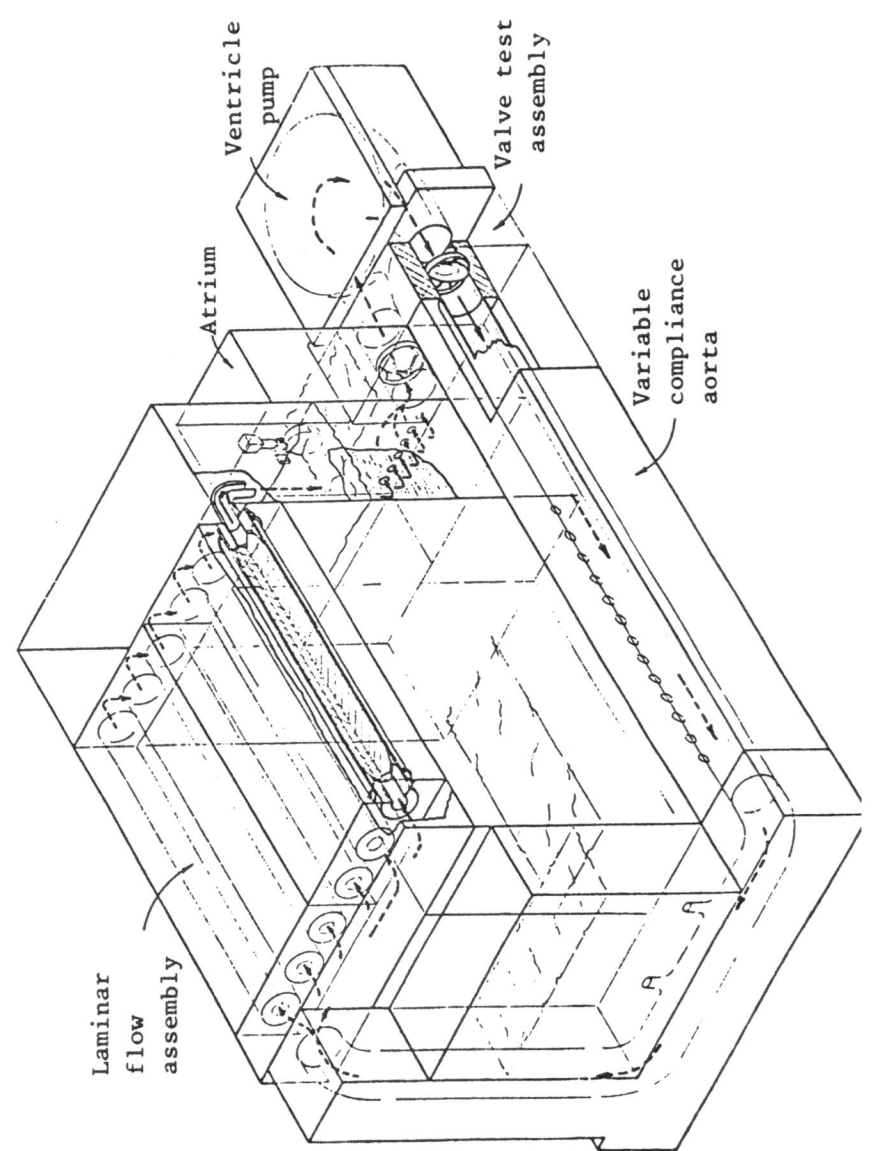

Ventricle pump

Valve test assembly

Atrium

Variable compliance aorta

Laminar flow assembly

Fig. 1. The mock circulatory system. (From Nardi et al.[6]. With permission.)

and hence blood as circulatory fluid, must be performed.

In this paper, some devices which have been designed at the Engineering School, University of Pisa, in order to test prosthetic heart valves in vitro are presented.

MOCK CIRCULATORY SYSTEM

It is a device able to evaluate the valve performance in conditions similar both to the physiological ones and to the most common cardiovascular pathologies. Such an apparatus has already been described in detail in a previous paper[6]. Besides testing prosthetic valves, this mock circulatory system can evaluate vascular prostheses, left heart replacement pumps, intraaortic balloon pumps and other circulatory assistance devices. Neither the anatomic complexity nor the physiological behaviour can be exactly reproduced, but the simplicity of use and the good results obtained with this device, show its efficiency.

The whole system, shown in Fig. 1, is built in Plexiglass. The circulatory fluid is ejected by an electropneumatically driven ventricular pump. Downstream of the pump, an aortic valve assembly is located: two different models have been built in order to offer lateral or frontal view of the prosthesis movements. Suitable stent adapters allow to test prostheses of different type and size. The aorta is a variable compliance rubber tube. Through a rigid conduit the fluid is conveyed to the laminar flow assembly which controls peripheral resistances. Aortic compliance and peripheral resistances are hydropneumatically controlled. The fluid, passing through a venous reservoir open to atmospheric pressure, reaches the left atrium. This is a rigid wall chamber in which a hydropneumatic system relates cardiac output to venous return, reproducing Frank--Starling's Law. Between atrium and ventricle there is another valve test assembly which allows to test mitral valves.

Output data from the mock circulatory system, during a cardiac cycle, are: the aortic and ventricular pressure diagrams, the flow rate through the valve diagram and the average values of these variables. As an example, Fig. 2 shows the output data obtained while testing a tilting disc Hall Kaster (HK) valve of 27 mm tissue anulus diameter (TAD). Nine different tests have been performed at three

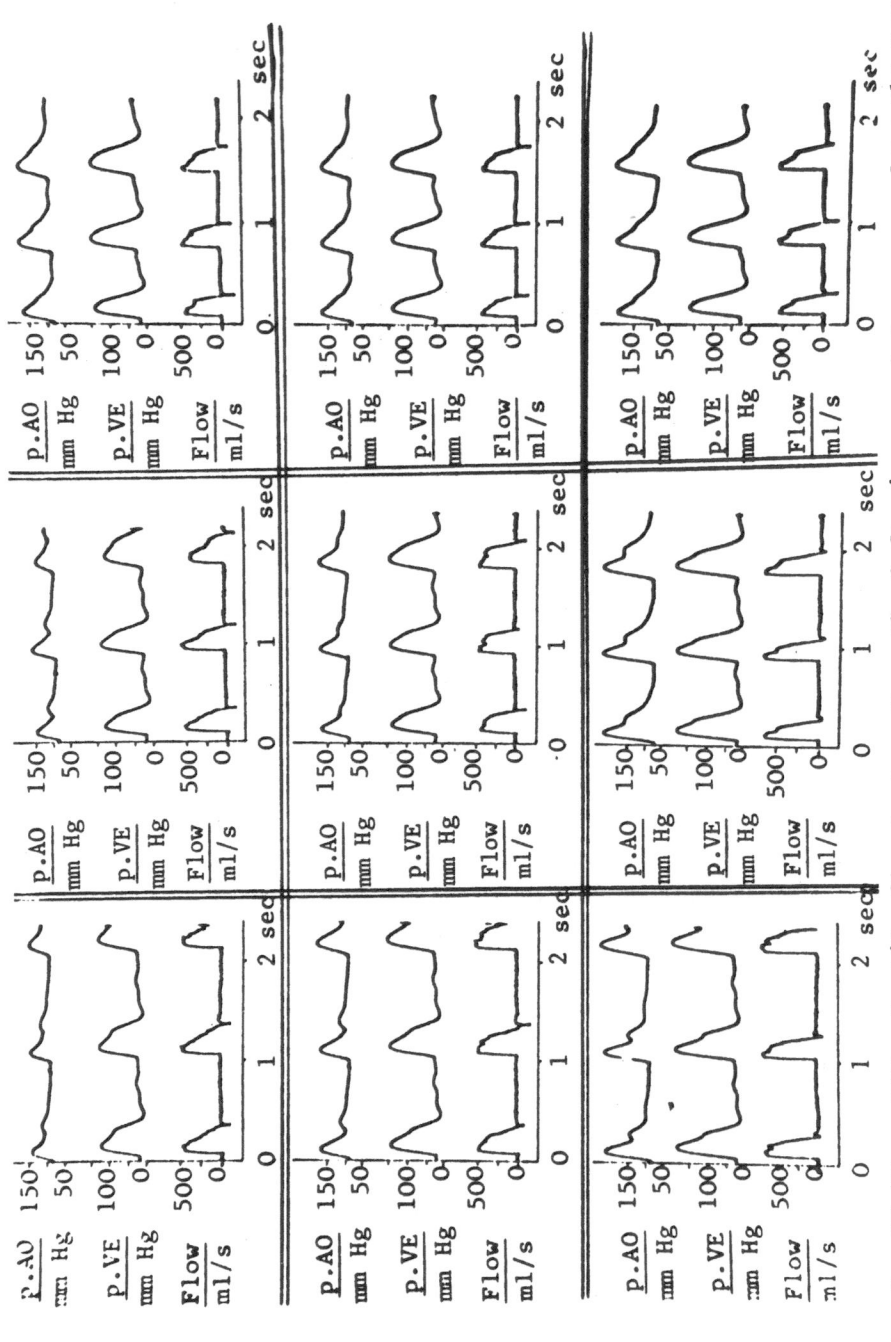

Fig. 2. HK 27 mm TAD pulsatile flow tests. From the left: increased frequency; from the top: increased average flow rate during the cycle.

Fig. 3. Some frames from a high speed film of the closing phase
         of a HK 27 mm TAD.

cardiac frequencies (60, 72, 85 beat/min) and at three ventricular
outputs (4, 5, 6.5 1/min). By varying the ventricular output and
hence the circulatory fluid volume, the device sensivity is especially shown up by the pressure value changes without morphological
alterations. The flow rate diagram, too, shows the change of the
circulatory volume, besides giving an evaluation of the closing
reflux volume.

By using a high speed camera it is possible to obtain some
films which show the valvular movements in detail. Some frames from
a high speed film of the closing phase of the HK prosthesis, 27 mm
TAD, are shown in Fig. 3. In this way it is possible to measure the
real area of the open valve and to compare it with the nominal area.
High speed films are very important for the bioprostheses because
their mechanics are not known, a priori. This is the only way to
examine the movements of the leaflets during the cardiac cycle.

TESTING DEVICE FOR THE MEASUREMENT OF THE BACK FLOW RATE

To evaluate the total regurgitation of a prosthetic heart valve
it is useful to consider its two parts: the first is the closing
reflux volume produced by the inertia of the closing valve; the
second is the leakage of the badly-fitting joint of the valve[7]. The
evaluation of the first part, shown by the natural valve too, needs
the use of the mock circulatory system. A really simple device
which permits to measure carefully the second cause of regurgitation
(steady retrograde flow) has been purposely designed and realized.
Fig. 4 shows a scheme of the device which consists, substantially,
in a chamber test in which suitable stent adapters allow to test
different prostheses. Upstream of the valve there is a water gauge
to provide a pressure of 110 mm Hg across the closed valve (this
value has already been used by other researchers[8]). The water-level
in the reservoir, and therefore the pressure across the valve, decreases during the test and so needs to be restored. An overflow
pipe, on the other hand, prevents exceeding the water-mark while
restoring. Downstream of the valve there is a vessel to measure the
steady retrograde flow. The leakage from the outside the stent
of the prosthesis and the choking of the very small orifice of the
closed valve, produced by tiny air bubbles or particles in the circulatory fluid, needed great care to be eliminated. In fact some
tests on the same prosthesis gave different results and only with

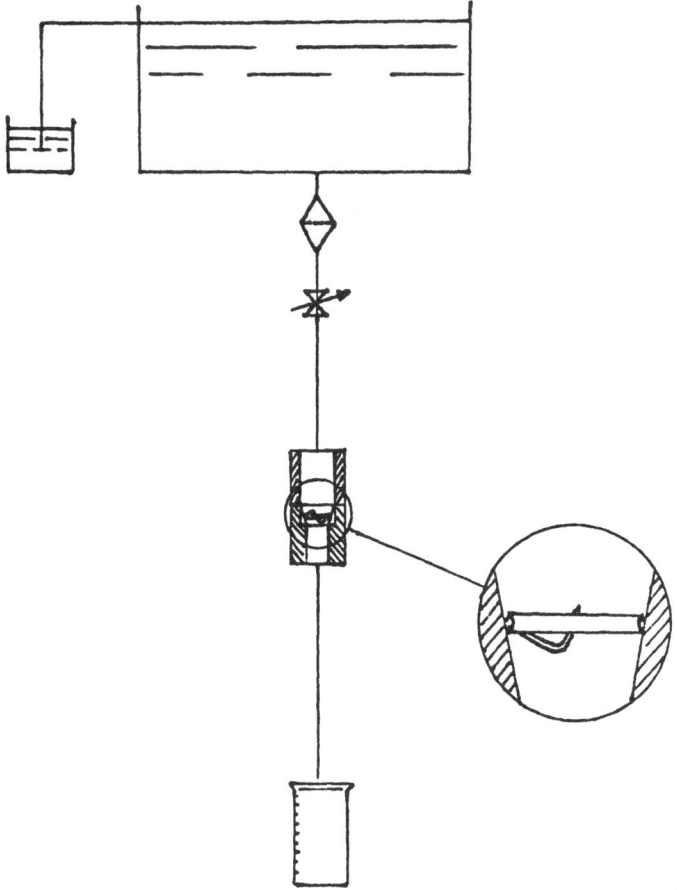

Fig. 4. The device for the evaluation of the steady retrograde flow.

the use of an o-ring packing outside the stent and of a filter upstream of the valve, it was possible to perform reliable measurements.

In Table 1 the results obtained with 5 prosthetic valves (HK 21,
23, 25, 27, 29 mm TAD) are shown. Water at room temperature was the
circulatory fluid. Steady retrograde flow is not an icreasing
function of the diameter, meaning that the working tolerances affect
very much the clearance between disc and stent. It is possible, by
knowing both pressure and flow rate across the valve, to calculate
a hydraulic equivalent diameter which shows the size of the disc-
-stent clearance. The ratio between such a diameter and the nominal
one provides a figure of merit for the comparative evaluation of
different prostheses.

TESTING DEVICE FOR STEADY FLOW EVALUATION

Systolic transvalvular pressure gradient and flow patterns are
very much important characteristics of prosthetic valves which the
purposely designed, steady flow apparatus, schematized in Fig. 5
is able to evaluate carefully. As in the previous devices, suitable
stent adapters allow to insert every kind of prosthesis in the
chamber test. Downstream of the valve the sinuses of Valsalva and
a rectilinear tract of the aortic wall are realized. The whole
system is built of Plexiglass to be transparent. Upstream of the
valve there is a flow straightener realized with many small diameter (2 mm) pipes, 100 mm of length. The test chamber and the reservoir are connected to a hydraulic circuit including a centrifugal
pump. It is possible to vary the flow rate through the valve by the
feed-cock located upstream of the valve, or by the pump by-pass
circuit. An electromagnetic turbine flowmeter permits to measure
the flow rate and a water differential manometer the pressure drop
across the prosthesis. A pressure measuring site is located upstream
of the valve; and two, one in the sinuses of Valsalva, the other in
the aorta, are provided downstream.

Table 1.   Dependence of steady retrograde flow from valve
           diameter.

Hall-Kaster valve diameter	(mm TAD)	21	23	25	27	29
Steady retrograde flow	(1/min)	0.306	0.547	0.335	0.670	0.650

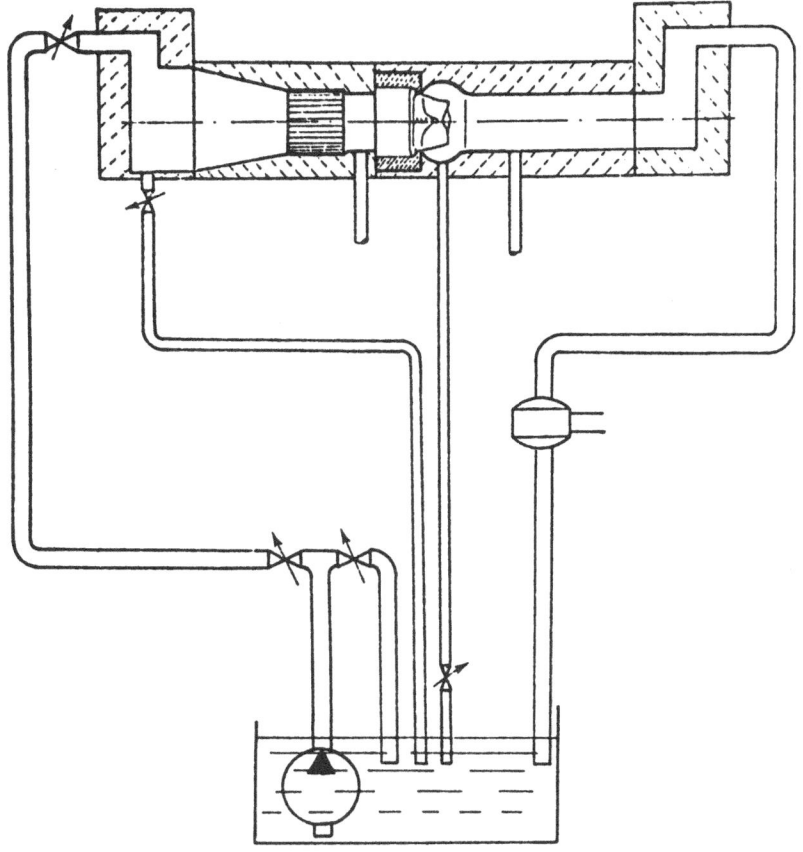

Fig. 5.   The steady flow device for the fluidodynamic evaluation.

The visualization of the flow patterns needed the building of
an apparatus able to light up only a diametral plane of the chamber
test. In such a way a clear view of the flow patterns without the
disadvantage of the simultaneous observation of the whole fluid circulating into the chamber test is obtained[9]. This apparatus consists
of a 1000 W lamp placed in the focus of a plano-convex lens which
permits that only a blade of light reaches the chamber test. The
whole lighting apparatus is contained in a black box to avoid the
leakage of light; an air blast cooling it has been built, but,
however, the risk of damaging the Plexiglass suggests not keeping
the light on more than the time necessary to photograph the flow
patterns. Microspheres of 400 μm of diameter are put in the circulatory fluid to show the flow patterns.

With the steady flow device it is possible to evaluate some
fluidodynamic characteristics of prosthetic valves. One of these is
the systolic transvalvular pressure gradient which gives a comparative evaluation between different kinds of prostheses. Regarding
the aortic valve as a hydraulic concentrated resistance, the equation of Darcy:

$$\Delta p = k \, \frac{Q^2}{d^4}$$

binds the pressure drop Δp both to the flow rate Q and to d, equivalent diameter. The steady flow device measure the pressure drop
across the valve for each flow rate permitted by the pump. Plotting
Δp against Q the parabole distinctive of the tested valve is obtained. It is possible to convert the parabole into a line using Q^2
instead of Q. The slope of this straight line shows the pressure
drop of the tested valve depending on the equivalent diameter d and
on the coefficient k which depends on the shape of the valve. The
same kind of valvular prosthesis shows a decrease of the slope of
the straight line while increasing the diameter. A better plot is
obtained using Δp against Q^2/d^4. In such a way a single line shows
the performance of each kind of prosthesis because its slope depends
only on the shape of the valve, and not on its diameter[10]. The coefficient k is used, as figure of merit, for a comparative evaluation
of different prostheses. The higher the value of k (the higher the
slope of the line), the worse the fluidodynamics of the valve. It
should be noted that steady flow instead of pulsatile flow means a
triple value of the flow rate, if the systole is equal to a third of

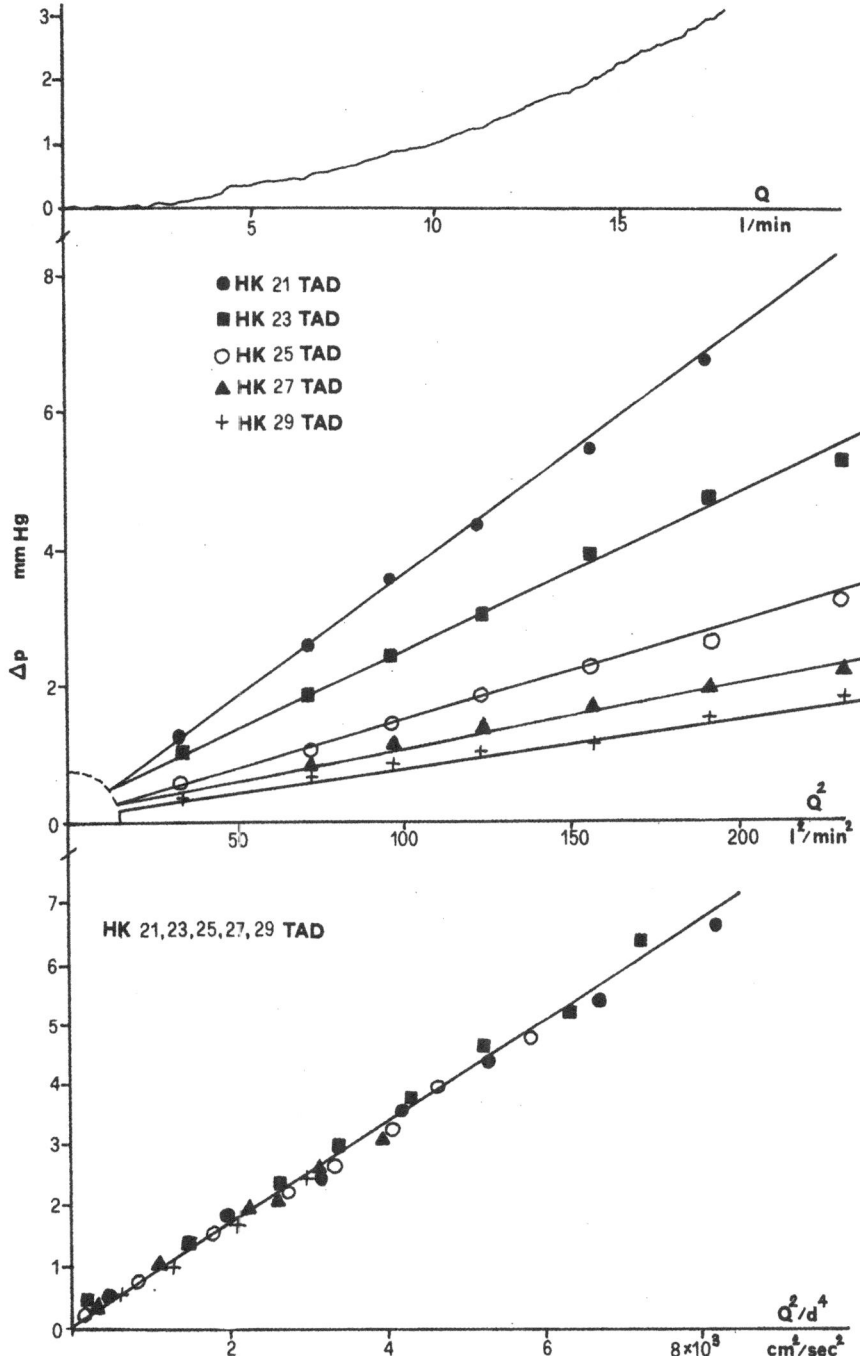

Fig. 6.   Evaluation of the pressure drops in condition of steady
         flow.

the cardiac cycle. In Fig. 6 the parabole for the HK 27 mm TAD, the
5 straight lines of 5 HK of different diameters and their condensing
in a single line are plotted.

The steady flow test gives the following indications: for the
bioprostheses an increased pressure gradient means an increased
stenosis and therefore, under the same conditions, an increased
cardiac work. An increased stenosis means an increased turbolent
flow and hence the possibility of hemolysis. For the mechanical
prostheses a high pressure gradient means that the disc or the ball
is very much occlusive in open position. For this reason there is
a turbolent flow downstream of the valve and a stagnant zone where
the fluid reaches the disc or the ball. In this way the prosthesis
may be damaged by the growth of thrombi or tissue.

The behaviour of the fluid around the valve can be studied
using the steady flow device for the visualization of the flow
patterns. With the steady flow device only the behaviour in condition of systolic peak is observable, but it is already a good indi-
cation. A pulsatile pump for the visualization of the flow patterns
in pulsatile flow condition is being designed.

In Fig. 7 some photographs of flow patterns are illustrated:
the first shows the flow patterns in the device without the valve
and the second the flow patterns of a bioprosthesis while the last
two show the flow patterns of a HK and a Sorin tilting disc valves.
Without the valve, the flow patterns are rectilinear and parallel to
each other and there is no turbulence. With the bioprosthesis
the zone of rectilinear flow is smaller, but central; the turbulent
flow appears only close to the aortic wall. The two mechanical
valves show a dissymmetrical flow with a great zone of turbulence
over the disc; hence their hemodynamics is worse than that of the
bioprosthesis. It is interesting to note the presence of vortexes
inside the sinuses of Valsalva in all the photographs. Such vortexes
are important for the closing mechanics of the natural leaflets[11].

FATIGUE TEST DEVICE

A pulsatile test device for the evaluation of the fatigue of
the valvular prostheses has been realized and is now being set up,
so that it is not possible to show any experimental results.

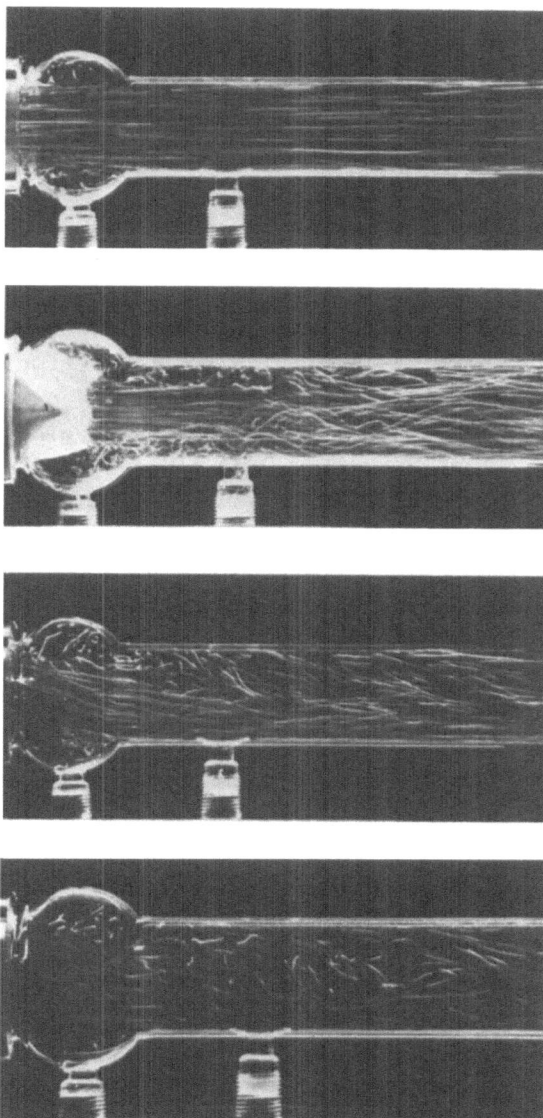

Fig. 7. Visualizations of flow patterns obtained by the steady
flow device. From the top: the device without the valve,
the biological valve, the HK 27 mm TAD and the Sorin
27 mm TAD.

With this device it is possible to increase the cardiac frequency
up to 10 times and to test 10 valves simultaneously. Anatomic shapes
and physiological behaviours are not reproduced, but valve working
and pressure values are guaranteed. Test chambers are entirely
made of Plexiglass and so, using a stroboscopic beam light, it is
possible to control the propagation of the fatigue failure during
the test.

CONCLUSIONS

For a systematic comparison between the different kinds of valvular prostheses already existing and for the development of new
prosthetic valves, a careful in vitro characterization has a great
importance. On the other hand it is not possible to measure exactly
an artificial valve. Within a project for the development of a new
polymeric valvular prosthesis, it seemed appropriate to support
the already existing mock circulatory system with some other devices
able to give more specific informations such as the steady retrograde
flow, the systolic transvalvular pressure gradient, the flow patterns
and the fatigue resistance. In this paper the main characteristics
of the devices designed and built to perform the above mentioned
tests have been described. The results of some representative tests
carried out in our laboratories using the illustrated devices have
also been reported. In spite of the fact that the results prove
the validity of the devices, some modifications to improve their
performances are under study.

REFERENCES

1. V. S. Luisi, E. Reginato, M. Bernabei, S. Eufrate, S. Cicconardi
    and G. Nardi, Metodiche di valutazione fluidodinamica di biopro-
    tesi valvolari aortiche, Rassegna Clinico-Scientifica, 11-12:75
    (1981).
2. C. R. Gentle, The role of simulation studies in cardiac valve
    prosthesis design, Eng. in Med., 7:101 (1978).
3. N. H. C. Hwang and H. Reul, In vitro characteristics of prosthe-
    tic valve testing, Proceeding of 34th ACEMB, 172 (1981).
4. H. Reul, W. Tillmann and G. Häussinger, New trends in artificial
    heart valve development and testing, Digest of the 2nd ICMMB,
    20 (1980).

5. W. H. Swanson and R. E. Clark, Testing of prosthetic heart valves, ASME Paper, 76-WA/BIO-3 (1976).

6. G. Nardi, S. Cicconardi, P. Dario and D. De Rossi, Modular hydro-pneumatic mock circulatory system for the evaluation of cardio-vascular prostheses, ASAIO J., 4:139 (1981).

7. T. M. Wright, The heart, its valves and their replacement, Bio--Med. Eng., 7:26 (1972).

8. L. N. Scotten, R. G. Racca, A. H. Nugent, D. K. Walker and R. T. Brownlee, New tilting disc cardiac valve prostheses, J. Thorac. Cardiovasc. Surg., 82:136 (1981).

9. J. T. M. Wright and L. J. Temple, A flow visualization study of prosthetic aortic and mitral heart valves in a model of the aorta and left heart, Eng. in Med., 6:31 (1977).

10. C. R. Gentle, A limit to hydraulic design of heart valve pros-theses, Eng. in Med., 6:17 (1977).

11. H. Reul and N. Talukder, Heart valve mechanics, in:"Quantitative cardiovascular studies - Clinical and research applications of engineering principles," N. H. C. Hwang, D. R. Gross and D. J. Patel, ed., University Park Press, Baltimore (1979).

DEVELOPMENT AND TESTING OF A NEW PROSTHETIC HEART VALVE

G. Nardi', S. Cicconardi', R. Pietrabissa", M. Lazzoni"
C. Migliaresi"' and P. Giusti"

'Istituto di Macchine, Fac. Ingegneria, Univ. Pisa, Italy
" Istituto di Chimica, Fac. Ingegneria, Univ. Pisa, Italy
"' Ist. Principi Ing. Chimica, Univ. Napoli, Italy

INTRODUCTION

The main problems so far found for the practical applications of the prosthetic heart valves in cardiovascular surgery seem to be related to the short life exhibited by the biological valves and to the unsatisfactory hemodynamic behaviour and thrombogenic properties shown by the mechanical ones. In attempt to alleviate the above problems, a study has been undertaken to design a new type of prosthetic leaflet heart valve entirely manufactured in polymeric materials.

The idea is to introduce, in this technological field too, the use of composite materials. In fact, many researchers who tried to realize a leaflet heart valve made of elastomers met with difficulties to supply fatigue strength. This property is the most difficult to give to a heart prosthesis, as flexibility and hemocompatibility are also required and, generally, these two characteristics can be obtained with materials not showing good fatigue strength. The leaflets of the new valve are made of composite materials to obtain a prosthesis able to provide both the fatigue strength by the inner component of the composite and the hemocompatibility by the outer component. The composite consists of:

1. a knotted Dacron fabric, able to impart mechanical
    strength

2. a thin layer of a polymer as coating of the fabric, able
    to provide both the hemocompatibility and the impermeabil-
    ity.

We directed our efforts towards the synthesis of the necessary
hemocompatible materials and, successively, their use in the composite. We have realized the prosthesis that has been tested both in
vitro and in vivo.

The study can be divided in two fundamental sections, the first
concerning the design of the appropriate geometrical shapes, the
second relating to the choice of materials.

MATERIALS

The materials used in this project can be divided in two
classes: conventional and commercial materials able to provide mechanical strength and new materials potentially hemocompatible made
in our laboratories.

The materials used for the stent and the core laeflets belong
to the first class while the materials used as coating agent of the
leaflets belong to the second class. The stent is made in Teflon
or in Delrin. Particularly the latter material shows sufficient
characteristics of fatigue strength and, in fact, it has already
been used in similar stents of some commercial biological heart
valves. The leaflets consist of a thin inner layer of a commercial
knotted Dacron fabric particularly light and of an outer coating
agent belonging to the following two classes:

1. hydrogels

2. polyurethanes.

Hydrogels represent an important area in the biomedical field,
because they simulate, in many ways, some hydrodynamical properties
of the inner lining of the vascular walls[1]. In fact, the presence
of large quantities of water chemically and physically bonded and
associated give these materials unique properties. So they are very
similar, for the same aspects, to some biological tissues that, as
is common knowledge, hold a lot of water.

The hydrogel used in this study has been the poly(2-hydroxy-ethylmethacrylate) obtained by a radical polymerization of the 2-hydroxyethylmethacrylate (HEMA) accomplished at 60°C for 2 hours using 0.5% by weight of ethyleneglycoldimethacrylate (EDMA) as cross-linking agent and 0.1% by weight of α,α'-bisisobutyronitrile (AIBN) initiator. The polymer shows perfect transparency and is able to absorb and desorb water reversibly[2].

The polyurethanes are the potentially hemocompatible polymers towards which we spent most of our efforts, as in several other laboratories in the world[3,4]. The synthesis of these polymers aims to produce an elastomer with a surface that could react as little as possible with the proteins of the blood. These polymers are characterized by the insertion of a lot of different functional groups in their main chains in such a way to create a surface able to interact with the proteins mostly by Van der Waals forces; in this way, probably, the main component of the resultant force field is of the dispersion type and its intensity fluctuates as a function of locus. Under these conditions, it can be expected that plasma proteins placed in such a force field will be in an energetically unstable state and so, probably, they are not inclined to interact with the foreign surface through an irreversible attachement[5].

When we speak of polyurethanes for biomedical uses, we refer to a large group of polymers characterized by several functional groups so that more precisely we can speak of polyether-urethanes, polyester-urethanes, polyether-urethanes-urea, polyisobutylen-urethanes-urea. They are block copolymers, whose physical properties depend on the presence of soft and hard segments[4]. The soft segments are formed by polyether residues and the hard segments by urethane and urea linkages. These segments aggregate into phase domains, with the hard segments acting as cross-links analogous to the covalent cross-links of conventional thermosetting elastomers. This structure may produce virtual or thermolabile cross-links, as the hard urethane domains act like joint-points below their glass transition temperature, but they allow polymer slipping at higher temperature. The results consist of a polymer acting as if it were vulcanized at usual temperatures, and if it were thermoplastic at the highest ones. Polyurethanes like these may be aromatic or aliphatic ones. We have considered aromatic ones, where diisocyanate consists of methylene bis(4-phenylisocyanate) (MDI). In our syntheses polyoxypropyleneglycol (PPG) with molecular weight of 400, 1200, 2000 and -OH terminated polyisobutylene (PIB) with molecular weight

of 3400, represent the soft part.   Long polymeric chains are obtained
with chain extenders.   In fact, prepolymers are chain-extended with
several extenders among which diamines and diols are the most
employed: only using the first one ureic groups are obtained.   We
used these extenders in our syntheses:

1.   ethyleneglycol

2.   1,4-cyclohexanedimethanol

3.   1,4-cyclohexanediol

4.   ethylenediamine

5.   4,4'-diamino-dicyclohexylmethan.

So we synthesized more than twenty different polymers all containing
MDI as a common element, and PPG or PIB plus one of the previous extenders.   It may be pointed out, as an exemple, that polyurethanes
having PPG 2000 or PIB 3400 as a soft segment sometimes show excessive elasticity; in this case it may be useful to use extenders
which are able to constitute hard blocks together with those
produced by MDI, in order to reduce the long-polymeric-chain effect.
This kind of problem does not exist using PPG 1200, while polymers
with PPG 400 as a soft segment have unsatisfactory elastomeric characteristics.   The prepared polymers are DMF and DMSO soluble, so
that it is possible to adopt dip-moulding and sprayed coating techniques as well as to prepare thin films used for their chemical and
physical characterization.   Structures of examinated polymers are
shown:

$$\left[\overset{O}{\overset{\|}{C}}NH\!\!\bigcirc\!\!CH_2\!\!\bigcirc\!\!NH\overset{O}{\overset{\|}{C}}OCH_2\!\!\bigcirc\!\!CH_2O\overset{O}{\overset{\|}{C}}NH\!\!\bigcirc\!\!CH_2\!\!\bigcirc\!\!NH\overset{O}{\overset{\|}{C}}O\!\!\left(\overset{CH_3}{\underset{CH_3}{C-CH_2}}\right)_x\right]_N$$

$$\left[\overset{O}{\overset{\|}{C}}NH\!\!\bigcirc\!\!CH_2\!\!\bigcirc\!\!NH\overset{O}{\overset{\|}{C}}NH\!\!\bigcirc\!\!CH_2\!\!\bigcirc\!\!NH\overset{O}{\overset{\|}{C}}NH\!\!\bigcirc\!\!CH_2\!\!\bigcirc\!\!NH\overset{O}{\overset{\|}{C}}O\!\!\left(\overset{CH_3}{\underset{}{CH-CH_2-O}}\right)_x\right]_N$$

In order to study the behaviour of these materials when in
contact with blood, some in vitro tests have been performed; they
show, in agreement with both Howell and Lee-White tests, and platelet adhesion measurement tests, that they are among the best hemo-

compatible materials which have ever been made.

GEOMETRIES

On realizing the valve a geometry similar to the natural one has been chosen. It is, in fact, a tri-leaflet valve, where the leaflets have been assembled on a stent similar to those used for bioprostheses. This stent has been provided with three prongs connected to the basic ring with very wide fillets in order to limit the notch effects. The used thicknesses guarantee a sufficient softness to the whole structure, without endangering its strength. At this first stage of research we have made this element with machine tools, but an injection moulding technique is far better. Such a typical stent is shown in Fig. 1.

As we have already said, leaflets consist of Dacron fabric. They are indeed constituted by a single fabric element conveniently sewn on the stent. The fabric is sewn on the stent from the outside

Fig. 1. The stent.

to avoid that only the suture thread (it is the usual surgical Dacron
thread) bears all the stresses, especially acting in the diastolic
phase. In this way, all the three strips of fabric are fairly el-
lipsoid-shaped[6]. During this first designing period we have mainly
been looking for hemocompatible materials and the leaflet geometries
have been chosen looking at some biological prostheses and looking
for a simple machining technology. On the other hand, we started a
research program that, by using stereophotogrammetry techniques as
H. M. Karara[7] suggests, enables us both to exactly verify the real
leaflet geometry and to make, with opportunately built moulds, the
best geometries, which are computed according to D. N. Ghista and
H. Reul[8].

Sewing being finished, the valve is spray-covered by a solution
of one of the previously described hemocompatible materials, dissolved in a suitable solvent. During the subsequent evaporation of the
solvent, leaflets can be given the definite desired shape. Before
covering, it is necessary to obtain a shell-shaped geometry. The
techinque we have used seems to be inadequate; in the future we are
going not to adopt a flat fabric, but one convenientely weaved along

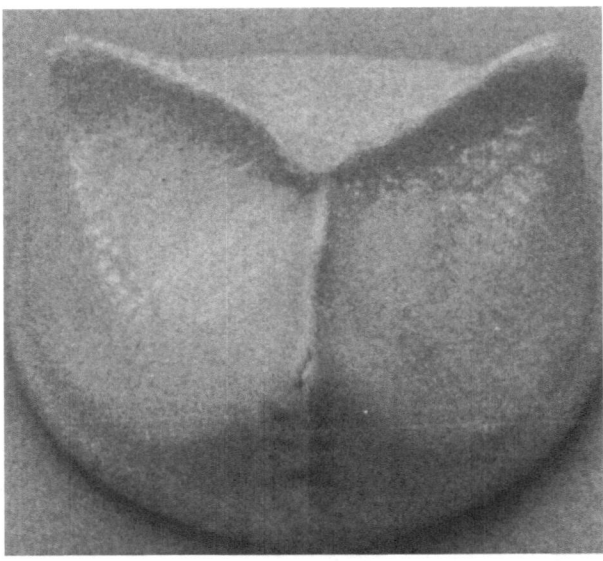

Fig. 2. The valve.

the definite bent surfaces. A prototype of the polymeric valve is shown in Fig. 2.

TESTS

The functionality of the valve has been tested with equipment which has been made in our laboratory . Pulsatile flow tests on a mock circulatory system, flow patterns, measurements of steady retrograde flow and pressure drop across the valve, in vivo tests have been carried out. By the pulsatile flow tests, the shapes of curves referring to ventricular and aortic pressure and to flow rate, similar to that shown by other kinds of valves, have been observed. A substantial regularity of opening and closing movements has been found by a high speed filming as shown in Fig. 3. The regurgitation in both pulsatile and steady flow are satisfactory, even if it must be pointed out that optimization studies on coaptation surfaces among the leaflets have still to be performed. Nowadays we believe, in fact, that the adopted solution for this valve portion is not yet satisfactory. By analizing the prosthesis with a proper device for the visualization of the flow patterns, it has been noted that they are very similar to those of the biological valves and therefore turbulent flow effects can be considered to be rather limited. Typical flow patterns of the polymeric valve are shown in Fig. 4. The prosthesis has been tested by experiments on living animals. Particularly, many valves differing in sizes have been used on pigs, at first in the descending aorta, in series with the natural aortic valve, later in proper aortic position, substituting the natural valve. Even if, because of long testing times, it is not possible yet to give a complete opinion on it, it must be pointed out that some of those animals survived a few months. Histological tests on the nature of natural tissues growing over these new prosthetic valves, after the permanence in the animals, are still being carried out.

CONCLUSIONS

A series of prosthetic leaflet aortic valves which, by performed trials, seem to show fluidodynamic behaviour that fits functional expected tasks has been developed using polymers. However it must be emphasized how much such prototypes can be im-

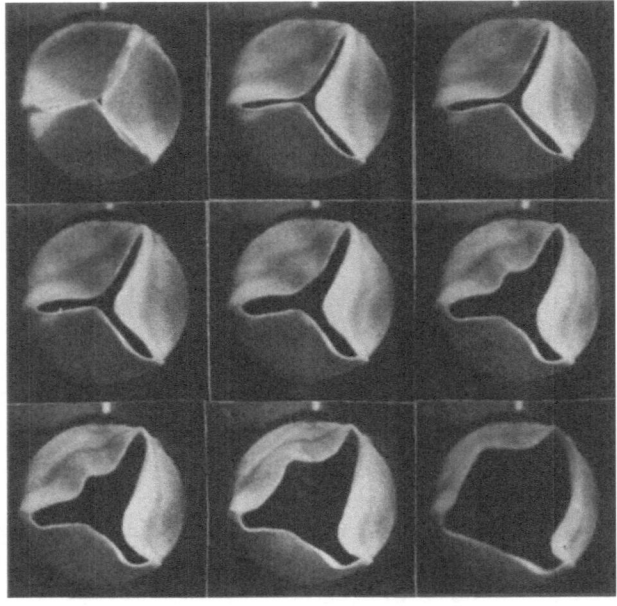

Fig. 3.   Some frames from a high speed film of the opening phase
          of the polymeric valve.

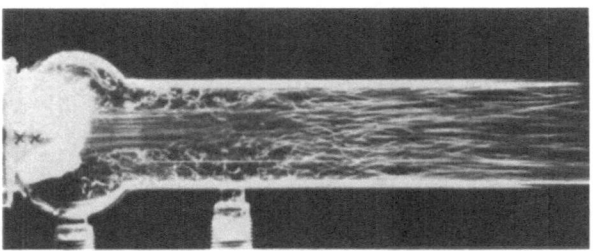

Fig. 4.   Visualization of the flow patterns of the polymeric valve.

proved both on the geometrical side, employing especially prepared cloths instead of commercial flat fabric, and on the material side using other promising potentially hemocompatible materials we are synthesizing in our laboratories. Satisfactory results which have been obtained from in vivo tests induce us to carry on with our research.

Work partially supported by CNR Special Project "Chimica Fine e Secondaria"

REFERENCES

1. S. D. Bruck, "Properties of biomaterials in physiological environment," C.R.C. Press, Boca Raton, Florida (1980).
2. C. Migliaresi and L. Nicolais, Composite materials for biomedical applications, Int. J. Artif. Organs, 3:114 (1980).
3. D. J. Lyman, Synthesis of polyurethanes, in:"Sintesi di polimeri," AIM (1982).
4. D. J. Lyman, Structural order and blood compatibility of polymeric prostheses, in:"Structural order in polymers," F. Ciardelli and P. Giusti, ed., Pergamon Press, Oxford (1981).
5. R. Ward and E. Nyilas, Production of biomedical polymers, in: "Organometallic polymers," C. E. Carraher Jr., ed., Academic Press, New York (1978).
6. H. Reul, V. Tillmann and G. Häussinger, New trends in artificial heart valve development and testing, Digest of 2nd ICMMB, 20 (1980).
7. H. M. Karara, Aortic heart valve geometry, Photogrammetry Eng. 1393 (1974).
8. D. Ghista and H. Reul, Optimal prosthetic aortic leaflet valve, J. Biomechanics 10:313 (1977).

HEPARIN-LIKE SUBSTANCES AND BLOOD-COMPATIBLE POLYMERS OBTAINED FROM CHITIN AND CHITOSAN

Riccardo A.A. MUZZARELLI

Institute of Biochemistry, Faculty of Medicine
University of Ancona, I-60100 Ancona, Italy

INTRODUCTION

The most widely distributed polysaccharides, cellulose, dextrans, pectins, alginic acid, agar-agar, agarose, starch and carrageenans, are either neutral or acidic substances. Chitin, |(1-4)-2-acetamido-2-deoxy-β-D-glucan|, the acetylated polymeric form of glucosamine, as well as chitosan, its deacetylated derivative, are the only abundantly available polysaccharides possessing sharply basic characteristics [1-3].

Chitin does not occur in vertebrates. Mammals have no chitinous tissues [4], however a number of glycoproteins and proteoglycans are present in their bodies, where they exert important functions [5]. Connective tissue glycoproteins, such as the collagens and proteoglycans of various animal species, are structural elements as are the cell wall glycoproteins of yeasts and green plants [6]. The body fluids of vertebrates are rich in glycoproteins secreted from various glands and organs: for instance, constituents of blood plasma which are glycoproteins include transferrin, ceruloplasmin, transcobalamin, immunoglobulins and all the clotting factors.

Among the connective tissue polysaccharides, hyaluronic acid, keratan sulfate and heparin possess repeating disaccharide units including N-acetylglucosamine. Many glycoproteins contain oligosaccharide chains linked N-glycosidically from N-acetylglucosamine to the amide nitrogen of asparagine in the peptide. The glycoproteins which contain N-acetylglucosaminyl-asparagine linkage as well as those known to contain the core structure of mannosyl-di-N-acetyl-chitobiose were recently reviewed [7].

Chitin, then, assumes importance because its monomers and dimers are present in certain human tissues, and also because it is indirectly related to our food of animal and fungine origin, and to medical aids and pharmaceuticals.

Among mammals, which are in general able to digest chitin, because they possess chitinases in their gastric mucosa, as it is the case for *Perodicticus potto* and *Cebus capucinus*, and sometimes in pancreas, *Homo sapiens* does not possess chitinases [8]. Chitin, however, is not unsafe to man: we bake our bread with chitin-bearing yeasts; we usually ingest some chitin when eating shrimps, lobsters or fermented foods. The production of canned crab and shrimp meat as well as the production of citric acid from *Aspergillus niger*, originates very large amounts of chitinous wastes, which not only represent a worry from the ecological standpoint, but also a destruction of renewable polysaccharidic resources [9, 10]. Moreover, insects destroy a substantial aliquot of our crops: thus, since insects possess a chitinous exoskeleton and their larvae are protected by a chitinous membrane, [11] an important approach to the crop protection and insect control is the inhibition of chitin biosynthesis [12 - 15]. We are also interested in protecting other vertebrates which represent our main supply of proteins, from aggression from chitinous organisms.

Many enemies of the human life rely on chitin for their body structure, food and reproduction [16]: parasites such as mites and ticks and infectious microorganisms such as *Aspergillus fumigatus* and *Candida albicans* are well known examples [17]. Mucormycosis can also be mentioned as a dramatic example of aggression by a chitinous organism *(Rhizopus oryzae, Mucor miehei, Absidia corymbifera, etc.)* to man, [18 - 20]. The bacterial cell wall itself can be interpreted as a chitin derivative, namely the ether of lactic acid and chitin at C3 of alternate units, currently named N-acetylmuramic acid units [1, 16].

It can be said that chitin is most relevant to our health and welfare, because of the very important interactions between human life and chitin-based life [21]. To explain the importance of the knowledge of the chemistry of chitin for the biomedical sciences, we could mention the well-known investigation on structure and function of lysozyme carried out with chitin oligomers: lysozyme is present in human body fluids as a defence against chitinous organisms on which it exerts hydrolytic action. A further example is the research on heparin and heparinoids: heparin is probably the poly-saccharide most closely related to chitin that can be found in the human body, where it reacts specifically with blood components to control blood fluidity and coagulation.

The possible applications of chitin in various fields including medicine, have been recently reviewed [22, 23]: they range from very simple applications such as powders for surgical gloves, to sophisticated ones, such as the immunoadjuvant effect of chitin which was found to impressively extend the survival of chitin-treated mice implanted with Ehrlich and sarcoma 180 ascites tumors or challenged with *Staphilococcus aureus* [24].

The present contribution is therefore restricted to only two possible applications of chitin in medicine, namely membranes and heparin-like substances. These applications appear to hold potential for becoming effective rather soon: one point in favor of this fore-cast is the progress done in the understanding of basic aspects of chitin. Too much empirism has affected research on chitin for a long time, thus preventing the utilization of chitin on a scientific basis: just recently it was realized that chitins are a class of substances whose differences are in terms of structural features, degree of acetylation, basicity, molecular size and polydispersity, presence of aminoacids, metal ions and inorganics. Isolation and production processes were put in scientific terms just recently: they can be tailor-made to suit the projected applications.

Similarly, it was not before the First International Conference on Chitin and Chitosan that it was realized that chitosan, the more deacetylated form of chitin, is a reactive substance on which all the reactions of the primary aliphatic amino group can be carried out, to obtain new polymeric products possessing the important $\beta(1-4)$ anhydroglycosidic backbone [25]. Today, fully characterized chitins are available in large amounts and at moderate costs and offer the choice of sets of characteristics according to the planned uses.

A NEW HEPARIN-LIKE SUBSTANCE: SULFATED N-CARBOXYMETHYL CHITOSAN

During the last few years, efforts have been made toward the introduction of sulfate and carboxyl groups in the chitosan macro-molecule. Early reports on this topic deal with the sulfation of chitosan [26], the selective formation of sulfamido groups [27], the sulfation and depolymerization of chitosan [28], and the preparation of formyl chitosan, a remarkably potent inhibitor of aldosterone production [29]. Among a variety of polysalts of chitosan, those with dextran sulfate have been reported to inhibit clot formation in vitro [30]. The antithrombogenicity of polysalts of chitosan and carboxymethyl dextrans was also studied [31]. In addition to these polysalt complexes, chitosan sulfate (27,000 dalton) was studied, and evidence of its anticoagulant activity was given [32]. Other polymers studied were sulfated aminodeoxycellulose [33], alginic acid [34], cellulose sulfate [35] and synthetic polymers [36, 37]. Sulfated pectins and dextrans are also known [38]. We believe that chitosan is particularly well suited as a raw material from which heparin-like substances can be obtained, because it has $\beta(1-4)$ anhydroglycosidic linkages and linear chains and already carries

amino and acetamido groups, whose introduction in a polysaccharide
chain has been found to be most difficult.

A number of heparin-like susbtances so far proposed lack
essential acetamido groups; some have been submitted to harsh treatments for the oxidation of primary alcohol groups to carboxyl groups.

Our study is therefore based on the following considerations:
(a) the heparin-like substances should possess simultaneously O-
sulfated, acetamido and carboxyl groups, which are recognized to be
effective in anticoagulant action; (b) carboxyl groups should preferably be in the form of carboxymethyl groups or, even better, in
the form of α-aminoacid groups. It has been recently reported [39]
that dextrans bearing carboxymethyl, sulfate, benzyl sulfate and
α-aminoacid groups possess high antithrombogenic activity. (c) The
introduction of the said groups should be as mild as possible, to
avoid side reactions and alterations whose consequences are difficult
to foresee and which prevent the correct description of the product.
(d) The ionic groups should be in optimum ratios to impart solubility
and (e) should be distributed regularly along the chain. (f) The
chain length should be tailored to impart optimum blood anticoagulant
properties.

For this purpose, we have used *Euphausia superba* chitosan [40]
to prepare N-carboxymethyl chitosan, and then we have submitted it
to sulfation [41]. Glyoxylic acid crystals were added to an aqueous
suspension of chitosan powder, to obtain a chitosan glyoxylate
solution (pH 3.2). Upon pH adjustement to 4.5 - 5.0, Schiff reaction
took place with formation of N-carboxymethylidene chitosan. The latter
was then reduced with sodium cyanoborohydride and isolated by
addition of acetone. The thus obtained N-carboxymethyl chitosan was
subsequently submitted to sulfation in N,N-dimethylformamide.

The infrared spectrum of a sample of sulfated N-carboxymethyl
chitosan insolubilized with acetone at pH 1.0 (hydrochloric acid)
as well as a spectrum of sulfated N-carboxymethyl chitosan obtained
by lyophilizing a solution of pH 8.3 show strong absorption bands
at 1230 and 800 cm^{-1}, assigned to the sulfate group: they do not
occur in the N-carboxymethyl chitosan spectrum (Fig. 2 of Ref.42).
In the spectrum of the polymer isolated at acidic pH value, the
1730 cm^{-1} band assumes evidence (undissociated carboxyl group). The
circular dichroism spectra taken on a number of samples showed a
negative Cotton band at 203 nm; molar ellipticity at pH 8.0 was
around 200,000.

The polymer was found to be insoluble at pH values below 3.5.
The titration of the amino groups required the calculated amount
of sodium hydroxide, and in comparison with the data for N-carboxy-

methyl chitosan (Fig. 6, Ref. 42) the titration interval for amino groups was narrower, this fact being in agreement with the increased average molecular weight of the repeating units. Sulfated N-carboxymethyl chitosan had a sulfur content of 11 %, after 3 and 15 hour sulfation. This indicates that a 3-hour sulfation period is enough for the preparation of the heparin-like substance.

Sulfated N-carboxymethyl chitosan was dissolved in acetate buffer (50 mg/ml), submitted to sonication, filtered on 0.45 μm Millipore membranes and submitted to chromatography on Bio-Gel P-100. Since it is known that the average molecular weight and poly-dispersity are important parameters to be taken into account when testing a blood anticoagulant, we have developed this approach in order to reduce the molecular weight of the sulfated N-carboxymethyl chitosan, which originally was over 1,000,000 dalton.

The detection of the eluates at 206 and 280 nm indicated that under the adopted conditions two fractions only were present. The elution peaks were centered at 39,000 and 80,000 dalton, and corresponded to about 25 % and 75 % of the sulfated N-carboxymethyl chitosan submitted to chromatography, respectively. Samples of sulfated N-carboxymethyl chitosan as prepared (without sonication, filtration and chromatography) and samples submitted to sonication and filtration were then tested for their blood anticoagulant activity.

The tests carried out for the evaluation of the blood anti-coagulant activity of sulfated N-carboxymethyl chitosan are the same as those for the determination of heparin, i.e. the anti-thrombin test for the thrombin inhibition and the heparin test for the factor Xa inhibition. Both of these tests were done by spectro-photometry at 405 nm on the p-nitroaniline liberated from a chromo-genic substrate.

The results confirmed that sulfated N-carboxymethyl chitosan bound to antithrombin inhibits the thrombin present according to the following reactions:

$$AT + SNCMC_{excess} \rightarrow |AT\text{-}SNCMC| + SNCMC_{remaining}$$

$$|AT\text{-}SNCMC| + Thrombin_{excess} \rightarrow |AT\text{-}SNCMC\text{-}Thrombin| + Thrombin_{remaining}$$

$$Chromogenic\ substrate \xrightarrow{Thrombin\ remaining} Peptide + p\text{-}nitroaniline$$

Similarly, sulfated N-carboxymethyl chitosan forms a complex with AT, and inhibits factor Xa, thus proportionally decreasing the splitting of p-nitroaniline from substrate:

$$SNCMC + AT_{excess} \rightarrow |SNCMC\text{-}AT|$$

$$|\text{SNCMC-AT}| + \text{Fxa}_{\text{excess}} \rightarrow |\text{SNCMC-AT-FXa}| + \text{FXa}_{\text{remaining}}$$

$$\text{Chromogenic substrate} \xrightarrow{\text{FXa remaining}} \text{Peptide} + \text{p-nitroaniline}$$

Sulfated N-carboxymethyl chitosan was therefore found to be a blood anticoagulant exerting its action for prolonged periods of time. Differences were observed between the sonicated and filtered samples and the untreated samples. Untreated sulfated N-carboxymethyl chitosan, which still contained high molecular weight fractions, when added to human blood originated a certain degree of hemolysis, platelet aggregation and adverse phenomena on the cellular structures. On the contrary, the sonicated and filtered sulfated N-carboxymethyl chitosan (a mixture of 39,000 and 80,000 dalton) did not produce any appreciable hemolysis; the lymphocytes observed at 400x, 24 hours after treatment, did not show any undesirable alteration and appeared like those in heparinized human blood. Erythrocytes were not altered in shape and volume, and no evidence of osmotic shock was detected. Their spontaneous sedimentation resulted quite analogous to that observed on samples treated with established products. The dose of the low molecular weight sulfated N-carboxymethyl chitosan necessary for complete and definite anticoagulant action was found to be 50 I.U./ml of human blood, i.e. only slightly higher than that currently used with heparin.

N-Carboxymethyl chitosan can therefore be considered as a suitable starting polyampholyte for the preparation of heparin-like products by sulfation. Favorable characteristics of N-carboxymethyl chitosan are linear $\beta(1-4)$ polyanhydroglycosidic chains, regular substitution with amino groups on each C2, part of which in form of glycino groups. It was recently realized that certain amino acid units favor the blood anticoagulant activity of artificial polymers, [43] and therefore the N-carboxymethyl group should not only be interpreted as a function carrying the essential carboxyl group, but also as an amino acid unit wich adds to the blood anticoagulant potency of these heparin-like substances.

The molecular size of sulfated N-carboxymethyl chitosan can be easily and conveniently reduced by sonication, without greatly depressing the overall yield of the preparation. The reduced molecular size of sulfated N-carboxymethyl chitosan permitted to lower its dosage for optimum and long-lasting anticoagulation effect on human blood and to preserve the cellular structures of lymphocytes and erythrocytes. Among the many artificial blood anticoagulants so far studied by other authors, sulfated N-carboxymethyl chitosan seems to be the heparin-like substance most similar to heparin from the chemical standpoint. At the present state of the research it can not be excluded that, during sulfation, some acetyl groups are replaced by sulfate, thus introducing a minor proportion of sulfamide groups.

While sulfated N-carboxymethyl chitosans differing in terms of degree of acetylation, N-carboxymethylation and sulfation can be indicated as a class of heparin-like substances of interest for applications in vitro, it is anticipated that further refinements could lead to the preparation of sulfated N-carboxymethyl chitosans suitable for use in vivo. The main advantage obtained in using sulfated N-carboxymethyl chitosan instead of sulfated chitosan is in fact the demonstrated lack of adverse phenomena on cellular structures.

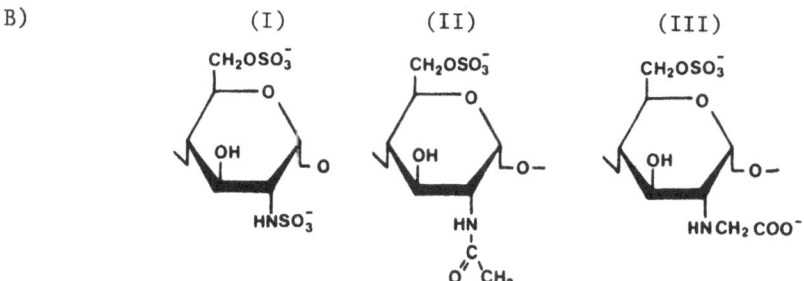

Figure 1. A) Fragment of heparin, illustrating some characteristic structural features. B) Repeating units of sulfated N-carboxymethyl chitosan: (I) present in minor amounts, (II) present at 42+4 %, (III) present at 58+4 %. Sulfur content, 11.0 %.

ANTITHROMBOGENIC SURFACES

Polytetrafluoroethylene, polyurethanes, polyethylene, silicones and acrylates have been proposed for replacement of both hard and soft tissues. These biomaterials must satisfy two important criteria to provide an useful function in a biological environment: they should possess the proper physical characteristics as replacement materials and should exhibit compatible interfacial properties with surrounding tissues and fluids. The interaction of blood with foreign surfaces resulting in thrombogenesis has received considerable attention, but still represents a problem.

Wettability, surface free energy, surface charge and compliance have been studied. Heparinized surfaces have exhibited considerable compatibility with blood. Typically, heparin is surface bound using cationic materials including tridodecyl methylammonium chloride, benzalkonium chloride, octadecylamine and quaternary ammonium polymers. Since polyelectrolyte complexes of heparin and chitosan have been prepared and exhibit either thrombogenic or non thrombogenic properties chitosan can be coated onto artificial polymeric supports and then reacted with heparin or with heparin-like substances [44].

Polyethylene catheters were surface primed either with chromic acid solution or with oxygen plasma treatment, then dipped into a 0.6 % chitosan solution in 1 % acetic acid. After drying, the catheters were exposed to ammonia and then soaked in a pH 7.0 phosphate buffer containing 1 % heparin. About 40 μg of chitosan were deposited per square cm. The heparinization procedure added 2 - 3 units of heparin per square cm. This heparin can be removed by soaking in 25 % sodium chloride, however, if the heparinization is performed in the presence of sodium cyanoborohydride, the brine removes only a minor part of heparin. Negligible amounts of the coating were removed by blood at 37°C after 24 hr contact.

The catheters were implanted for 30 min in carotid and femoral arteries. ^{125}I-Fibrinogen was injected prior to implantation; the radioactivity of the entire catheter and extruded thrombus was determined by γ-ray spectrometry. Table I presents the data on chitosan--heparin coated polyethylene catheters. The tridodecyl methylammonium chloride-heparin surface performs poorly in comparison to the chitosan--heparin cyanoborohydride surface.

Chitosan-heparin coated polymers display excellent thrombo-resistance properties. The lifetime of the thromboresistance can be extended by covalently binding the heparin to chitosan with the aid of sodium cyanoborohydride. This surface treatment is useful for bio-medical applications requiring blood compatibility for periods as long as four days.

Table I

Antithrombogenicity of coated polyethylene catheters.

Material	Fraction of catheter with clots	Radioactivity, cpm mean ± std dev.
Catheter assay in dogs		
Polyethylene	12/12	27,600 ± 5630
Chitosan + Heparin treated Polyethylene, sterilized with ethylene oxide	1/12	861 ± 80
Chitosan + Heparin treated Polyethylene, sterilized with gamma radiations	2/12	807 ± 123
Catheter assay in pigs		
Polyethylene	6/12	8,280 ± 3,790
Chitosan + Heparin + $NaCNBH_3$ leached with 25 % NaCl	0/12	431 ± 89
Tridodecylmethylammonium.HCl + Heparin treated Polyethylene	7/12	17,680 ± 7,050

CHITOSAN MEMBRANES

The film-forming ability of chitosan is well documented. Many articles have dealt with the use of chitosan membranes, for the removal of toxic metal ions, hemodialysis, treatment of brines, immobilization of enzymes and other purposes [1, 45 - 47].

Membranes are usually cast from chitosan solutions in acetic acid or other suitable acids. Threads, filaments and hollow fibers can be easily spun; these manufacts are neutralized with sodium hydroxide or ammonia before drying.

We have studied the chelation of copper on membranes of chitosan by spectrophotometry, electron spin resonance and infrared spectrometry. The chelates at pH values 4.0 and 5.0 involve one or two nitrogen atoms per copper ion, resulting from protonation of the amino groups of the biopolymer. At higher pH values, the coordination number increased and hydroxyl groups were also involved in the chelates. Ammonia molecules could enter the chelate and be retained on the polymer by dative bonds to the cupric ion. While sodium ions easily permeate the chitosan membrane separating the sodium sulfate solution from distilled water, cupric ions had to undergo chelation equilibria and their passage through the membranes was slowed down. Over a 70 hour period, with a 8-μm thick membrane, concentration equilibrium was reached by sodium, while the copper concentration in the distilled water remained below detection limits.

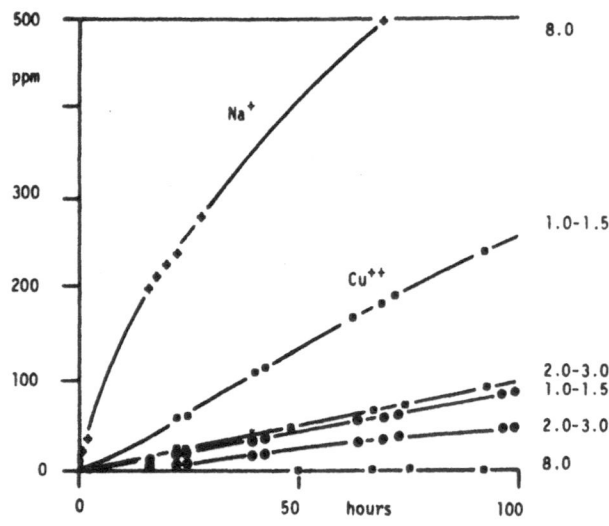

Figure 2. Permeation of chitosan membranes by copper and sodium ions: the figure shows the amount of metal ion in μg/g, appearing in the distilled water with time. The membrane thickness is indicated on the right side of the drawing. (+) Sodium ion; (■) Cupric ion; (●) Copper-amino complex. Sodium passes through 8.0 μm thick chitosan membrane from a 1000 μg/ml sodium sulfate solution into distilled water and reaches equilibrium after 70 hours, while cupric ion (points on the abscissa) does not pass through.

Ion-exchange membranes containing amino groups, insoluble in acidic and alkaline solutions were prepared from chitosan, poly-(vinyl alcohol) and glutaraldehyde. When using these membranes in a diaphragm cell, one side being adjusted to be acidic and the other alkaline, inorganic anions such as chloride, bromide and iodide and organic anions such as benzenesulfonate and amino acids were actively transported from the acidic side to the alkaline side across the membranes against the concentration gradient of anions between both sides of the membranes. Transport fractions and trans-port rate of anions through the membrane were significantly

influenced by pH gradients and electric potential differences. The
mechanism of the active transport of an amino acid through a chitosan
membrane is explained in Fig. 3.

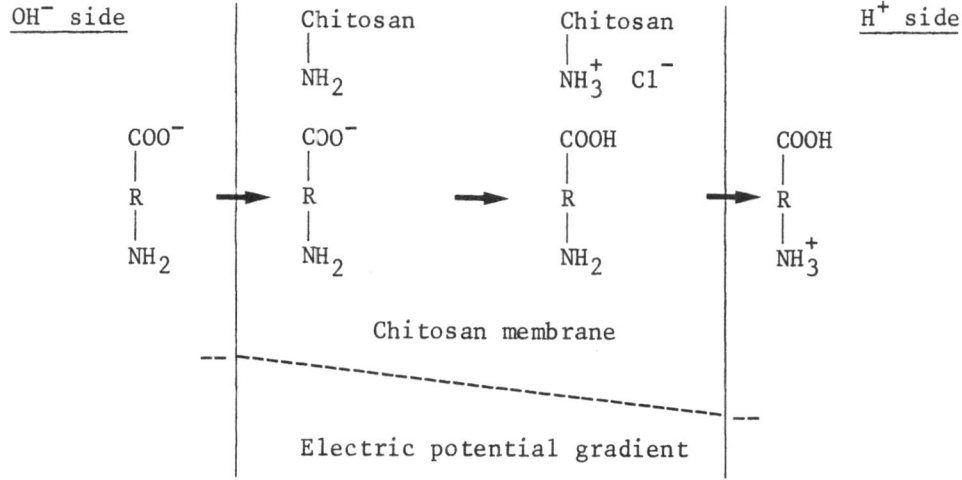

Figure 3. The L-phenylalanine molecule is charged negatively
on the alkaline side and positively on the acidic side. The
chitosan membrane is kept neutral on the alkaline side and
positive on the acidic side. Therefore, it is very difficult
that L-phenylalanine enters the membrane at the acidic side,
because of electrostatic repulsion. Negatively charged
L-phenylalanine is incorporated into the membrane on the
left hand side (OH$^-$ side) and transferred to the membrane due
to the electric potential difference between both sides of
the membrane. When negatively charged L-phenylalanine reaches
the right hand side (H$^+$ side), it is released by H$^+$ ions.
Therefore, L-phenylalanine is actively transported through
the chitosan membrane from the alkaline side to the acidic
side.

The film-forming ability is a characteristic property of chitosan derivatives: when a chitosan membrane is acetylated, a regenerated
chitin (N-acetylchitosan) membrane is obtained. The flow-rate of water
through N-acetylchitosan membranes was 10.0 -23.6 x 10^{-3} ml/cm^{2}min,
under a pressure of 3 kg/cm^2, and was unaffected by membrane thickness in the range 12 - 60 μm. The increase of chain length in the
N-acyl groups caused a slight decrease of the flow-rate. Low-molecular
weight compounds such as D-glucose, maltose, urea, calcium chloride,
sodium chloride, cyclohepta-amylose and maltodextrin (m. w. 2900)
passed though N-acetylchitosan membranes, whilst high-molecular weight

compounds, such as cytochrome c (m. w. 13,000) did not pass through.

A variety of N-acyl chitosan membranes are available today. They seem to offer advantages over Cuprophane, the cellulosic product which is commonly used for artificial kidneys, in terms of more efficient ultrafiltration of middle molecular size compounds (m. w. 1000 - 2000). The manufacture of Cuprophane membranes is also complicated by the need of removing copper from the product.

Commercial chitosans, however, contain great quantities of metal ions, which are collected on chitosan from well or tap water during the final washing, after the deacetylation step in the production process. To demonstrate that metals do not come from the shrimp or crab shells, we have carried out determination by atomic absorption spectrometry of a number of metals on a chitosan sample obtained from a production process where demineralized water was used. The results in Table II demonstrate that the total concentration of nine metals can be kept below 5.0 µg/g.

In this respect, we have demonstrated that chitosan and chitosan derivatives well qualify for use as medical aids to be brought in contact with blood. It is known that stringent specifications are imposed as far as the metal content of the membranes is concerned. With the exclusion of iron, which is considered to be non-toxic, the maximum allowed limit is 5 µg metals per g of membrane.

Table II

Determinations of metals in chitosan prepared from crab shells with use of demineralized water [48].

Metal	µg/g	Metal	µg/g
Vanadium	0.12	Silver	0.02
Chromium	0.36	Cadmium	0.22
Manganese	0.09	Mercury	0.025
Nickel	2.03	Lead	0.57
Copper	1.03		
		TOTAL	4.465

On the other hand, the chelating ability of chitosan seems to guarantee the absence of release of metal ions to blood.

A series of N-alkyl chitosans has been recently reported [49]: their potential usefulness for medical applications has still to be evaluated.

CONCLUSIONS

    The data presented in this lecture permit to conclude that the chemistry of chitin and chitosan offers the opportunity to exploit an abundant natural resource for the preparation of such sophisticated medical aids as heparin-like substances and blood-compatible devices.

    At the present time, chitin finds a number of applications; in Japan alone, about 500 tons/year are used for water treatment.

    More refined chitosan-based products, like those dealt with in this lecture, or like the modified chitosans used for the immobilization of enzymes and whole cells [50], open new perspectives for the utilization of chitin in the near future.

Acknowledgements. The original results presented in this lecture have been obtained with the financial contribution of the Consiglio Nazionale Ricerche, Roma (CT.81.01687.03).

REFERENCES

1.   R.A.A. Muzzarelli, "Chitin", Pergamon Press, Oxford (1977).
2.   R.A.A. Muzzarelli and E.R. Pariser (Eds.) "Proc. First Intl. Conference on Chitin & Chitosan", M.I.T. Sea Grant, Cambridge, USA (1978).
3.   S. Hirano and S. Tokura (Eds.) "Proc. Second Intl. Conference on Chitin & Chitosan" Japan Society for Chitin, Sapporo, Hokkaido, Japan (1982).
4.   C.H. Brown, "Structural Materials in Animals", Pitman, London (1975).
5.   W.J. Lennarz (Ed.), "The Biochemistry of Glycoproteins and Proteoglycans", Plenum Press, London (1980).
6.   R.D. Preston, "The Physical Biology of Plant Cell Walls", Chapman & Hall, London (1974).
7.   R. Kornfeld and S. Kornfeld, Comparative aspects of glycoprotein structure, Annu. Rev. Biochem. 45, 217 (1976).
8.   C. Jeuniaux and C. Cornelius, Distribution and activity of chitinolytic enzymes in the digestive tract of birds and mammals, in "Proc. First Intl. Conference on Chitin & Chitosan", R.A.A. Muzzarelli and E.R. Pariser (Eds.) M.I.T. Sea Grant, Cambridge, USA (1978).
9.   R.A.A. Muzzarelli, F. Tanfani and G. Scarpini, Chelating, film-forming and coagulating ability of the chitosan-glucan complex from Aspergillus niger industrial wastes. Biotechnol. Bioengin. 22:885 (1980).
10. G. Reed, "Prescott & Dunn's Industrial Microbiology" MacMillan Publ. Co.,New York (1982).
11. R.A.A. Muzzarelli, Biochemical Modifications of Chitin, in

"The Insect Integument", H.R. Hepburn (Ed.), Elsevier, Amsterdam (1976).

12. A.C. Grosscurt, "Some physiological aspects of the insecticidal action of diflubenzuron, an inhibitor of chitin synthesis". Doctoral Thesis, Agricultural University of Wageningen, Netherlands (1980).

13. U. Schlüter, Ultrastructural evidence for inhibition of chitin synthesis by nikkomycin. Wilhelm Roux's Arch. 191: 205 (1982).

14. S.R. Hansen and R.R. Garton, The effects of diflubenzuron on a complex laboratory stream community. Arch. Env. Cont. Toxicol. 11:1 (1982).

15. H.P. Fielder, R. Kurth, J. Langharian, J. Delzer and H. Zahner, Nikkomycins, microbial inhibitors of chitin synthase, J. Chem. Tech. 32:271 (1982)

16. R.C.W. Berkeley, G.W. Gooday and D.C. Ellwood (Eds.)"Microbial Polysaccharides and Polysaccharases", Academic Press, London (1979).

17. J.E. Smith and D.R. Berry (Eds.) "The Filamentous Fungi", Edward Arnold, London (1976).

18. J. Santamaria Cano, R. Pallarés Giner and F. Graus Ribas, Rhino-cerebral mucormycosis: a review. Medicina Clinica (Madrid) 78:453 (1982).

19. R.T. Waldo, Rhinocerebral mucormycosis: guidelines for therapy. Tex. Med. 78: 50 (1982).

20. E.C. Lazzaro and B. Sloan, Mucormycosis: case presentation and discussion. Ann. Ophthalm. 14:660 (1982).

21. P. Bernfeld (Ed.) "Biogenesis of Natural Compounds", Pergamon Press, Oxford (1963).

22. R.A.A. Muzzarelli, F. Tanfani and M.G. Muzzarelli, La chitina e i suoi derivati. Chim. Ind. (Milano) 64:18 (1982).

23. R.A.A. Muzzarelli, Chitosan and its derivatives: new trends of applied research. Carbohydr. Polymers 3:   (1983).

24. S. Suzuki, Y. Okawa, Y. Okura, K. Hashimoto and M. Suzuki, Immunoadjuvant effect of chitin and chitosan, in "Proc. Second Intl. Conference on Chitin & Chitosan" S. Hirano and S. Tokura (Eds.), Japan Society for Chitin, Sapporo Hokkaido, Japan (1982).

25. R.A.A. Muzzarelli, "Chairman's address: Chitin, an important natural polymer", in "Proc. First Intl. Conference on Chitin & Chitosan", R.A.A. Muzzarelli and E.R. Pariser (Eds.), M.I.T. Sea Grant, Cambridge, USA (1978).

26. M.L. Wolfrom and T.M. Shen Han, The sulfonation of chitosan, J. Am. Chem. Soc. 81:1764 (1959).

27. D.T. Warner and L.L. Coleman, Selective sulfonation of amino groups in amino alcohols. J. Org. Chem. 23:1133 (1958).

28. K. Nagasawa and N. Tanoura, Depolymerization and sulfation of chitosan by sulfuric acid. Chem. Pharm. Bull. 20:157 (1972).

29. E. Glaz and P. Vecsei, "Aldosterone", Pergamon Press, Oxford (1975).

30. H. Fukuda and Y. Kikuchi, In vitro clot formation on the poly-
         electrolyte complexes of sodium dextran sulfate with
         chitosan. J. Biomed. Mat. Res. 12:531 (1972).
31. Y. Kikuchi and A. Noda, Polyelectrolyte complexes of heparin
         with chitosan. J. Appl. Polymer Sci. 20:2561 (1976).
32. K. Nagasawa, Anticoagulant sulfonated chitosan. Japan 76.06720
         (1976).
33. T. Teshirogi, H. Yamamoto, M. Sakamoto and H. Tonami, Prepara-
         tion of sulfated aminodeoxycelluloses. Sen-I Gakkaishi
         36:78 (1980).
34. O. Larm, K. Larsson, E. Scholander, L.O. Andersson, E. Holmer
         and G. Söderström, The preparation of a heparin ana-
         logue from alginic acid. Carbohydr. Res. 73:332 (1979).
35. G. Kindness, W.F. Long and F.B. Williamson, Evidence for anti-
         thrombin III involvement in the anticoagulant activity
         of cellulose sulfate. Br. J. Pharmacol. 68:645 (1980).
36. M. Okada, H. Sumimoto, M. Hasegawa and H. Komada, Sulfated
         synthetic polysaccharides having physiological activity.
         Makromol. Chem. 180:813 (1979).
37. H. Komada, M. Okeda and H. Sumimoto, Synthetic polysaccharides
         containing amino groups. Makromol. Chem. 181:2305 (1980).
38. L.B. Jaques, Heparins: anionic polyelectrolyte drugs. Pharm. Rev.
         31:99 (1979).

39. M. Jozefowicz, J. Jozefowicz, C. Fougnot and D. LaBarre, New
         heparin-like insoluble materials, in "Chemistry and
         Biology of Heparin", R.L. Lundblad, W.V. Brown, K.G.
         Mann and H.R. Roberts (Eds.), Elsevier/North Holland,
         Amsterdam (1981).
40. R.A.A. Muzzarelli, F. Tanfani, M. Emanuelli, M.G. Muzzarelli and
         G. Celia, The production of chitosans of superior
         quality. J. Appl. Biochem. 3:316 (1981).
41. R.A.A. Muzzarelli and F. Tanfani, N-Carboxymethyl chitosans and
         N-carboxybenzyl chitosans: novel chelating polyampho-
         lytes, in "Proc. Second Intl. Conference on Chitin &
         Chitosan" S. Hirano and S. Tokura (Eds.) Japan Society
         for Chitin, Sapporo, Hokkaido, Japan (1982).
42. R.A.A. Muzzarelli, F. Tanfani, M. Emanuelli and S. Mariotti,
         N-Carboxymethylidene chitosans and N-carboxymethyl
         chitosans: novel chelating polyampholytes obtained from
         chitosan. Carbohydr. Res. 107:199 (1982).
43. R.A.A. Muzzarelli, F. Tanfani, M. Emanuelli and S. Mariotti,
         Sulfated N-carboxymethyl chitosans: novel blood anti-
         coagulants, to be published.
44. W.J. Hammar, H.V. Mendenhall, R.L. Vigdahl, R.H. Ferber and L.C.
         Haddad, Chitosan-heparin as a thromboresistant surface.
         in "Proc. Second Intl. Conference on Chitin & Chitosan",
         S. Hirano and S. Tokura (Eds.), Japan Society for Chitin,
         Sapporo, Hokkaido, Japan (1982).

45. R.A.A. Muzzarelli, F. Tanfani, M. Emanuelli and S. Gentile, The
        chelation of cupric ions by chitosan membranes. J. Appl.
        Biochem. 2:380 (1980).
46. S. Hirano, K. Tobetto, M. Hasegawa and N. Matsuda, Permeability
        properties of gels and membranes derived from chitosan.
        J. Biomed. Mat. Res. 14: 477 (1980).

47. S. Hirano, K. Tobetto and Y. Noishiki, SEM ultrastructure studies
        of N-acyl- and N-benzylidene chitosan and chitosan
        membranes. J. Biomed. Mat. Res. 15:903 (1981).
48. R.A.A. Muzzarelli, original results.
49. R.A.A. Muzzarelli, F. Tanfani, M. Emanuelli and S. Mariotti,
        The characterization of N-methyl, N-ethyl, N-propyl,
        N-buthyl and N-hexyl chitosans, novel film-forming
        polymers. J. Membr. Sci., in press.
50. R.A.A. Muzzarelli, Immobilization of enzymes on chitin and
        chitosan. Enz. Microb. Technol. 2:177 (1980).

ADVANCES IN THE DEVELOPMENT OF

EXTRACTION RESISTANT FLEXIBLE PVC COMPOUNDS

M.S.Biggs and D.Robson

BIP Vinyls Limited,(now Norsk Hydro Polymers Limited)
Newton Aycliffe,
Co. Durham,   United Kingdom

INTRODUCTION

The established use of flexible PVC in the manufacture of
medical devices is increasingly threatened by doubts about the
toxicity of phthalate plasticisers normally used in the PVC compounds
concerned. For many years BIP Vinyls Limited have been working to
develop alternative flexible PVC compounds using plasticiser systems
which are more acceptable than standard phthalates.

This paper summarises this development work which has resulted
in a comprehensive range of extraction resistant flexible PVC
compounds and discusses application areas ranging from stomach feeding tubes and nutritional fluid bags to blood contact devices such as
dialysis sets and blood bags.

EXTRACTION RESISTANCE

Flexible or soft PVC compounds obtain their flexibility by
incorporation of plasticiser which is not normally chemically bonded
to the PVC polymer. Under certain conditions of use such as contact
with fatty substances or oils for example, the plasticiser is able to
migrate from the flexible PVC product into the contact material.
When plasticiser is extracted in this way the PVC becomes less
flexible and can eventually assume a permanent set and possibly crack
or fail under stress. Many types of plasticiser are available and
some of these have good resistance to extraction from PVC compounds
because of their chemical structure and molecular size. Use of this
type of plasticiser enables an extraction resistant flexible PVC
compound to be produced.

IMPORTANCE IN MEDICAL APPLICATIONS

Because they possess many desirable properties, PVC products are in widespread use throughout the world today for the manufacture of medical devices and the majority of these utilise flexible PVC compounds. Whenever a device is used in contact with fatty tissues and fluids from the human body, the opportunity exists for extraction of plasticiser from the PVC. Some applications are more likely than others to result in plasticiser extraction, examples being any long term use of devices such as gastric feeding tubes or blood bags. In both cases a prolonged contact time of up to three weeks is possible between PVC and blood or stomach fluids such that significant and measurable plasticiser extraction can take place. This can give rise to several problems in medical applications.

In extreme cases the resultant changes in the properties of the PVC such as stiffening of a flexible tube may prevent a device from continuing to function properly and if originally inserted in a patient, removal of a device with much reduced flexibility can be problematical. This situation has actually occurred in practice with stomach feeding tubes (1).

However, a more serious general consequence of plasticiser extraction by body fluids can be the transference of plasticiser to a patient with attendant concern about the possible toxic effects and the additional biological burden suffered by the patient whose metabolism or ability to deal with the foreign substance may already be impaired.

CURRENT SITUATION WITH DEHP

The plasticiser used in the majority of flexible PVC currently used in medical applications is di-octyl phthalate (DOP) or more correctly di-(2 ethyl hexyl) phthalate (DEHP) and this is a prime example of a plasticiser which can be extracted under appropriate conditions.

Because DEHP plasticised PVC has been used in medical applications for nearly twenty-five years, a considerable amount of data has been provided from studies of its biological effects (mainly on animals) and often conflicting conclusions have been reached concerning its desirability and possible toxic effects. Largely because of its record of in-use performance for such a long period of time, DEHP is currently almost the only plasticiser permitted wherever positive lists exist defining additives that can be used in PVC compounds for medical applications. Examples are the European Pharmacopoeia monograph and DIN standard concerning plastic blood storage bags (2) (3). In view of the apparent low acute toxicity of the plasticiser it seems reasonable to permit use of DEHP in short contact time devices such as transfusion sets or any single-use disposable devices for treatment of

patients. However, it is a known and accepted fact that DEHP is
extracted into blood in long term contact (up to three weeks) with
flexible PVC such as in blood storage bag applications.

In the past eighteen months or so renewed anxiety has been expressed about the use of DEHP - especially in medical applications -
following the publication of results of US National Cancer Institute
studies on various commonly used plasticisers including DEHP (4).
These rat feeding studies which used high feeding dosages conclude
that this plasticiser is a possible carcinogen for humans. A debate
currently exists about the merits of extrapolating results of these
animal tests to make conclusions about effects on humans and this has
initiated new studies in both Europe and the USA to try and resolve
this point. Although results of the new studies are not expected
before 1983, national regulatory health authorities are known to be
closely observing the situation with respect to the use of DEHP. It
also seems certain that some authorities will sooner or later place
severe restrictions on the use of this plasticiser in all PVC used in
medical applications. Apart from the possible toxic or carcinogenic
effects, a factor having increasing influence on any such decision is
that of biological burden to patients. Patients particularly exposed
to risk in this way are those receiving frequent treatment over long
periods of time, such as haemodialysis patients. It is likely that
the metabolism of such patients is already unable to operate normally
when treatment is commenced and it is the fact that a continuous
additional biological burden is placed on the patient as a result of
the regular treatment required that gives rise to concern. It has been
shown that a patient can receive up to 150mg of DEHP in a five hour
dialysis session (5) and on average a patient receives 100 sessions
per year.

NEED FOR IMPROVEMENT

Flexible PVC has achieved a leading position in the range of
plastic materials used in the manufacture of medical devices due to
the many beneficial chemical and physical properties that it provides.
PVC devices have been shown to be highly reliable and have a very low
risk to benefit ratio in service and it is obviously desirable that
this type of product continues to be available to the medical device
industry but preferably without some of the plasticiser extraction
short-comings already described. Recognising this situation BIPVL
started development work in consultation with the UK DHSS as long ago
as 1975, seeking to develop a flexible PVC compound having improved
extraction resistance to body fluids. Initially this development was
specifically aimed at stomach feeding tube applications and the work
has been summarised in an earlier paper (6). The development reached
fruition in 1979 when the newly developed extraction resistant PVC was
first used commercially for the manufacture of feeding tubes. However,
work has continued in order to offer a comprehensive range of

compounds and the previously described concern about DEHP extraction
in many blood contact applications has more recently caused increased
interest by the medical device industry about the possibilities offered by the DEHP-free flexible PVC compounds.

DEVELOPMENT OF EXTRACTION RESISTANT PVC

     In the early stages BIPVL screened many materials as potential
alternatives to DEHP and a vegetable oil immersion test was shown to
 ive a good measure of the extraction resistance of the PVC compound
(6). Figure 1 compares the performance of several plasticisers in
this test showing significant differences in performance. Two types
of polymeric adipate give very good resistance to extraction in
comparison with DEHP and it is interesting to note that a trimellitate
plasticiser of the type currently being recommended for use in medical
PVC compounds in the USA performs no better than DEHP in this test.
Figure 2 shows the comparison of PVC based on DEHP and polymeric
adipate plasticisers in a commercially available synthetic feeding
fluid Intralipid (7). This is a 20% emulsion of oil in water and
weight loss measurements due to plasticiser extraction were affected
by a certain amount of water absorption by the PVC at the test temperature (37°C). Nevertheless after an initial increase in weight by
both materials the adipate based PVC maintained a constant weight
change in contrast with the DEHP based material which thereafter
showed an increasing loss in weight associated with extraction of
plasticiser into the fatty constituents of the test fluid.

     While it is relatively easy to select alternatives to DEHP that
have good resistance to extraction in oils and fatty fluids, it is
much more difficult to find materials which at the same time have as
good resistance to extraction by water as DEHP. This can be seen in
Table 1 which summarises some of the typical aqueous extraction tests
that are required by most national standards or pharmacopoeia as a
means of demonstrating freedom from chemical toxicity.

Table 1. Aqueous Extraction Tests (Chemical Toxicity)

Test	Typical Specifications	DEHP	Polymeric Adipate A	B
Oxidisable Matter	≯ 2ml of 0.1 N Na thiosulphate	0.8	5.8	1.8
pH	4.5 to 7.0	5.0	3.85	4.5
Evaporation residue	≯ 3 mg	1.0	27.4	2.2
UV absorption	≯ 3 at 220 nm	0.130	0.470	0.156

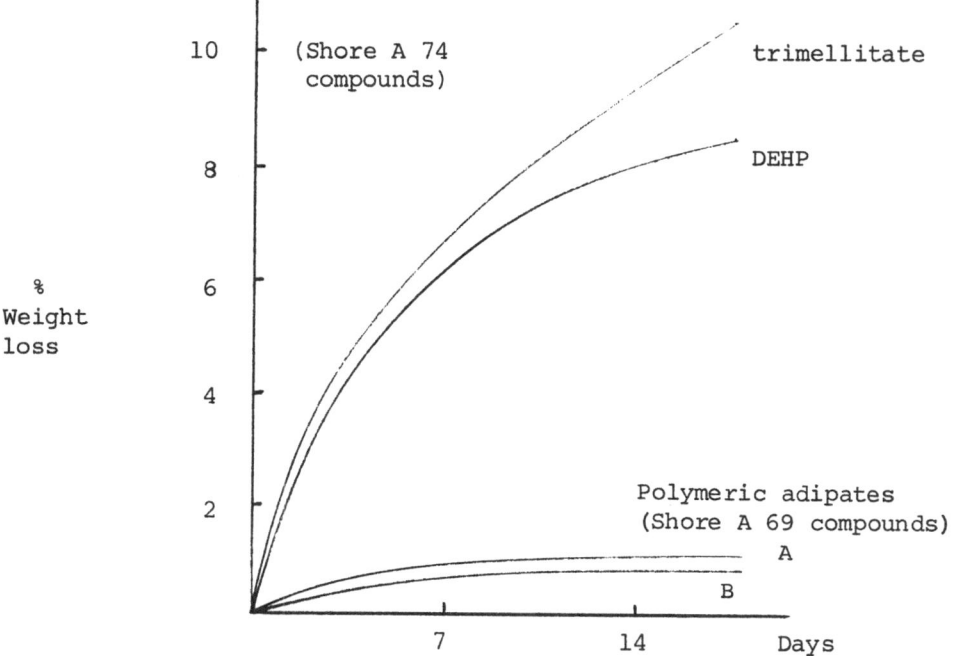

Figure 1.   Immersion in vegetable oil

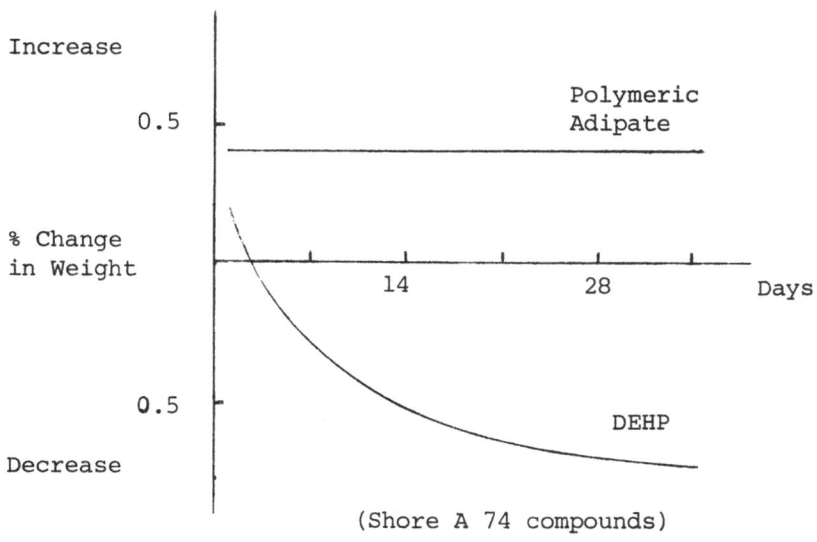

Figure 2.   Immersion in Intralipid.

In this case the PVC was extracted in water for thirty minutes at 110°C using a surface area to water volume ratio of 50 cm² per 10 ml. These conditions and the test limits quoted for the first three tests are those given in BS 2463 (17). Other specifications require different extraction and use slightly different test limits and normally require measurement of UV absorption in addition, but the test limits shown are fairly typical requirements.

A marked difference is observed between the two polymeric adipates and the specially selected adipate (B) used in BIPVL extraction resistant PVC is seen to have almost equal performance to DEHP in these important aqueous extraction tests.

It can be recognised, therefore, that the total extraction resistance requirements in both oily fluids and water imposed great limitations on the choice of alternative plasticisers. It was only after a large number of materials were screened that a suitable candidate was found for use in more extensive development work.

OTHER REQUIREMENTS

As well as achieving good extraction resistance it is necessary that any new PVC compound should retain the desirable properties of existing DEHP based compounds used in medical applications. Table 2 lists typical properties of this type, some of which relate to production and processing of the PVC compound, others being requirements of the PVC in product form. Any basic formulation changes in a flexible PVC compound which is a mixture of at least four additives (PVC resin, plasticisers and stabilisers) will have a direct effect on one or more of the listed properties. Simply the preservation of good colour and clarity means that many possible formulation components are rejected and this alone restricts the choice of alternative plasticisers to DEHP. PVC compounds based on the polymeric adipate selected for its extraction resistance performance satisfy the majority of the physical and chemical properties, although inevitably some differences are seen with DEHP based PVC. For example, at a given Shore A hardness the new compounds will have slightly higher density, tensile strength and elongation, but slightly poorer low temperature resistance. The adipate based compound is not processed quite so easily to a high standard of appearance as DEHP based material normally, but careful selection and combination of all formulation ingredients including PVC polymer has enabled BIPVL to offer compounds which can be processed on existing equipment with the need for only minor adjustments. For example, a higher extrusion melt temperature is desirable in order to obtain best results with the extraction resistant compounds and use of water cooled screws of high compression ratio (greater than 3/1) is also most advisable for optimum performance.

Table 2.   Important Properties of Flexible PVC
used in Medical Applications

A)   Compounds

Physical Properties - Flexibility, Tensile Strength, Elongation
Ease of Processing - Extrusion, Injection Moulding, Calendering

B)   Products

Appearance	- Colour, Clarity, Gel Speck Level
Sterilisation Resistance	- Steam, Ethylene Oxide, Irradiation
Weldability	- Solvent, Radio Frequency Radiation
Gas Permeability	)
Low Temperature Resistance	)
Particulate Matter Generation	) Film/Packaging Applications
Moisture Vapour Transmission	)

The behaviour of the extraction resistant compounds under various methods of sterilisation has been shown to be similar to DEHP plasticised PVC and the same is true with respect to weldability.

TOXICOLOGICAL STATUS

The preceeding paragraphs have shown that many factors have to be considered in formulating an extraction resistant PVC for medical applications in order to achieve acceptable technical performance. Having met this target in spite of many conflicting demands the final and perhaps most demanding requirement is that any new material must be toxicologically acceptable. Basic guidelines now exist for evaluation of plastic materials and devices for biological hazards (8) (9) and Table 3 summarises the types of test recommended in order to provide initial toxicity data. Generation of even this amount of information has been very costly and time consuming but BIPVL have been able to provide information on the majority of these items for the new extraction resistant compounds with satisfactory results so far.

The main need has been to provide information on the biological effects of use of the particular polymeric adipate which has not previously been used in medical applications, although plasticisers of this type are permitted for use in flexible PVC for food contact applications (10) (11) (12). Depending on the type of test, results have been obtained on either the plasticiser alone or more usually on the PVC compound itself or extracts from the compound.

Table 3. Toxicological Data Requirements

Physical/Chemical Toxicity	–	Plasticiser, PVC Compound, Extracts	
Cytotoxicity	–	Plasticiser, PVC Compound	
Acute Oral Toxicity	–	Plasticiser	
Acute Systemic Toxicity	–		Extracts
Intracutaneous Toxicity	–		Extracts
Intramuscular Implant	–	PVC Compound	
Blood Compatibility	–	PVC Compound	
Mucous Membrane Irritation	–	PVC Compound	
Mutagenicity	–		Extracts
Carcinogenicity	–	PVC Compound	

The physical/chemical toxicity tests are of the type specified in many national pharmacopoeia and standards (2),(3),(13) to (19) inclusive, mainly involving tests on aqueous extracts including those already referred to in Table 1. Other typical tests required on the extracts are taste, odour, colour, clarity, chloride, ammonia, organic phosphorus and heavy metal content, and in addition ignition residue and heavy metal tests are made directly on the PVC compound. Specified limits for some of these tests vary so it would be misleading to state that all of these tests are met in every case but at present we do know that the new extraction resistant compounds have been shown to satisfy BS 2463 (17), DHSS Specifications (18) and DIN 58361 Part 4 (3) for example.

In vitro cell culture tests with plasticiser and PVC compound showed these to be non-toxic with L929 mouse fibroblast cells. Similar tests on the PVC compound and an eluate using human lung fibroblast cells also showed no evidence of cytotoxicity.

Acute oral toxicity studies for seven days on rats using the plasticiser emulsified in vegetable oil showed no clinical signs of toxicity with dose levels 0.2, 1.0 and 5.0 ml per kg body weight and no significant histopathological changes were observed.

The acute systemic toxicity has been successfully tested according to the USP method (13) using four extractants - saline, alcoholic saline, vegetable oil and PEG 400, extraction taking place over sixty minutes at 121°C followed by intraveneous or intraperitoneal injection into mice.

Intracutaneous toxicity of the same four extracts was assessed by injection into rabbits, again following the USP test method (13). A sample of extraction resistant PVC compound tested in this way has passed satisfactorily without showing frequently observed effects found with DEHP plasticised PVC extracted in PEG 400.

The intramuscular implantation test of the USP assessed toxicity arising from implantation in rabbits for seven days and the new PVC compound proved satisfactory in this test. A further test of this type compared extraction resistant and DEHP medical type PVC over a three week implantation period. While microscopic examination showed no significant difference in reaction, histopathological evidence suggested the DEHP based compound produced more severe reactions than those produced by the new PVC.

Successful results in the implantation, acute systemic and intra-cutaneous toxicity tests, enabled the extraction resistant compound to gain a USP Class VI (121°C) rating.

For initial assessment of blood compatibility, tubing made from extraction resistant and DEHP compounds was compared in tests using fresh dog blood which is generally more susceptible to chemical haemolysis than any other species. No evidence of haemolysis was observed within the limits of the test.

Mucosal irritation was studied for both extraction resistant and standard PVC compounds by implantation in hamster cheek pouches for three weeks with subsequent macroscopic and histopathological exam-ination. No significant differences were observed between the compounds.

Mutagenicity data as provided by the Ames test using an extract from the PVC compound is in the course of being obtained.

ACCEPTANCE OF EXTRACTION RESISTANT PVC

Approval or acceptance for use by medical regulatory authorities is chiefly dependant on the toxicological evidence that can be provided not only for the basic plastic such as extraction resistant PVC in this case but also for a finished product or medical device. As a manufacturer of PVC compounds BIPVL is only able to provide information on the extraction resistant PVC and customers of BIPVL - the medical device manufacturers - are required to provide the additional information demonstrating the suitability of a device for its intended application. Limited clinical trials may however be permitted at an intermediate stage.

The toxicity tests already described were chiefly designed to show the extraction resistant compound is satisfactory for use in the original target application - that of stomach feeding tubes. In 1979 the Danish Health Ministry accepted the material for just such an application followed in 1980 by the UK DHSS who raised no objection to use in a slightly wider range of application areas including wound drainage and blood contact. BIPVL extraction resistant compounds have therefore been in commercial use since 1979 and they are

Table 4.   Status for use of Extraction Resistant
PVC in medical Applications

UK   - Filed with DHSS.
            Use permitted in nasogastric tubing, wound
            drainage tubing and blood contact applications.

DENMARK - Filed with Health Ministry.
            Use permitted for stomach tubing and catheters.

USA   - Filed with F & DA (Drug Master File).
            Plasticiser type permitted for food contact use.

ITALY   - Filed with Health Ministry.
            Plasticiser permitted for food contact use.

currently being used in evaluations and development work for many
medical device applications throughout Europe. In order to assist
customers in obtaining clearance for medical devices it is normal
practice for BIPVL to make available all toxicological data and appropriate formulation details to national regulatory authorities and
examples of this are shown in Table 4.

FURTHER APPLICATIONS

     Wider application of these compounds depends on several factors
one being the need for additional toxicity and biological contact
information particularly with respect to blood compatibility and
extraction data in contact with blood and other fluids such as IV
solutions. One of the major difficulties in generating such information in the past has been the problem of detecting and measuring low
levels of the polymeric adipate plasticiser. Methods of measurement
for small quantities of DEHP which is a monomeric substance cannot be
applied to the adipate which being polymeric is not easily character-
ised like DEHP. Also due to its much lower extractability only
extremely low levels of the polymeric plasticiser are likely to be
present in contact fluids such as blood and no satisfactory methods
have yet been developed for direct measurement especially in the
presence of such a chemically complex fluid as blood. Work has been
commissioned seeking to develop an appropriate chemical method for
quantitative assay of the plasticiser in a series of biological and
synthetic fluids.

     As an alternative approach to the problem, the use of radio
labelling techniques is being considered for assessing extraction of
plasticiser into a similar series of fluids including blood. It is
important in this case to ensure levels of radio activity are sufficient to enable meaningful measurements to be made on any plasticiser
extracted into contact fluids, and this will directly affect the cost

of such work.   In addition of course many device manufacturers are
carrying out their own additional tests appropriate to the intended
applications.

Another factor that cannot be overlooked is the cost of any new
PVC compound.   It is a fact of life that use of almost any plasticiser
other than DEHP will result in increased raw material costs.   The
extent of the increase will be greater as the plasticiser type becomes
more specialised as in the case of the extraction resistant compounds.
There is therefore very little incentive for the industry to change
from use of DEHP plasticised PVC for many short term contact medical
devices even if technical and toxicological advantages are demonstrated
for a new but more expensive PVC.   This situation would be different of
course if any moves are made to restrict the use of DEHP in medical
applications and pressure to this end may well be exerted in the
future by regulatory authorities on the one hand and from the end-users
on the other.   In the meantime in the critical application areas of
storage of blood and blood components and haemodialysis applications
for example where cost may not be paramount, the BIPVL extraction
resistant compounds are readily available as an alternative to DEHP
based PVC and are actively under evaluation. Development work is also
in progress at several companies to examine the suitability of the new
PVC for intravenous and nutritional fluid storage bags. Wherever
problems currently exist with DEHP extraction the extraction resistant
compounds should be seriously considered.

Because of the interest shown by the medical device industry and
demands for different compounds a wide range of extraction resistant
flexible PVC compounds has now been formulated by BIPVL and this is
illustrated in Table 5. While basically formulated for extrusion
applications, some of these compounds have also been shown to be
perfectly satisfactory for injection moulding of small components for
medical devices, and also for film production by calendering and
blown lay-flat film processes.

SUMMARY

BIPVL view the present compounds as a first step towards establishing DEHP free flexible PVC which may be eventually required for
wider application than in the medical industry alone. However it is
accepted that even further improvements to the current extraction
resistant PVC may be desirable in the most demanding areas where for
example all round resistance to extraction in aqueous and oily fluids
or tissues is required combined with very good low temperature properties and other potential systems are continually being screened. The
trimellitate based material referred to earlier appears to show
promise with low extractability in blood and while there may be strong
political pressures to rapidly establish a replacement for DEHP particularly in the USA, there remain some uncertainties about the use of
this particular material. For example expert opinion is critical of

Table 5.   Extraction Resistant Flexible PVC Compounds
           for Medical Applications

BIPVL REFERENCE	BS SOFTNESS	SHORE A HARDNESS
VHW-68584	80	57
VHW-68902	65	63
VLW-67733	55	68
VHW-68576	45	74
VHW-68623	35	80
VHW-68169	30	83
VHW-68710	20	90

the claims made concerning the absolute level of extraction into
human plasma because of the low levels of radio-activity that were
used in the radio-labelling studies employed to investigate plasticiser extraction.

It may also be argued that the chemical structure of the trimellitate is very similar to that of DEHP. As it seems quite possible
that any toxic effects are caused by metabolites or breakdown products
of the plasticiser it can be speculated that the same chemicals may be
produced from both DEHP and trimellitate.

This paper has shown the many often conflicting requirements
demanded of any new flexible PVC compound for use in medical applications. Even when the physical and technical properties are satisfied
the new material has to meet toxicological requirements which are
becoming increasingly demanding and complex. This inevitably means
that development to the point where a new material can be offered to
the medical device industry is a very time-consuming and costly
operation. BIPVL therefore believes the newly developed extraction
resistant flexible PVC merits serious attention for use in the
manufacture of medical devices in the immediate future.

REFERENCES

1. E. G. Hayhurst and M. Wyman, Am Jnl Dis Children, 129:72
       (1975).
2. European Pharmacopoeia, VI 1.2 Plastic Materials and VI 2.2
       Plastic Containers.
3. Plastic Containers for Blood DIN 58361 Part 4 (September 1980).
4. National Toxicology Programme/Inter Agency Regulatory Liaison
       Group, Conference on phthalates Washington D.C., U.S.A.
       (June 1981).
5. T. P. Gibson, W. A. Briggs and B. T. Boone, J. Lab. Clin. Med.
       87:519 (1976).

6.   M. S. Biggs and J.Baldwin, Plast. and Rub. Matls. and Appctns.
     5(4) : 187 (1980)

7.   Intralipid (20% emulsion of soya bean oil in water),
     Manufactured by Kabi Vitrum AB, Sweden.

8.   Evaluation of medical devices for biological hazards,
     BS 5736 Part 1 (1979)

9.   Guideline for evaluating the safety of materials used in
     medical devices, US HIMA Document No 10 Vol 3 (1981).

10. West German BGA Recommendation 38.

11. USA Code of Federal Regulations, Title 21, Pt 178, 3740.

12. Italian Ministerial Decree 1973 (as amended).

13. Biological and Physico Chemical Tests - Plastics,
     US Pharmacopeia XX (1980).

14. Italian Farmacopea Ufficiale VIII, Vol 1.

15. The Pharmacopoeia of Japan IX, Revised (8/79).

16. Pharmacopée francaise 9th edition.

17. Transfusion Equipment for medical use, BS 2463.

18. UK DHSS Specifications TSS/B/320/005, 008 and 020.

19. European Agreement on the exchange of therapeutic substances
     of Human origin, Council of Europe, Strasbourg (1981)

IMMOBILIZATION OF ENZYMES ON HOLLOW FIBERS ASSEMBLED AFTER CHEMICAL MODIFICATION

Piergiorgio Pietta, Alma Calatroni and Dario Agnellini

Istituto di Chimica Organica ed Analitica

Via Celoria, 2 - MILANO

INTRODUCTION

The use of enzymes in therapy[1,2] is limited by several collateral effects, consisting mainly in the antibody formation[3], hypersensitivity reactions and toxicity in different organs[4,5].

To reduce these side-effects the extracorporeal use of enzymes has been developed by linking enzymes to the different biocompatible polymers[6-8]. Enzymes have been, as an alternative immobilized to hollow fibers of commercial hemodialyzer without altering the dialytical properties of the fibers and their hemocompatibility[9-12].

Nevertheless to avoid the risk of serious damaging the chemical modification on hollow fibers hemodialyzers has to be performed in the absence of organic solvents and consequently the choice of the activation procedures is very limited.

To overcome this problem, we developed a simple technique, which involves two steps: 1. chemical activation of hollow fibers and their characterization; 2. assessment of the functionalized hollow fibers in conventional hemodialyzers and their use for coupling several nucleophiles, such as aminoacids, peptides and enzymes.

MATERIALS

Cuprophan Hollow fibers were supplied from SORIN Biomedica, S.p.A. (Saluggia, Italy). 3-Aminopropyltriethoxysilane, o-hydroxynaphtaldehyde dicyclohexylcarbodiimide (DCC), n-hydroxysuccinimide (HOSu) were obtained from Fluka. N-ethyl-N'-(3-dimethylaminopropyl)carbodiimide·HCl (WSC) was from Merck.

Val-Leu-Ser-Pro,Val-His-Leu and Eledoisin were synthesized in our
laboratory according to the literature.[13-15] Trypsin,asparaginase and
urease were from Sigma;other chemicals were standard analytical grade
reagents.

METHODS

Hollow fibers activation

1. Reaction with 3-aminopropyltriethoxysilane

    After washing with acetone,hollow fibers(5g,30cm^2 surface) were
reacted with 2% 3-aminopropyltriethoxysilane in acetone for 2 min at
room temperature. Then the fibers were thoroughly washed with acetone and dried at 45°C for 2 h. The capacity of the resulting aminofi-
bers,expressed as ,umol of aminogroup/g of fiber,was assayed by the
method of Esko et.al.[16] The stability of the bond between the cellulosic fibers and the silyl arm was checked by determining the capacity
after circulating water,KCl 50mM and 0.1 M phosphate buffer(pH 7) for
different times at 37°C.

2. Succinylation

    The aminofibers(2.5 g) were reacted with succinic anhydride(2.5 g)
and NaOH(2 g) in water at pH 6 for 3 h. The reaction was followed
assaying an aliquot of the fiber(10 mg) with 1% trinitrobenzensulpho-
nic acid(TNBS) in 0.1 M borate buffer.[17]

Coupling of aminoacids,peptides and enzymes

a. Hydroxysuccinimide ester.
    The succinylated fibers(1.5 g) were reacted with HOSu(400 mg) in
dioxane(40 ml) and DCC(1.2 g) for 3 h at room temperature. After washing with dioxane and methanol,the fibers were assembled in the
hemodialyzer form(50 fibers each,31 mm^2) and stored in desiccator
until used.
    Each hemodialyzer was allowed to react with 10 ,umol of Gly,Ala,Arg,
Phe,Gly-Gly,Gly-Gly-Gly,Val-Leu-Ser-Pro,Val-His-Leu and Eledoisin
in 0.1 M phosphate buffer pH 7 or with 5 ml of 1 mg/ml enzymes solu-
tion(Asparaginase,Urease,Trypsin) at pH 7 for 2 h at 4°C. Then the
filters were washed with the same buffer and stored at 4°C.

b. N-Ethyl-N'-(3-dimethylaminopropyl)carbodiimide.HCl(WSC).
    Assembled succinylated fibers(50 fibers) were equilibrated with
phosphate buffer pH 7.5,added of the equal amount of each nucleophile

reacted with WSC(100 mg) for 2 h at 4°C and washed with the same buffer.

After each coupling procedure, the degree of substitution, expressed as μmol/g of fiber for the aminoacids and the peptides was evaluated with automatic aminoacid analyzer Beckman mod. 120B.

The amount of the enzymes immobilized on the filter(mg active protein/g of fiber) was determined by assaying the bound activity.

RESULT AND DISCUSSION

The chemical activation of the cellulose hollow fibers and the coupling of nucleophiles(RNH$_2$) were performed according to the following scheme:

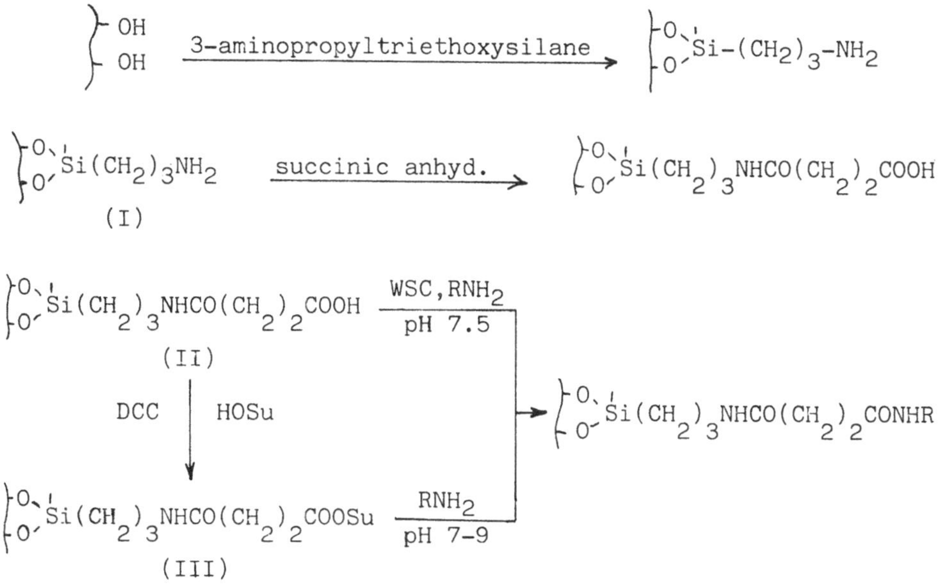

$$\text{OH}, \text{OH} \xrightarrow{\text{3-aminopropyltriethoxysilane}} \text{O}_2\text{Si-(CH}_2)_3\text{-NH}_2$$

$$\underset{(I)}{\text{O}_2\text{Si(CH}_2)_3\text{NH}_2} \xrightarrow{\text{succinic anhyd.}} \text{O}_2\text{Si(CH}_2)_3\text{NHCO(CH}_2)_2\text{COOH}$$

$$\underset{(II)}{\text{O}_2\text{Si(CH}_2)_3\text{NHCO(CH}_2)_2\text{COOH}} \xrightarrow[\text{pH 7.5}]{\text{WSC, RNH}_2}$$

DCC | HOSu

$$\underset{(III)}{\text{O}_2\text{Si(CH}_2)_3\text{NHCO(CH}_2)_2\text{COOSu}} \xrightarrow[\text{pH 7-9}]{\text{RNH}_2}$$

$$\longrightarrow \text{O}_2\text{Si(CH}_2)_3\text{NHCO(CH}_2)_2\text{CONHR}$$

To maintain the dialytical properties of the fibers, the reaction with 3-aminopropyltriethoxysilane was carried out for short time (up to 10 min) using increasing concentration of the reagents.

The best results referring to both the capacity of amino fibers (I) and their physical properties, were reached using 2% reagent in acetone for 2 min. In these conditions, the capacity of the resulting I ranged between 30 to 60 μmol/g of fiber. Reaction of I with large excess of succinic anhydride at pH 6 allowed to obtained II free of any residual aminogroup, as supported by the negative response to the TNBS test.

II was converted to the corresponding active ester support III, that was assembled in the hemodialyzer form and reacted with

aminoacids,peptides and enzymes. Alternatively,II was assembled and
used to immobilize the same nucleophiles in the presence of WSC.

The results obtained by these two procedures are compared in
table 1.

Table 1. Degrees of substitution obtained by the HOSu and WSC
procedures.

AMINOACIDS PEPTIDES	μmol/g HOSu	WSC	ENZYMES	mg prot.act./g HOSu	WSC
Ala,Gly,Phe,Arg	49	60	TRYPSIN	0.2	0.3
Gly-Gly	45	55	UREASE	0.75	1.2
Gly-Gly-Gly	41	45	ASPARAGINASE	0.3	1.0
Val-Leu-Ser-Pro	44	40			
Val-His-Pro	37	45			
Eledoisin	30	40			

From these data,it appears clearly that the active ester procedure yielded degrees of substitution lower than those obtained by
the WSC method.

The physical properties of the hollow fibers were evaluated after
each activation step,following the diffusion coefficient of standard
potassium nitrate and riboflavin. As shown in table 2,the activation
did not adversely change the peculiarity of semipermeable membrane.

Table 2. Physical properties of hollow fibers after each
activation step.

TREATMENTS	Diffusion coefficient cm^{-1}min^{-1} KNO_3	Riboflavin
Hollow fibers without treatment	$7.41 \cdot 10^{-6}$	$2.25 \cdot 10^{-6}$
Acetone	$4,57 \cdot 10^{-6}$	$1.70 \cdot 10^{-6}$
Silylpropylamine	$8,56 \cdot 10^{-6}$	$2.90 \cdot 10^{-6}$
Succinic anhydride, hydroxysuccinimide	$4.80 \cdot 10^{-6}$	$1.76 \cdot 10^{-6}$

Furthermore,the bond between the cellulosic fiber and the silyl
arm proved to be stable in the condition described under Methods.

Further evidences for the stability of the coupled nucleophiles were obtained by determining the degree of substitution after circulating phosphate buffer pH 7 for 3 h at 37°C.

In conclusion, the proposed method based on the chemical activation of the hollow fibers followed by their assessment in the bioreactor form,may be a solution to the problems normally encountered during the activation of commercial hemodialyzers.

REFERENCES

1. S.J.Bach and D. Swaine, Br. J. Cancer,19:379(1965)
2. E.Harris,J.Whitecar,G.P.Bodey and E.J.Freireich,
   Proc. Am. Cancer Res.,10:35(1969).
3. R.G.Peterson,R.E.Handschmacher and M.S.Mitchel,
   J. Clin. Invest.,50:1080(1971).
4. G.P.Whitecar;J.E.Harris,G.P.Bodey and E.J.Freireich,
   Clin. Res.,17:408(1969).
5. G.Celle,M.Dodero and J.Panaciulli, Europ. J. Cancer,9:55(1973).
6. J.J.Marshall,TIBS,April:79(1978).
8. D.A.Cooney,H.H.Weetal and E. Long, Biochem. Pharm.,24:503(1975).
9. H.Hyden, Biomat. Med. Dev.,Art. Org.,8:1(1980).
10. P.G.Pietta,D.Agnellini,G.Mazzola,G.Vecchio,S.Colombi and
    G.Bianchi, Enzyme Engineering vol.5,Plenum Press,New York(1980).
11. G.Mazzola,G.Vecchio, Int. J. Artif. Organs,4:102(1981).
12. V.Rossi,A.Malinverni and L.Callegaro, Int. J. Artif. Organs,
    4:102(1981).
13. A.Corbellini,P.G.Pietta and L.Rossi-Bernardi,
    Gazz. Chim. Ital.,97:1138(1967).
14. F.Weygand,w.Steglich and P.G.Pietta, Chem. Ber.,100:3841(1969).
15. P.G.Pietta,P.F.Cavallo,K.Takahashi and G.R.Marshall,
    J. Org.Chem.,39:44(1974).
16. K.Esko,S.Karlsson and J.Porath, Acta Chem. Scan.,22:3342(1968).
17. P.G.Pietta,D.Agnellini and M.Pace, J.Pharm. Sci.,68:1565(1979).

POLYMER COMPOSITE MATERIALS IN ORTHOPAEDIC SURGERY

G. W. Hastings

Bio-Medical Engineering Unit
North Staffordshire Polytechnic and North Staffordshire
Health District, c/o The Medical Institute,
Hartshill, Stoke on Trent, Staffordshire, ST4 7NY

IMPLANTS AND BONE

Polymeric materials have achieved regular use in orthopaedic
surgery since the introduction of room temperature hardening
acrylic cement and ultra high molecular weight polyethylene
(UHMWPE) and the various applications have been reviewed [1,2,3].

This combination of materials has had a profound effect upon
the development of surgery for the treatment of joint disease and
prostheses are readily available, not just for the hip joint but
also for the knee, ankle, shoulder and other joints. It is the
good biological acceptability of UHMWPE coupled with its mechanical
properties that has led to this widespread acceptance. In
particular, the tribological characteristics appear to be the most
satisfactory for use in a metal-polymer combination. Prosthesis
designs utilising alumina ceramic also incorporate an acetabular
component of UHMWPE in consequence of the low rate of wear observed.

However, when the interaction with the living bone is
considered the situation is somewhat different from the consideration
of one implant material with another. The effects of wear debris
and of corrosion/degradation products is a separate issue not to
be considered in this chapter. Attention will be confined to the
relationship between the mechanical properties of bone and an implant
material which together form a new compound system.

Mechanical forces applied to a living bone are known to affect
the pattern of growth provided that they exceed a critical
physiological minimum value and are of sufficient duration. Growth
can be redirected, enhanced, or inhibited and this fact is used

clinically in treatment of congenital effects and correction of
skeletal malalignments[4] . The actual mechanism by which this occurs
is not understood. It is observed though, that a piezo-electrical
response is produced by bone in response to mechanical strain.
This may have no consequence in vivo but the observation of this
phenomenon has led to methods of clinical treatment of bone
healing [5,6] . The potentials produced may be of several milli-volts
and are precisely associated with microscopical structural units
of bone (osteons) so that it is not unreasonable to suppose some
biological effect arising from them. It is known from blood
compatibility studies that blood components are attracted to
charged surfaces and in a similar way these strain generated
potentials could influence macromolecular orientations in growing
bone[6] .

    The control of growth processes in bone by means of an effect
directly operable at cellular level and activated by implantable
materials or devices is a development yet awaited. Macromolecular
chemistry may offer such a method for control as it does for
cancer therapy. For the present, the control is somewhat less
specific and has arisen from a clinical observation of bone
healing in the presence of stiff metallic implants.

    The objective in any operative method to achieve fixation of
fractures should be stability at the fracture site, good
anatomical reconstruction and unrestricted use of neighbouring
joints without the need for external splinting. This has usually
involved the use of a metal plate or other device intended to
provide stability and to have adequate strength to maintain the
situation during healing.

    Stainless steel is widely used as a component of joint
prostheses and also as bone plates used for internal fixation of
bone fractures. When a fracture is stabilised by means of a plate
attached to it by screws it is observed that the bone underlying
the plate is reduced in thickness and therefore weakened[7,8,9].
Although this is not universally observed[10] it seems to be
sufficiently accepted as stress induced osteopoenia. Refracture
of bone after plate removal may occur up to two years later. Less
rigid means of fixation have been sought as a means of over-
coming this problem whilst retaining stability. The unknown
parameter is the amount of movement that can be permitted at a
fracture site during healing without inhibiting the effectiveness
of natural healing processes.

    Normal bone healing[11] occurs by growth and organisation of
calcified tissue initiated from the periosteal membrane that
surrounds the bone. This is referred to as periosteal callus and
bridges the fracture gap to encircle it with a band of new bone
tissue. In time this callus is resorbed as the main bone structure

is repaired and strengthened. A rigid metal plate hinders this
process and callus may never be seen. Healing may then take place
by direct cell growth across the divided bone sections at points
where they are in contact. Primary bone healing or remodelling
union is the name applied to this. A final healing process
occurring rather later is by endosteal callus arising from the
inner surfaces of the medullary canal.

Excessive movement can destroy all these processes but a small
undefined amount can actually promote callus formation. Research
is being directed towards definition of the factors necessary for
good bone healing but they can only be defined qualitatively so
far. Figure 1 is a representation, schematically, of the relationship between rigidity of fixation and strain (deformation) at the
fracture site and the three clinical effects resulting. Super-
imposed on this is an optimum stability. As stability decreases
there can come a stage when only non-union of the fracture is
possible. The situation is, of course, much more complex than
can be attempted to be shown in a single diagram.

The aim, therefore, has been to use the advantages or rigid
internal fixation but to reduce the dependency of the bone on the
implant by considering less rigid fixation systems.

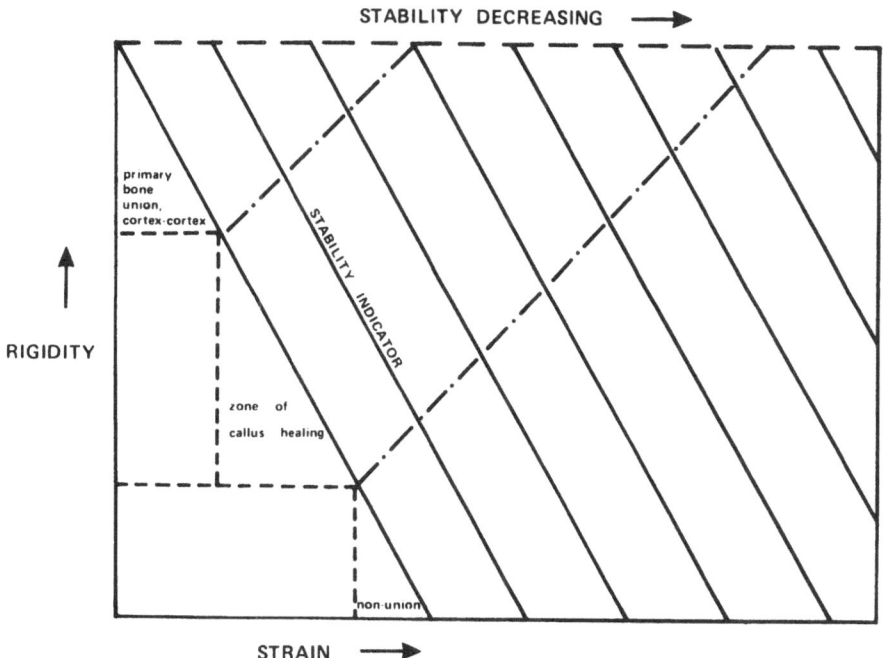

Fig. 1. Representation of the inter-relationship of stability,
         regidity and type of result in bone healing.

POLYMER COMPOSITES

In order to reduce the rigidity of a bone-plate system a carbon fibre epoxy resin (cfr epoxy) composite has been developed [12,13]. This arose out of studies into water emulsifiable epoxy resins which seemed to offer a possible use as a bone substitute material. Reinforcement was needed and carbon fibres were investigated and shown to be suitable for the purpose. The further study of the water based epoxy system was deferred in favour of conventional resins and these have formed the basis of all subsequent work. From the outset it was intended that prosthesis developments would be the aim [12] but the main emphasis has been on bone plates as providing a suitable device for product development.

Carbon fibre composite materials have several advantages for use as surgical implants. (Table 1).

Table 1. Advantages of Carbon Fibre Composites as Surgical Implants

Carbon fibre biocompatible
Strength/Weight ratic   high
Design according to principal stress magnitude and direction
Good fatigue properties
Moisture resistance
Cost effective

Epoxy resins have had widespread clinical use as encapsulants for heart pacemakers and have been used for prosthesis fixation without adverse effects [14]. This use was therefore feasible and laboratory studies on fatigue and in vivo trials were begun.

EXPERIMENTAL

In Vivo Trials

A series of blocks of the epoxy-carbon composite chosen were implanted into rat muscle and the reaction was satisfactory in all cases. The samples were surrounded by a thin fibrous sheath as shown in Figure 2.

Trials were begun in two groups of sheep. The first group had a simple transverse osteotomy performed in the midshaft of one tibia. This was reduced and secured with a cfr epoxy plate similar in design to an 8 hole broad dynamic compression plate. In the second group a double osteotomy was performed to give a 1cm long avascular bone segment. This was also reduced and fixed with a similar plate.

Animals were killed sequentially from the 5th post-operative week to the 25th and the intact tibiae and the strength of the

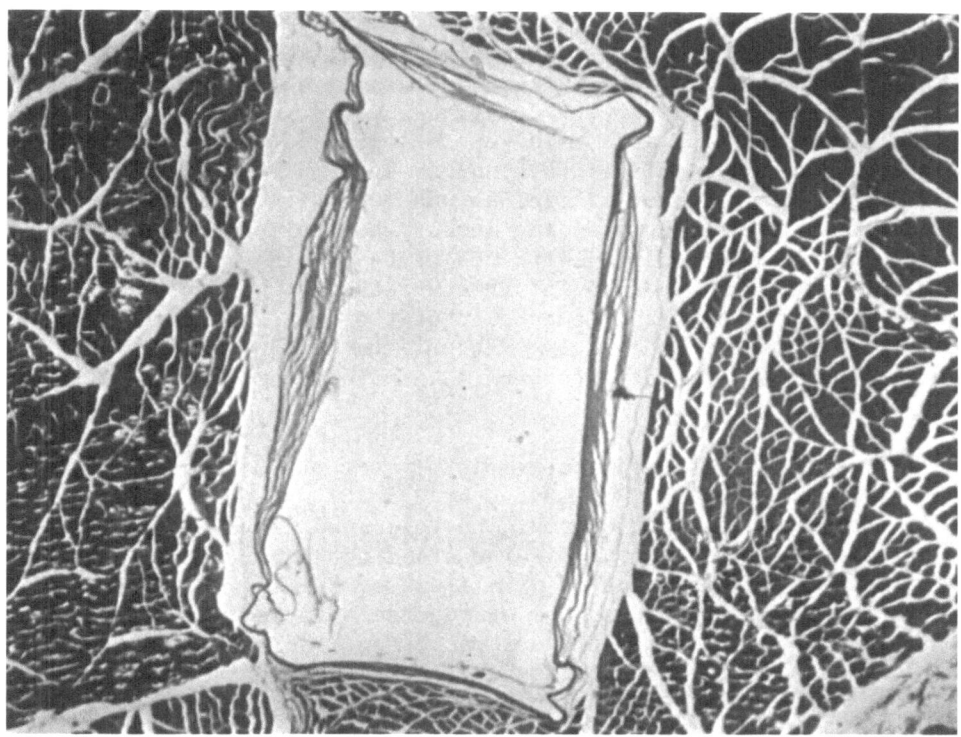

Fig. 2. Histological section from cfr epoxy implanted into rat
         muscle.

healing bone was compared with that of the intact bone in the
opposite leg.

    The first group of osteotomies healed with the production of an
abundant amount of periosteal callus with normal bone strength
achieved in ca. 20 weeks. A similar result was obtained in the
second group with normal strength being achieved in ca. 25 weeks[15].

EXPERIMENTAL

In Vitro Laboratory Trials

    In addition to inter-laminar shear strength tests to evaluate
the static mechanical performance of the composites, fatigue
testing was carried out, since not only must the plate carry gross
loads, but must also be able to withstand cyclic loads during the
working cycle. Mechanical test results have been reported[13] and
the fatigue studies are to be reported in detail elsewhere[16].

    All fatigue tests were done using an Avery reversed plane
bending Dynamic Fatigue Testing machine modified to accept bone
plate specimens. Samples were enclosed in a rubber envelope
through which Ringer-Lactate solution was circulated at a rate of
1 litre per hour.   Plates used were stainless steel 6 hole dynamic
compression plates (ASIF, DCP) and cfr epoxy plates of a similar
pattern. The frequency of cycling was 10Hz with a fixed strain
amplitude. The maximum bending moment was applied for the part of
the cycle which corresponded to the plate bending in a way that
would compress the ends of the fractured bone together. Half that
value was applied to the other part of the cycle corresponding to
distraction of the bone ends. The maximum bending moments applied,
determined from static tests were 12.5Nm for steel and 13.0Nm for
cfr epoxy.

    The steel plates failed completely except for that one cycled
at the lowest loading (3.5Nm) and at 2 x 10^6 cycles it was estimated
that the maximum load that could be tolerated was 6Nm. In the
case of the composite plates failure did not occur and 10.7 Nm was
tolerated at 2 x 10^6 cycles. The limit of cycling was chosen as
being approximately the number of reversals in one year of average
walking. Reduction in stiffness was taken as the criterion of
change in the composite plates and Figure 4 shows these results.
The comparison with steel plates is shown in Figure 3.

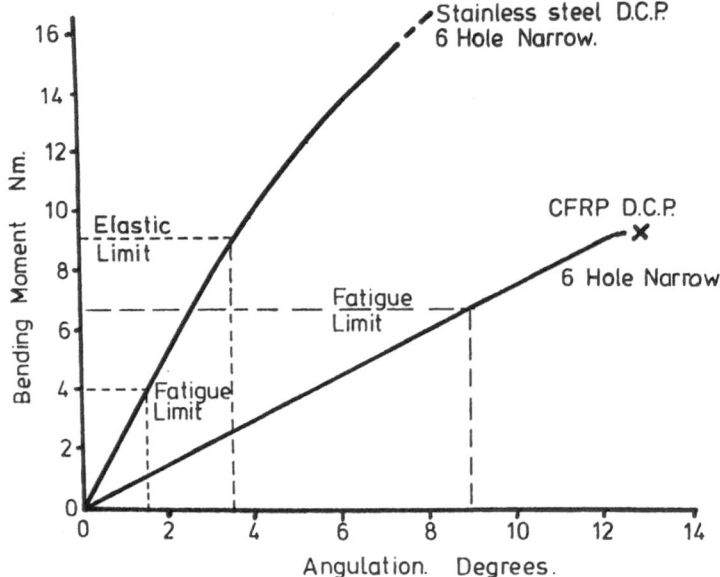

Fig. 3.   Comparative properties of comparable steel and cfr epoxy
          plates. (Adapted from J.S. Bradley, PhD Thesis, (17)).

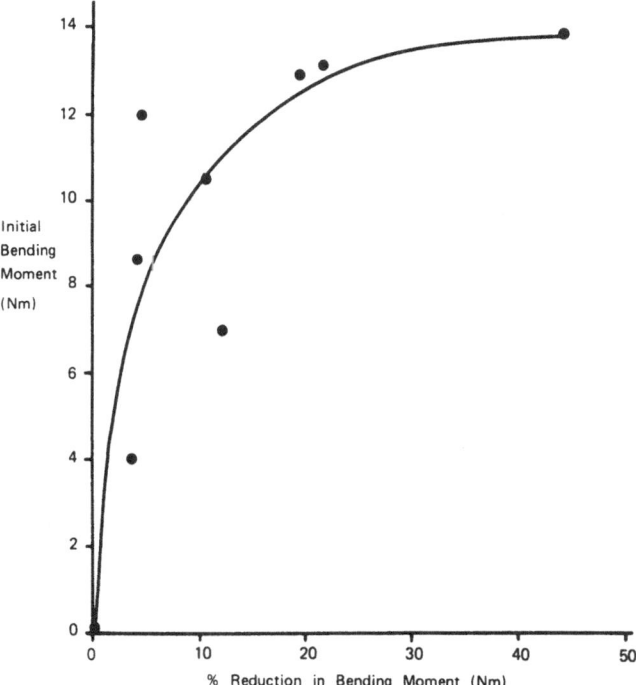

Fig. 4. Fatigue studies of cfr epoxy. Interpolated "S-N" curve
showing reduction of stiffness in comparison with original
stiffness. (Adapted from J.S. Bradley, PhD thesis (17)).

EXPERIMENTAL

Clinical Trials

The fatigue studies were confirmed in vivo by the sheep trials
in which the plates withstood the rigours of full-weight bearing
without failure. Since it was felt that little further benefit
would be gained from prolonged experiments of similar nature, a
limited clinical trial was begun on transverse and short oblique
fractures of the tibial mid-shaft including those with butterfly
fragments. Full details of this trial on 20 patients has been
reported[15].

Surgery was performed as soon as possible after injury and
between the third and tenth post-operative day mobilisation and
weight bearing were commenced depending on the comfort of the
patient. Regular clinical and radiological examination was carried
out. All plates were removed when the fracture was judged to be
clinically sound and unrestricted weight bearing allowed. The first
seven cases were fixed with a plate having oval holes similar to
the DCP type but subsequent plates had cylindrical holes with
conical countersinks.

The final result was satisfactory in all cases, even when infection was present and plates were removed at around 40 weeks post operation. All patients were permitted totally unrestricted activity in less than 14 months from the date of injury, the fractures healing by means of external bridging callus which gave early rapid consolidation of the fracture. There was no evidence of stress protection effects.

Following this experience, two further clinical trials are in progress in the forearm (radius, ulna) using two slightly different designs of plate and methods of fixation. To date about twenty fractures in total have been treated but it is still early to draw firm conclusions. A typical result is shown in Figure 5.

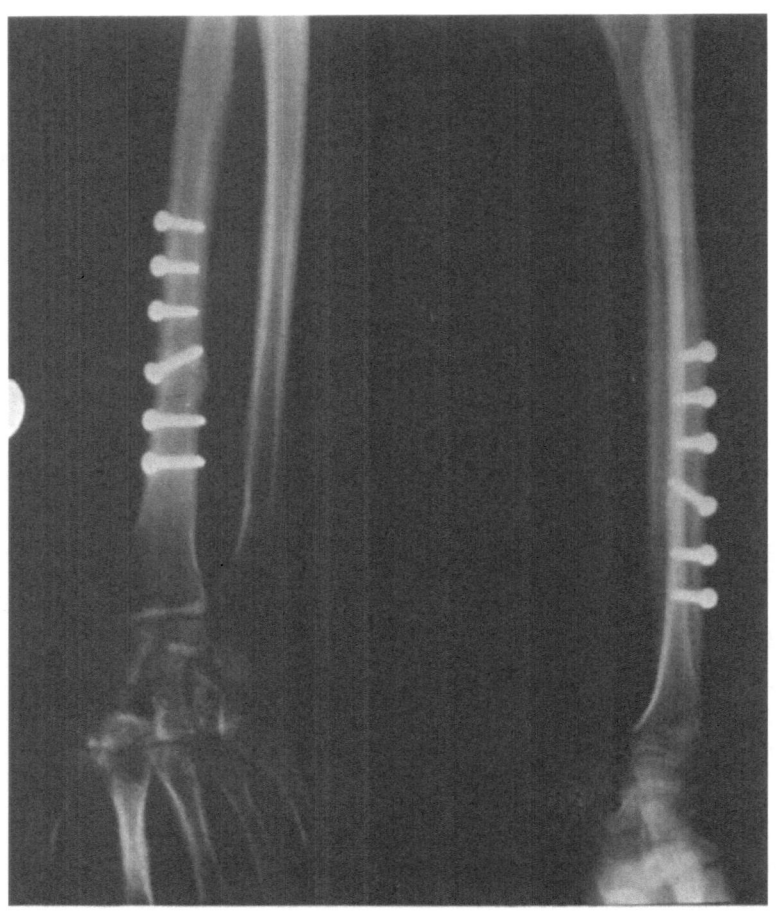

Fig. 5.   Radiograph of forearm fracture treated with cfr epoxy plate. Note the radio-transparency of the plate.

DISCUSSION

The advantages of a composite material in this orthopaedic application are clearly shown. Fatigue life is better than stainless steel and this is important because:

a)   greater movement is permissible at the fracture site, thus stimulating periosteal callus formation;
b)   early weight bearing is encouraged with beneficial effects in healing and in mobility in the adjacent joints;
c)   stability is provided for a longer period than usual if necessary for some reason e.g. infection, or gapping at the fracture site.

Direct comparisons between cfr epoxy plates and stainless steel plates may be rather misleading unless due account is taken of dimensional factors. In the tibial series reported earlier a broad composite plate was used whereas the normal steel plate recommended for this site is narrower. The elastic modulus E of the composite is approximately one third the value of stainless steel ($65 \times 10^9 \text{N/m}^2$; $210 \times 10^9 \text{N/m}^2$) and calculation of Flexural Rigidity (EI where I is the 2nd moment of area) gives corresponding values of 45.9 Nm^2 and 53.7 Nm^2 for the solid section of the plate, 32.4 Nm^2 and 31.3 Nm^2 for the hole section.

This reveals that the plate used on the tibia was of comparable stiffness to the recommended plate and that the advantage gained was not, therefore, in terms of reduced rigidity. It was the improved fatigue life which gave successful stabilisation under conditions which may not always have been ideal, but there is an additional factor. The composite plates deform elastically over the whole range of applied load up to the point of failure and the range of elastic deformation is greater than for steel. Hence, a range of movement is possible which will still retain the function whereas a deforming steel plate, because it yields permanently, will progressively hold the bone ends separated.

It is necessary to evaluate the plates fully over a range of clinical applications and this draws attention to need for an adequate philosophy for testing new implant materials.

It has been evident for some time that passive implantation for a short time cannot be sufficient by itself to establish biological acceptance. Long term studies are essential since effects are possibly only realisable after tissues and implant have been in intimate contact for periods in excess of one year[18]. These are presently in progress. Functional implantation is also necessary to study the effects of load-bearing or movement in the same way that vascular implantation is necessary for evaluation of thrombo-genetic activity. In this context, it may be noted that the

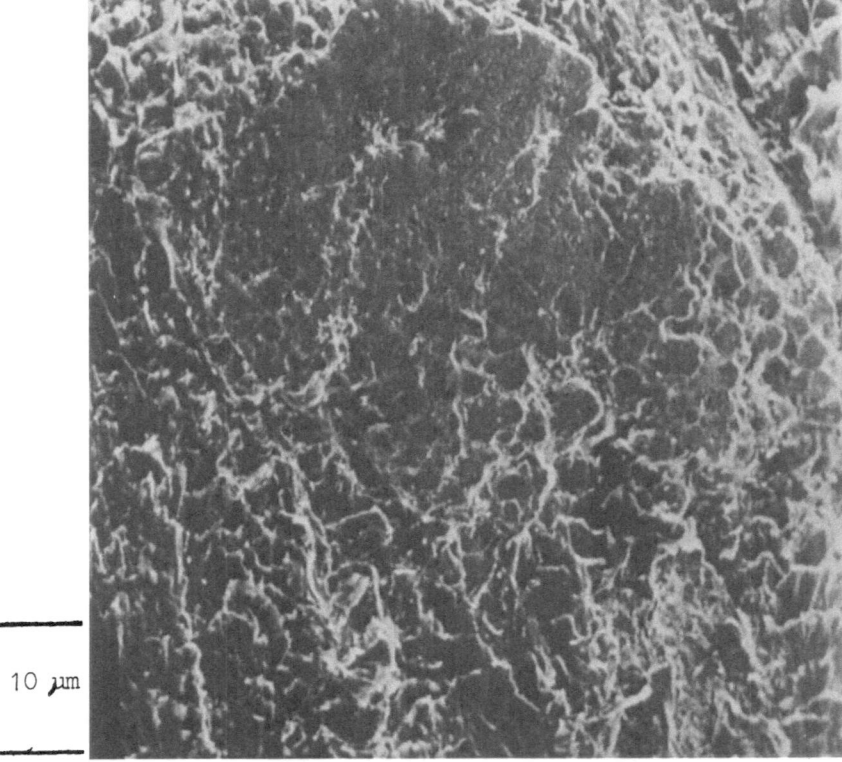

10 μm

Fig. 6.   Wear patterns of UHMWPE from examination of retrieved knee
          joint prostheses.   SEM Examination.

examination of retrieved implants is one of the most valuable means
for evaluation of implant behaviour, and yet has not been established
as a routine procedure.   In cases where this has been done,
information can be gained on in vivo effects on plastics materials[19].
Figure 6.   In the study, various patterns of wear in the polyethylene
were observed, including abrasion from embedded acrylic cement
particles, creep and galling processes in which regions of differing
mechanical properties from the bulk polymer were removed from the
surface.   A surface delamination was also seen.   Creep appears to
be one of the longer-term effects seen in prostheses but in
circumstances of abnormal loads, may be seen in the earlier stages
of implantation.

     Another application of carbon fibres has been realised by an
attempt to overcome creep effects in polyethylene by random
incorporation of chopped carbon fibres into the polymer.   However,
when the proportion of carbon fibre is optimised with respect to
the mechanical properties (creep) the rate of wear against a
stainless steel counterface in a pin on disc machine is increased.

The reinforced polymer was the pin in this case.   When wear
properties were optimised by reducing the amount of carbon fibre
present, then the creep was increased[13,17].

    Simulation of clinical use is one of the important problems
facing biomaterials scientists. Laboratory tests will evaluate
important properties and are essential preliminaries. Tissue
culture will eliminate undesirably toxic materials and animal studies
take this a stage further. However, gait and load-bearing
characteristics in four footed animals, small and large, are very
far removed from those in the human. A certain degree of functional
assessment is possible in laboratory simulators but an adequate
testing machine would be prohibitively expensive. This impels one
to look more closely at retrieved implant examination, realising,
of course, that these may represent out of date materials and
designs. Evaluation will always lag behind development.

POSTSCRIPT

    The functional development of an implant material and its
introduction into clinical practice have been described. In one
sense, this process continues since long-term in vivo evaluation
is essential and the application to other sites and other devices
is desirable. Joint prostheses are presently in the prototype
stage and a range of different plates are now being made. The over-
riding interest is in the interaction between tissues and implants,
and this extends beyond the present materials.

    The author acknowledges with gratitude the work of several
postgraduate students in Australia and the U.K. and the collaboration
of scientific and clinical colleagues, in particular Mr. K. J.
Tayton (Cardiff) and Mr. C. Wynn Jones (Stoke).

    The Department of Health and Social Security (UK) are thanked
for the award of a grant by which this work was largely carried
out.

REFERENCES

1.   G. W. Hastings. Load bearing polymers in orthopaedic surgery.
         Brit. Pol. J. 1978, 10, 251-255.
2.   G. W. Hastings. Biomedical engineering and materials for
         orthopaedic implants. J. Physics (E), 1980, 13, 599-607.
3.   B. Bloch, G. W. Hastings. Plastics materials in surgery.
         2nd edition. 1972. Springfield, Charles C. Thomas.
4.   F. Pauwels. Biomechanics of the normal and diseased hip.
         1976. Berlin, Springer-Verlag.
5.   C. T. Brighton, J. Black, S. R. Pollack (Editors).
         Electrical properties of bone and cartilage.
         1979. New York, Grune & Stratton.

6.   G. W. Hastings. Structural and mechanistic considerations
         in the strain generated electrical behaviour of bone
         in G. W. Hastings & P. Ducheyne (Eds). Structure
         property relationships in biomaterials. Vol. 3.
         Living materials of the musculo skeletal system. 1983.
         Boca Raton, CRC Press Inc.
7.   Z. Stromberg, N. Dalan, P. Laftman, F. Sirgurdson.
         Atrophy of cortical bone caused by rigid plates and
         its recovery in Uhthoff, H. K. (Editor). Current
         concepts of internal fixation of fractures. Berlin:
         Springer-Verlag, 1981: 289-90.
8.   P. Statis, P. Paavolainen, E. Karaharju, T. Holmstrom.
         Structural and biomechanical changes in bone after
         rigid plate fixation in Uhthoff, H. K. (Editor).
         Springer-Verlag. 1981; 291.
9.   A. J. Tonino, C. L. Davidson, P. J. Klopper, L. A. Linclau.
         Protection from stress in bone and its effects.
         J. Bone Jt.Surg. 1976, 58B, 107-113.
10.   S. G. Steinmann.
         Characteristics of an ideal implant material for stable
         fixation in Uhthoff, H. K. (Editor). Current concepts
         of internal fixation of fractures. Berlin; Springer-
         Verlag, 1981; 93-99.
11.   B. McKibbin. The biology of fracture healing in long bones.
         J. Bone Jt. Surg. 1978, 60B, 150-162.
12.   G. W. Hastings. Carbon fibre resin composites for surgical
         implants. Composites. 1978, 9, 193-208.
13.   J. S. Bradley, G. W. Hastings, C. Johnson-Nurse.
         Carbon fibre reinforced epoxy as a high strength, low
         modulus material for internal fixation plates.
         Biomaterials, 1980, 1, 38-40.
14.   B. Bloch, G. W. Hastings, G. R. Wallwork. Major prosthetic
         replacement of bone for chondrosarcoma. Med. J. Australia.
         1970, 1, 938-940.
15.   K. J. Tayton, C. Johnson-Nurse, B. McKibbin, J. S. Bradley,
         G. W. Hastings. The use of semi rigid carbon fibre
         reinforced plastic plates for fixation of human fractures.
         J. Bone Jt. Surg. 1982, 64B, 105-111.
16.   G. W. Hastings, J. S. Bradley, P. Bell. Fatigue studies of
         carbon fibre composites intended for less rigid fracture
         fixation. In preparation.
17.   J. S. Bradley. PhD thesis. North Staffordshire Polytechnic,
         Stoke on Trent, England. 1980.
18.   L. L. Hench. Future developments and applications of bio-
         materials - an over-view. Biomat. Med. Dev. Artif.
         Organ. 1979, 7, 339-350.
19.   G. W. Hastings. In vivo wear studies of polyethylene used
         in the knee joint. Wear, 1979, 55, 1-9.